Books are to be returned on or before
the last date below.

510.2462

8/07

D0302029

To Rebecca, Charlotte and Elinor

National Engineering Mathematics

Volume 2

J.C. Yates
Senior Mathematics Lecturer
Wigan and Leigh College

MACMILLAN

First published 1994 by
THE MACMILLAN PRESS LTD
Houndmills, Basingstoke, Hampshire RG21 2XS
and London
Companies and representatives
throughout the world

ISBN 0-333-54855-8

A catalogue record for this book is available
from the British Library.

Copy-edited and typeset by Povey–Edmondson
Okehampton and Rochdale, England

Printed in Hong Kong

10	9	8	7	6	5	4	3	2	1
03	02	01	00	99	98	97	96	95	94

Acknowledgements

For the help I have received whilst writing this volume I am truly grateful. I wish to
thank Alan Skinner for his constructive and thoughtful comments. My deepest thanks
are reserved for my wife, Veronica, for her continued encouragement and patience.

The author and publishers would also like to acknowledge the following illustration
sources: British Alcan Aluminium plc; Harold E. Egerton, MIT, Cambridge, Mass.,
USA; Elonex plc; D. W. Molyneux; Gordon Roberts; Sefton Photo Library; Wella of
Great Britain.

Contents

Contents of Volume 1

Contents of Volume 3

Author's Note

This is the second volume covering the Mathematics required by a National Certificate or Diploma or Advanced GNVQ student in engineering and science. It aims to complete the coverage of the objectives of standard BTEC and college-devised syllabi with an eye on practical applications of the subject. This means some readers may neither wish nor need to study all the chapters. Readers aiming to continue into higher education will find most, if not all, the chapters essential.

Some colleges split the subject matter into 'Algebra' and 'Calculus'. The 'Algebra' half-unit can include Chapters 1–6 and the majority of Chapter 12 covering partial fractions. The 'Calculus' half-unit can include a selection of the remaining chapters. The order of chapters is a suggested overall order of study. However, Chapters 11–18 are almost independent, using Chapter 10 as a basis. Chapter 15 uses some of the ideas of Chapter 4, which in turn is linked to Chapter 3.

Each chapter is introduced with an Assignment. It is a specimen example of how the mathematics within the chapter can be applied. The Assignment provides a common thread for the chapter, linking together the theory and techniques as they develop. At appropriate stages each Assignment is revisited for further attention.

An electronic calculator is used throughout the text. You may find a graphical calculator useful, but not essential. Also you may find computer software packages useful to assist with the graphical work and the checking of your answers.

Greek letters used

Mathematics needs more letters than those provided by the alphabet. This is why the following Greek letters have been used:

α	alpha	π	pi
β	beta	ρ	rho
γ	gamma	ϕ	phi
δ	delta	ω	omega
θ	theta	$\sum$	sigma – this is capital sigma
λ	lambda	Ω	omega – this is capital omega
μ	mu		

1 Determinants and Matrices

The objectives of this chapter are to:

1 Evaluate a third order (i.e. 3×3) determinant.

2 Revise the rules for determinants.

3 Remove a common factor from a row (or column).

4 Know that addition/subtraction of rows (or columns) leaves a determinant unchanged.

5 Solve simultaneous linear equations with three unknowns using determinants.

6 Calculate the sum and difference of two 3 x 3 matrices.

7 Calculate the product of two 3×3 matrices.

8 Understand that it is possible to add/subtract/multiply more than two matrices.

9 Obtain the inverse of a 3×3 matrix.

10 Solve simultaneous linear equations with three unknowns using matrices.

11 Relate the use of matrices to simple technical problems.

Introduction

We have two main sections in this chapter, similar to the first chapter in Volume 1. The first one looks at **determinants** and the second at **matrices**. We are extending the work on second order determinants and matrices to third order. You may find it useful to re-read the first chapter of Volume 1 or refer to it as you work through this one. As you work through the matrices section you will notice some similarities with the determinants section. The different notations are important, though they both appear as types of array. Determinants and matrices are not the same. They have different mathematical meanings. However, there is a connection. You can find the determinant of a matrix, but not the other way round.

A major use for both determinants and matrices is the solution of simultaneous linear equations.

◼◼◼ ASSIGNMENT ◼◼◼

This assignment looks at a company assembling personal computers. It uses three assembly teams of differing skill levels to produce its three models; the PC/SD, PC/DD and PC/HD. The following table shows the hourly assembly rates for each team for each model.

Hourly Production	PC/SD	PC/DD	PC/HD
Assembly Team North	3	4	5
Assembly Team East	3.5	3	2.5
Assembly Team West	5.5	6	5.5

The company has a large order book for these relatively new models. A long-standing customer has placed a valuable, urgent order which will need to be met by overtime working. It is for 28 of the PC/SD, 30 of the PC/DD and 29 of the PC/HD. We want to know how to schedule this order using all the assembly teams.

Suppose the order will be met by the teams working x, y and z hours of overtime respectively. Combining the hourly production rates with these hours of overtime gives

Order Production	PC/SD	PC/DD	PC/HD
Assembly Team North	$3x$	$4x$	$5x$
Assembly Team East	$3.5y$	$3y$	$2.5y$
Assembly Team West	$5.5z$	$6z$	$5.5z$
Total no. of PCs	28	30	29

Having set out the problem we will return to it once we have looked at some relevant Mathematics.

Determinants

Remember that a **determinant** is an array of numbers (or letters) in rows and columns. Rows are horizontal, $\updownarrow$, and columns are vertical, $\longleftrightarrow$. Determinant is often shortened to either **det** or Δ. All determinants are **square**, i.e. they have the same number of rows as columns.

In this chapter we will look at third order determinants, having 3 rows, and hence 3 columns. A 3×3 ('3 by 3') determinant is another way of writing a third order determinant.

$\begin{vmatrix} a & b & c \\ e & f & g \\ k & l & m \end{vmatrix}$ is a general example of a third order determinant.

We need to use 2×2 determinants to find the value of a 3×3 determinant. In turn we strike out selected rows and columns. We strike out the row and first column where a is the common element. This leaves us with the minor $\begin{vmatrix} f & g \\ l & m \end{vmatrix}$.

Next we strike out the row and second column where b is the common element. This leaves us with the minor $\begin{vmatrix} e & g \\ k & m \end{vmatrix}$.

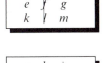

Finally we strike out the row and third column where c is the common element. This leaves us with the minor $\begin{vmatrix} e & f \\ k & l \end{vmatrix}$.

The next stage is to multiply each chosen element with its minor. This gives us $a\begin{vmatrix} f & g \\ l & m \end{vmatrix}$, $b\begin{vmatrix} e & g \\ k & m \end{vmatrix}$ and $c\begin{vmatrix} e & f \\ k & l \end{vmatrix}$.

Finally we gather together all these parts with **alternate $+/-$ signs** to get

$$\begin{vmatrix} a & b & c \\ e & f & g \\ k & l & m \end{vmatrix} = a\begin{vmatrix} f & g \\ l & m \end{vmatrix} - b\begin{vmatrix} e & g \\ k & m \end{vmatrix} + c\begin{vmatrix} e & f \\ k & l \end{vmatrix}.$$

Examples 1.1

Find the values of the third order determinants

i) $\begin{vmatrix} 12 & 6 & 2 \\ 5 & 4 & 9 \\ 3 & 1 & 7 \end{vmatrix}$, ii) $\begin{vmatrix} 1 & 7 & 0 \\ -6 & 21 & 2 \\ 3 & 10 & 5 \end{vmatrix}$.

We simply follow the order of operations we have shown before these examples. In the first example we will work out the stages slowly, giving you time to check them.

i) We strike out the row and first column where 12 is the common element. This leaves us with the minor $\begin{vmatrix} 4 & 9 \\ 1 & 7 \end{vmatrix}$.

Next we strike out the row and second column where 6 is the common element. This leaves us with the minor $\begin{vmatrix} 5 & 9 \\ 3 & 7 \end{vmatrix}$.

Finally we strike out the row and third column where 2 is the common element. This leaves us with the minor $\begin{vmatrix} 5 & 4 \\ 3 & 1 \end{vmatrix}$.

The next stage is to multiply each chosen element with its minor. This gives us $12\begin{vmatrix} 4 & 9 \\ 1 & 7 \end{vmatrix}$, $6\begin{vmatrix} 5 & 9 \\ 3 & 7 \end{vmatrix}$ and $2\begin{vmatrix} 5 & 4 \\ 3 & 1 \end{vmatrix}$.

Finally we gather together all these parts with alternate $+/-$ signs to get

$$\begin{vmatrix} 12 & 6 & 2 \\ 5 & 4 & 9 \\ 3 & 1 & 7 \end{vmatrix} = 12\begin{vmatrix} 4 & 9 \\ 1 & 7 \end{vmatrix} - 6\begin{vmatrix} 5 & 9 \\ 3 & 7 \end{vmatrix} + 2\begin{vmatrix} 5 & 4 \\ 3 & 1 \end{vmatrix}$$

$$= 12(4 \times 7 - 9 \times 1) - 6(5 \times 7 - 9 \times 3) + 2(5 \times 1 - 4 \times 3)$$
$$= 12(28 - 9) - 6(35 - 27) + 2(5 - 12)$$
$$= 12(19) - 6(8) + 2(-7)$$
$$= 228 - 48 - 14$$
$$= 166.$$

ii) For the second example we will link all the stages together. Notice that the last minor is multiplied by the element 0. This eases the calculation to give

$$\begin{vmatrix} 1 & 7 & 0 \\ -6 & 21 & 2 \\ 3 & 10 & 5 \end{vmatrix} = 1 \begin{vmatrix} 21 & 2 \\ 10 & 5 \end{vmatrix} - 7 \begin{vmatrix} -6 & 2 \\ 3 & 5 \end{vmatrix} + 0 \begin{vmatrix} -6 & 21 \\ 3 & 10 \end{vmatrix}$$

$$= 1(21 \times 5 - 2 \times 10) - 7(-6 \times 5 - 2 \times 3) + 0(-6 \times 10 - 21 \times 3)$$
$$= 1(105 - 20) - 7(-30 - 6) + 0$$
$$= 1(85) - 7(-36)$$
$$= 85 + 252$$
$$= 337.$$

The order of writing the rows and columns is important. In the next set of examples we look at some interchanges. We restate the general rule from Volume 1.

If any two rows (or two columns) are interchanged then the sign of the determinant changes $+/-$.

There are other ways of finding the value of a determinant. Where we use the first row, they are based on any row or column.

▓▓▓▓▓ **Examples 1.2** ▓▓▓▓▓

Find the values of the third order determinants

i) $\begin{vmatrix} 3 & 1 & 7 \\ 5 & 4 & 9 \\ 12 & 6 & 2 \end{vmatrix}$,   ii) $\begin{vmatrix} 12 & 2 & 6 \\ 5 & 9 & 4 \\ 3 & 7 & 1 \end{vmatrix}$,   iii) $\begin{vmatrix} 3 & 7 & 1 \\ 5 & 9 & 4 \\ 12 & 2 & 6 \end{vmatrix}$.

i) In this first example we have interchanged rows ① and ③ of Examples 1.1(i)

$$\begin{vmatrix} 3 & 1 & 7 \\ 5 & 4 & 9 \\ 12 & 6 & 2 \end{vmatrix} = 3 \begin{vmatrix} 4 & 9 \\ 6 & 2 \end{vmatrix} - 1 \begin{vmatrix} 5 & 9 \\ 12 & 2 \end{vmatrix} + 7 \begin{vmatrix} 5 & 4 \\ 12 & 6 \end{vmatrix}$$

$$= 3(4 \times 2 - 9 \times 6) - 1(5 \times 2 - 9 \times 12) + 7(5 \times 6 - 4 \times 12)$$
$$= 3(8 - 54) - 1(10 - 108) + 7(30 - 48)$$
$$= 3(-46) - 1(-98) + 7(-18)$$
$$= -138 + 98 - 126$$
$$= -166.$$

As we predicted the interchange of the rows has altered the $+/-$ sign of the determinant.

ii) In this second example we have interchanged columns ② and ③ of Examples 1.1(i).

$$\begin{vmatrix} 12 & 2 & 6 \\ 5 & 9 & 4 \\ 3 & 7 & 1 \end{vmatrix} = 12\begin{vmatrix} 9 & 4 \\ 7 & 1 \end{vmatrix} - 2\begin{vmatrix} 5 & 4 \\ 3 & 1 \end{vmatrix} + 6\begin{vmatrix} 5 & 9 \\ 3 & 7 \end{vmatrix}$$

$$= 12(9 - 28) - 2(5 - 12) + 6(35 - 27)$$
$$= 12(-19) - 2(-7) + 6(8)$$
$$= -228 + 14 + 48$$
$$= -166.$$

As we predicted the interchange of the columns has altered the $+/-$ sign of the determinant.

iii) In this third example we have combined the previous interchange of rows and columns.

$$\begin{vmatrix} 3 & 7 & 1 \\ 5 & 9 & 4 \\ 12 & 2 & 6 \end{vmatrix} = 3\begin{vmatrix} 9 & 4 \\ 2 & 6 \end{vmatrix} - 7\begin{vmatrix} 5 & 4 \\ 12 & 6 \end{vmatrix} + 1\begin{vmatrix} 5 & 9 \\ 12 & 2 \end{vmatrix}$$

$$= 3(54 - 8) - 7(30 - 48) + 1(10 - 108)$$
$$= 3(46) - 7(-18) + 1(-98)$$
$$= 138 + 126 - 98$$
$$= 166.$$

The first interchange has altered the $+$ sign of the determinant to a $-$ sign. The second interchange has altered that sign, returning it to a $+$ sign. Remember $(-)(-) = +$.

If any rows (or columns) are the same then the determinant has a value of 0.

▓▓▓▓▓ **Examples 1.3** ▓▓▓▓▓▓▓▓▓▓▓▓▓▓▓▓▓▓▓▓▓▓▓▓▓▓▓▓▓▓▓▓▓▓▓▓

Find the values of the third order determinants

i) $\begin{vmatrix} 2 & 2 & 3 \\ 7 & 7 & 5 \\ 6 & 6 & 0 \end{vmatrix}$, ii) $\begin{vmatrix} 2 & -3 & 4 \\ 6 & 11 & \frac{1}{2} \\ 2 & -3 & 4 \end{vmatrix}$.

i) $\begin{vmatrix} 2 & 2 & 3 \\ 7 & 7 & 5 \\ 6 & 6 & 0 \end{vmatrix} = 2\begin{vmatrix} 7 & 5 \\ 6 & 0 \end{vmatrix} - 2\begin{vmatrix} 7 & 5 \\ 6 & 0 \end{vmatrix} + 3\begin{vmatrix} 7 & 7 \\ 6 & 6 \end{vmatrix}$

$$= 2(0 - 30) - 2(0 - 30) + 3(42 - 42)$$
$$= 2(-30) - 2(-30) + 3(0)$$
$$= 0,$$

because columns ① and ② are the same.

ii) $\begin{vmatrix} 2 & -3 & 4 \\ 6 & 11 & \frac{1}{2} \\ 2 & -3 & 4 \end{vmatrix} = 2\begin{vmatrix} 11 & \frac{1}{2} \\ -3 & 4 \end{vmatrix} - -3\begin{vmatrix} 6 & \frac{1}{2} \\ 2 & 4 \end{vmatrix} + 4\begin{vmatrix} 6 & 11 \\ 2 & -3 \end{vmatrix}$

$$= 2(44 - -1.5) + 3(24 - 1) + 4(-18 - 22)$$
$$= 2(45.5) + 3(23) + 4(-40)$$
$$= 91 + 69 - 160$$
$$= 0,$$

because rows ① and ③ are the same.

Writing the rows as columns and the columns as rows leaves the value of the determinant unchanged.

▓▓▓▓▓▓ Examples 1.4 ▓▓▓▓▓▓

Find the values of the third order determinants

i) $\begin{vmatrix} 5 & 9 & 7 \\ -2 & 6 & 1 \\ 10 & 0 & 4 \end{vmatrix}$, ii) $\begin{vmatrix} 5 & -2 & 10 \\ 9 & 6 & 0 \\ 7 & 1 & 4 \end{vmatrix}$.

i) $\begin{vmatrix} 5 & 9 & 7 \\ -2 & 6 & 1 \\ 10 & 0 & 4 \end{vmatrix} = 5\begin{vmatrix} 6 & 1 \\ 0 & 4 \end{vmatrix} - 9\begin{vmatrix} -2 & 1 \\ 10 & 4 \end{vmatrix} + 7\begin{vmatrix} -2 & 6 \\ 10 & 0 \end{vmatrix}$

$$= 5(24 - 0) - 9(-8 - 10) + 7(0 - 60)$$
$$= 5(24) - 9(-18) + 7(-60)$$
$$= 120 + 162 - 420$$
$$= -138.$$

ii) $\begin{vmatrix} 5 & -2 & 10 \\ 9 & 6 & 0 \\ 7 & 1 & 4 \end{vmatrix} = 5\begin{vmatrix} 6 & 0 \\ 1 & 4 \end{vmatrix} - -2\begin{vmatrix} 9 & 0 \\ 7 & 4 \end{vmatrix} + 10\begin{vmatrix} 9 & 6 \\ 7 & 1 \end{vmatrix}$

$$= -138, \text{ as before.}$$

This is the expected answer because we have interchanged the rows and columns.

▮▮▮▮▮▮ **EXERCISE 1.1** ▮▮▮▮▮▮

Find the values of the third order determinants.

1 $\begin{vmatrix} 2 & 1 & 17 \\ 3 & 4 & 10 \\ 0 & 5 & 9 \end{vmatrix}$ **3** $\begin{vmatrix} -6 & 3 & 0 \\ 2 & 5 & 2 \\ 4 & 10 & 4 \end{vmatrix}$

2 $\begin{vmatrix} 1 & 5 & 20 \\ 9 & 2 & 3 \\ -1 & 0 & 6 \end{vmatrix}$ **4** $\begin{vmatrix} 1 & 1 & 5 \\ -2 & 3 & 0 \\ 5 & 1 & 1 \end{vmatrix}$

$$
\mathbf{5} \quad \begin{vmatrix} 0 & 4 & -4 \\ -7 & 0 & 2 \\ 5 & 3 & 1 \end{vmatrix}
\qquad
\mathbf{8} \quad \begin{vmatrix} 56 & -9 & 0.5 \\ -12 & 3 & 0 \\ 4 & 1 & 2 \end{vmatrix}
$$

$$
\mathbf{6} \quad \begin{vmatrix} \frac{1}{2} & \frac{1}{3} & \frac{2}{3} \\ -\frac{1}{2} & 0 & \frac{1}{2} \\ 1 & 3 & 0 \end{vmatrix}
\qquad
\mathbf{9} \quad \begin{vmatrix} 8 & 25 & 55 \\ 1 & 0 & 2 \\ 2 & 32 & 44 \end{vmatrix}
$$

$$
\mathbf{7} \quad \begin{vmatrix} 0 & -7 & 5 \\ 4 & 0 & 3 \\ -4 & 2 & 1 \end{vmatrix}
\qquad
\mathbf{10} \quad \begin{vmatrix} 0.7 & 0.5 & 1.2 \\ 0.6 & -1 & 0 \\ -0.25 & 0.4 & 0.3 \end{vmatrix}
$$

Common factors can be removed from rows (or columns).

In the next set of examples we demonstrate removing common factors from determinants. The idea is to make the arithmetic easier if possible. It is unlikely that you will be able to find common factors in all your determinants.

Examples 1.5

Find the value of the third order determinant $\begin{vmatrix} 10 & 20 & 30 \\ 0 & 4 & 1 \\ 27 & 6 & 18 \end{vmatrix}$.

i) If you work out this determinant in the usual way you will find its value is -2040.

ii) Alternatively you may spot there is a common factor of 10 in the first row. We may write

$$
\begin{vmatrix} 10 & 20 & 30 \\ 0 & 4 & 1 \\ 27 & 6 & 18 \end{vmatrix} = 10 \begin{vmatrix} 1 & 2 & 3 \\ 0 & 4 & 1 \\ 27 & 6 & 18 \end{vmatrix}
$$

$$
= 10 \left\{ 1 \begin{vmatrix} 4 & 1 \\ 6 & 18 \end{vmatrix} - 2 \begin{vmatrix} 0 & 1 \\ 27 & 18 \end{vmatrix} + 3 \begin{vmatrix} 0 & 4 \\ 27 & 6 \end{vmatrix} \right\}
$$

$$
= 10\{1(72 - 6) - 2(0 - 27) + 3(0 - 108)\}
$$

$$
= 10\{66 + 54 - 324\}
$$

$$
= -2040.
$$

iii) Yet again you may spot a common factor of 3 in the third row so that

$$
\begin{vmatrix} 10 & 20 & 30 \\ 0 & 4 & 1 \\ 27 & 6 & 18 \end{vmatrix} = 3 \begin{vmatrix} 10 & 20 & 30 \\ 0 & 4 & 1 \\ 9 & 2 & 6 \end{vmatrix}.
$$

You can work this out for yourself to check the answer is also -2040.

iv) Another option is to remove a common factor of 2 from the second column, so that

$$\begin{vmatrix} 10 & 20 & 30 \\ 0 & 4 & 1 \\ 27 & 6 & 18 \end{vmatrix} = 2\begin{vmatrix} 10 & 10 & 30 \\ 0 & 2 & 1 \\ 27 & 3 & 18 \end{vmatrix}.$$

This will also give you an answer of -2040.

v) You may remove as many or as few common factors as you wish. Finally the factors we have spotted we can remove in stages, as

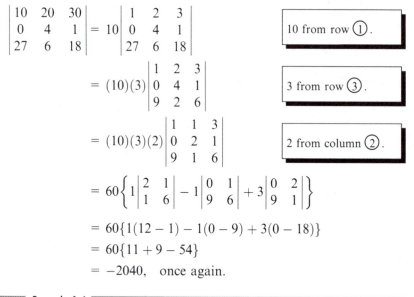

$$\begin{vmatrix} 10 & 20 & 30 \\ 0 & 4 & 1 \\ 27 & 6 & 18 \end{vmatrix} = 10\begin{vmatrix} 1 & 2 & 3 \\ 0 & 4 & 1 \\ 27 & 6 & 18 \end{vmatrix}$$ 10 from row ①.

$$= (10)(3)\begin{vmatrix} 1 & 2 & 3 \\ 0 & 4 & 1 \\ 9 & 2 & 6 \end{vmatrix}$$ 3 from row ③.

$$= (10)(3)(2)\begin{vmatrix} 1 & 1 & 3 \\ 0 & 2 & 1 \\ 9 & 1 & 6 \end{vmatrix}$$ 2 from column ②.

$$= 60\left\{1\begin{vmatrix} 2 & 1 \\ 1 & 6 \end{vmatrix} - 1\begin{vmatrix} 0 & 1 \\ 9 & 6 \end{vmatrix} + 3\begin{vmatrix} 0 & 2 \\ 9 & 1 \end{vmatrix}\right\}$$

$$= 60\{1(12 - 1) - 1(0 - 9) + 3(0 - 18)\}$$

$$= 60\{11 + 9 - 54\}$$

$$= -2040, \quad \text{once again.}$$

Example 1.6

Find the value of the third order determinant $\begin{vmatrix} 2 & 3 & 1 \\ 12 & 18 & 6 \\ 2 & 4 & 5 \end{vmatrix}$.

We can find the value in the usual way. Alternatively we can spot the common factor of 6 in the second row to give

$$\begin{vmatrix} 2 & 3 & 1 \\ 12 & 18 & 6 \\ 2 & 4 & 5 \end{vmatrix} = 6\begin{vmatrix} 2 & 3 & 1 \\ 2 & 3 & 1 \\ 2 & 4 & 5 \end{vmatrix}.$$

It is immediately obvious that the first and second rows are the same. This means the determinant has a value of 0. We will see this pattern of multiples later in some simultaneous linear equations.

Addition (or subtraction) of rows (or columns) leaves a determinant unchanged.

In the next set of examples we demonstrate adding and subtracting rows or columns within determinants. The idea is to make the arithmetic easier if possible. It is unlikely that you will be able to do this in all your determinants.

Example 1.7

Use addition/subtraction to ease the working out of $\begin{vmatrix} 21 & 11 & 31 \\ -38 & 15 & -61 \\ 47 & -17 & 78 \end{vmatrix}$.

In each determinant you need to compare the rows and then compare the columns. Ask yourself whether any addition (or subtraction) would make it easier to work out the determinant. Generally work from the largest number. In this example look at rows ③ and ② and think about addition. Suppose we add row ② to row ③

$$\begin{vmatrix} 21 & 11 & 31 \\ -38 & 15 & -61 \\ 47-38 & -17+15 & 78-61 \end{vmatrix} = \begin{vmatrix} 21 & 11 & 31 \\ -38 & 15 & -61 \\ 9 & -2 & 17 \end{vmatrix}. \qquad \boxed{\text{row ③ +} \atop \text{row ②.}}$$

Notice how we have reduced the elements in the third row. Now look at columns ① and ③ and think about subtraction. Suppose we subtract column ① from column ③ ,

$$\begin{vmatrix} 21 & 11 & 31-21 \\ -38 & 15 & -61+38 \\ 9 & -2 & 17-9 \end{vmatrix} = \begin{vmatrix} 21 & 11 & 10 \\ -38 & 15 & -23 \\ 9 & -2 & 8 \end{vmatrix}. \qquad \boxed{\text{col ③ − col ①.}}$$

Though not quite the largest elements in the determinant, we concentrate on the first row.

We subtract column ② from column ①.

$$\begin{vmatrix} 21-11 & 11 & 10 \\ -38-15 & 15 & -23 \\ 9+2 & -2 & 8 \end{vmatrix} = \begin{vmatrix} 10 & 11 & 10 \\ -53 & 15 & -23 \\ 11 & -2 & 8 \end{vmatrix}. \qquad \boxed{\text{col ① − col ②.}}$$

Notice the creation of 0 as we subtract column ③ from column ① .

$$\begin{vmatrix} 10-10 & 11 & 10 \\ -53+23 & 15 & -23 \\ 11-8 & -2 & 8 \end{vmatrix} = \begin{vmatrix} 0 & 11 & 10 \\ -30 & 15 & -23 \\ 3 & -2 & 8 \end{vmatrix}. \qquad \boxed{\text{col ① − col ③.}}$$

We do a final subtraction before finding the value of our determinant.

$$\begin{vmatrix} 0 & 11-10 & 10 \\ -30 & 15+23 & -23 \\ 3 & -2-8 & 8 \end{vmatrix} = \begin{vmatrix} 0 & 1 & 10 \\ -30 & 38 & -23 \\ 3 & -10 & 8 \end{vmatrix} \qquad \boxed{\text{col ② − col ③.}}$$

$$= 0 - 1(-240+69) + 10(300-114)$$
$$= 171 + 1860$$
$$= 2031.$$

■■■■■■■ **EXERCISE 1.2** ■■■■■■■■■■■■■■■■■■■■■■

Find the values of the third order determinants. Where possible try to remove common factors. Also see if you can ease the arithmetic by adding or subtracting rows or columns.

$$
\mathbf{1} \quad \begin{vmatrix} 22 & 21 & 1 \\ 3 & 4 & 5 \\ 6 & 6 & 7 \end{vmatrix}
\qquad\qquad
\mathbf{4} \quad \begin{vmatrix} 1 & 1 & -1 \\ -2 & 3 & -2 \\ 0 & 4 & 5 \end{vmatrix}
$$

$$
\mathbf{2} \quad \begin{vmatrix} 1 & 5 & 1 \\ 9 & 12 & 3 \\ 4 & -3 & 0 \end{vmatrix}
\qquad\qquad
\mathbf{5} \quad \begin{vmatrix} 17 & 9 & 14 \\ 3 & -1 & 2 \\ 15 & 8 & 14 \end{vmatrix}
$$

$$
\mathbf{3} \quad \begin{vmatrix} -6 & 3 & 9 \\ 2 & 5 & 6 \\ 4 & 0 & 3 \end{vmatrix}
$$

Simultaneous equations 1

We learned how to solve simultaneous linear equations in Volume 1. The next example, using second order determinants, should refresh your memory.

▓▓▓▓▓▓ **Example 1.8** ▓▓▓▓▓▓▓▓▓▓▓▓▓▓▓▓▓▓▓▓▓▓▓▓▓▓▓▓▓▓▓▓

Solve the pair of simultaneous equations $5x - 3y = 21$

and $7x + 8y = 5$.

Our first step is to gather all terms in each equation to one side of the "=" sign,

i.e. $5x - 3y - 21 = 0$

and $7x + 8y - 5 = 0$.

For the determinant under x we ignore those numbers associated with x and highlight $\begin{vmatrix} -3 & -21 \\ 8 & -5 \end{vmatrix}$.

For the determinant under y we ignore those numbers associated with y and highlight $\begin{vmatrix} 5 & -21 \\ 7 & -5 \end{vmatrix}$.

For the determinant under the 1 we ignore the pure numbers and highlight $\begin{vmatrix} 5 & -3 \\ 7 & 8 \end{vmatrix}$.

Linking these all together we have

$$\frac{x}{\begin{vmatrix} -3 & -21 \\ 8 & -5 \end{vmatrix}} = \frac{-y}{\begin{vmatrix} 5 & -21 \\ 7 & -5 \end{vmatrix}} = \frac{1}{\begin{vmatrix} 5 & -3 \\ 7 & 8 \end{vmatrix}}$$

i.e.

$$\frac{x}{15 - -168} = \frac{-y}{-25 - -147} = \frac{1}{40 - -21}$$

$$\frac{x}{183} = \frac{-y}{122} = \frac{1}{61}$$

We have 3 fractions, all equal to each other. Using the first and last ones we have

$$\frac{x}{183} = \frac{1}{61}$$

i.e.

$$x = 183 \times \frac{1}{61} = 3.$$

Using the middle and last fractions we have

$$\frac{-y}{122} = \frac{1}{61}$$

i.e.

$$y = -122 \times \frac{1}{61} = -2.$$

Our complete solution is $x = 3$, $y = -2$.

We check our solution with one of the original equations. Substituting both values into the left hand side of the first equation gives

$$5x - 3y - 21 = 5(3) - 3(-2) - 21 = 15 + 6 - 21 = 0.$$

These values are consistent, confirming our solution.

In Volume 1 we built up the theory for linear simultaneous equations in 2 unknowns, x and y. Our solutions involved second-order determinants. We will not go through the theory again here, but will just extend those ideas in the next example. This time we will have linear simultaneous equations in 3 unknowns, x, y and z. This means our solutions will involve third-order determinants.

Example 1.9

Solve the simultaneous equations
$$4x - 2y + z = 3,$$
$$3x - 8y - 4z = 10$$
and
$$7x + 5z = -1.$$

Our first step is to gather all terms in each equation to one side of the "=" sign,

i.e.
$$4x - 2y + z - 3 = 0,$$
$$3x - 8y - 4z - 10 = 0$$
and
$$7x + 5z + 1 = 0.$$

For the determinant under x we ignore those numbers associated with x
and highlight $\begin{vmatrix} -2 & 1 & -3 \\ -8 & -4 & -10 \\ 0 & 5 & 1 \end{vmatrix}$.

For the determinant under y we ignore those numbers associated with y
and highlight $\begin{vmatrix} 4 & 1 & -3 \\ 3 & -4 & -10 \\ 7 & 5 & 1 \end{vmatrix}$.

For the determinant under z we ignore those numbers associated with z
and highlight $\begin{vmatrix} 4 & -2 & -3 \\ 3 & -8 & -10 \\ 7 & 0 & 1 \end{vmatrix}$.

For the determinant under the 1 we ignore the pure numbers and highlight
$\begin{vmatrix} 4 & -2 & 1 \\ 3 & -8 & -4 \\ 7 & 0 & 5 \end{vmatrix}$.

We link these all together, alternating the $+/-$ signs as we did in Volume 1.
On the top line it is important to notice we get x, $-y$, z and -1 as

$$\frac{x}{\begin{vmatrix} -2 & 1 & -3 \\ -8 & -4 & -10 \\ 0 & 5 & 1 \end{vmatrix}} = \frac{-y}{\begin{vmatrix} 4 & 1 & -3 \\ 3 & -4 & -10 \\ 7 & 5 & 1 \end{vmatrix}} = \frac{z}{\begin{vmatrix} 4 & -2 & -3 \\ 3 & -8 & -10 \\ 7 & 0 & 1 \end{vmatrix}} = \frac{-1}{\begin{vmatrix} 4 & -2 & 1 \\ 3 & -8 & -4 \\ 7 & 0 & 5 \end{vmatrix}}$$

We work out these determinants in the usual way to get

$$\frac{x}{36} = \frac{-y}{-18} = \frac{z}{-54} = \frac{-1}{-18}$$

We have 4 fractions all equal to each other. Using the first and last ones
we have

$$\frac{x}{36} = \frac{-1}{-18}$$

i.e. $x = 36 \times \dfrac{-1}{-18} = 2.$

Using the second and last fractions we have

$$\frac{-y}{-18} = \frac{-1}{-18}$$

i.e. $y = 18 \times \dfrac{-1}{-18} = 1.$

Using the third and last fractions we have

$$\frac{z}{-54} = \frac{-1}{-18}$$

i.e. $\qquad z = -54 \times \dfrac{-1}{-18} = -3.$

Our complete solution is $x = 2$, $y = 1$ and $z = -3$.
We check our solution with one of the original equations. Substituting the values into the left hand side of the first equation gives

$$4x - 2y + z - 3 = 4(2) - 2(1) + -3 - 3 = 8 - 2 - 3 - 3 = 0.$$

These values are consistent, confirming our solution.

Examples 1.10

Solve the simultaneous equations

i) $\qquad 2x + 3y - z + 11 = 0,$ ii) $\qquad 2x + 3y - z + 11 = 0,$
$\qquad\qquad x - y + z + 4 = 0$ $\qquad\qquad\qquad 6x + 9y - 3z + 33 = 0$
and $2x + 3y - z - 9 = 0;$ $\qquad$ and $\quad x - y + z + 4 = 0.$

i) Let us look closely at the first and third equations. They have the same terms for x, for y and for z. However the pure numbers are different. Remember that the solutions for x, y and z must apply to equations simultaneously. How can $2x + 3y - z$ give the value of 0 when combined with the different 11 and -9? This should not happen. These equations are **inconsistent**. They have no simultaneous solution.

In determinant form we have

$$\frac{x}{\begin{vmatrix} 3 & -1 & 11 \\ -1 & 1 & 4 \\ 3 & -1 & -9 \end{vmatrix}} = \frac{-y}{\begin{vmatrix} 2 & -1 & 11 \\ 1 & 1 & 4 \\ 2 & -1 & -9 \end{vmatrix}} = \frac{z}{\begin{vmatrix} 2 & 3 & 11 \\ 1 & -1 & 4 \\ 2 & 3 & -9 \end{vmatrix}} = \frac{-1}{\begin{vmatrix} 2 & 3 & -1 \\ 1 & -1 & 1 \\ 2 & 3 & -1 \end{vmatrix}}$$

It is the last determinant that causes the problem. The first and third rows are the same. This causes the determinant to have a value of 0 and we know that $\dfrac{1}{0}$ is *not* allowed in Mathematics.

ii) Look closely at the first and second equations. In fact they are only one equation, because $6x + 9y - 3z + 33 = 0$ is just three times $2x + 3y - z + 11 = 0$. These 2 equations are **dependent** on each other. This means that, overall, the problem reduces to only 2 different equations in 3 unknowns. Thus we are unable to solve them simultaneously. Remember, because we have 3 unknowns, x, y and z, we need at least 3 different equations.
In determinant form we have

$$\frac{x}{\begin{vmatrix} 3 & -1 & 11 \\ 9 & -3 & 33 \\ -1 & 1 & 4 \end{vmatrix}} = \frac{-y}{\begin{vmatrix} 2 & -1 & 11 \\ 6 & -3 & 33 \\ 1 & 1 & 4 \end{vmatrix}} = \frac{z}{\begin{vmatrix} 2 & 3 & 11 \\ 6 & 9 & 33 \\ 1 & -1 & 4 \end{vmatrix}} = \frac{-1}{\begin{vmatrix} 2 & 3 & -1 \\ 6 & 9 & -3 \\ 1 & -1 & 1 \end{vmatrix}}$$

Now look at the second rows of all these determinants. From each second row we can remove a common factor of 3 to get

$$\frac{x}{3\begin{vmatrix} 3 & -1 & 11 \\ 3 & -1 & 11 \\ -1 & 1 & 4 \end{vmatrix}} = \frac{-y}{3\begin{vmatrix} 2 & -1 & 11 \\ 2 & -1 & 11 \\ 1 & 1 & 4 \end{vmatrix}} = \frac{z}{3\begin{vmatrix} 2 & 3 & 11 \\ 2 & 3 & 11 \\ 1 & -1 & 4 \end{vmatrix}} = \frac{-1}{3\begin{vmatrix} 2 & 3 & -1 \\ 2 & 3 & -1 \\ 1 & -1 & 1 \end{vmatrix}}$$

The first and second rows are the same in each determinant. This time all 4 determinants have values of 0. (You can check this for yourself as an exercise.) Again $\frac{1}{0}$ is *not* allowed in Mathematics.

From Example 1.10 we can define two rules where there is no solution.

If only the determinant under the 1 is zero, the equations are inconsistent. If all 4 determinants are zero, the equations are dependent.

■■■■ ASSIGNMENT ■■■■

Earlier in this chapter we set out the order production figures in total and for all the assembly teams. We may combine them to create three linear equations

$$3x + 3.5y + 5.5z = 28,$$
$$4x + 3y + 6z = 30$$

and $5x + 2.5y + 5.5z = 29.$

We may bring all the terms to the left-hand side in order to use our determinant method. These changes give

$$3x + 3.5y + 5.5z - 28 = 0,$$
$$4x + 3y + 6z - 30 = 0$$

and $5x + 2.5y + 5.5z - 29 = 0.$

In determinant form we can write

$$\frac{x}{\begin{vmatrix} 3.5 & 5.5 & -28 \\ 3 & 6 & -30 \\ 2.5 & 5.5 & -29 \end{vmatrix}} = \frac{-y}{\begin{vmatrix} 3 & 5.5 & -28 \\ 4 & 6 & -30 \\ 5 & 5.5 & -29 \end{vmatrix}} = \frac{z}{\begin{vmatrix} 3 & 3.5 & -28 \\ 4 & 3 & -30 \\ 5 & 2.5 & -29 \end{vmatrix}} = \frac{-1}{\begin{vmatrix} 3 & 3.5 & 5.5 \\ 4 & 3 & 6 \\ 5 & 2.5 & 5.5 \end{vmatrix}}$$

We can work out each determinant to get

$$\frac{x}{-7.5} = \frac{-y}{10} = \frac{z}{-15} = \frac{-1}{5}$$

Using the first and last fractions we have

$$\frac{x}{-7.5} = \frac{-1}{5}$$

i.e. $x = -7.5 \times \dfrac{-1}{5} = 1.5.$

Using the second and last fractions we have

$$\frac{-y}{10} = \frac{-1}{5}$$

i.e. $y = -10 \times \dfrac{-1}{5} = 2.$

Using the third and last fractions we have

$$\frac{z}{-15} = \frac{-1}{5}$$

i.e. $z = -15 \times \dfrac{-1}{5} = 3.$

Now we have a complete solution. Together the assembly teams can meet the order. The Assembly Team North has 1.5 hours of overtime, the Assembly Team East has 2 hours and the Assembly Team West has 3 hours.

■ EXERCISE 1.3 ■

Using determinants, solve the following simultaneous linear equations.

1 $x - y - 2z = 4,$
   $x + y - z = -6$ and
   $3x + 2y - z = 7$

2 $3x + y - z = 1,$
   $x + y - 2z = 0$ and
   $2x + 2y - 3z = 5$

3 $2x + y - 3z = 7,$
   $x - y = -1$
   $x + y - z = 6$

4 $x - 2y + z = 2,$
   $x + 2y - 5z = 0$ and
   $-x + y + z = 2$

5 $3x - y + z = 5,$
   $5x + 2y - 12z = 4$ and
   $-x + y - z = 3$

6 $-3x + 8y + z = 9,$
   $x - 5y + z = 7$ and
   $-x + y - z = 1$

7 $3x + 2y - z = 4,$
   $x + 3y + 3z = 0.5$ and
   $-x + 4y + 2z = 7$

8 $x + \frac{1}{2}y - 4z = 6,$
   $\frac{1}{3}x - y + z = 9$ and
   $x - 3y - z = 7$

9 $2x - 5y - 2z = 2,$
   $7x - y - 4z = 1$ and
   $-x + 3y + 2z = 2$

10 $\frac{1}{2}x - 2y + z = 4,$
   $\frac{1}{3}x + y + 2z = 3.5$ and
   $x + 2y + z = 3.5$

Matrices

Remember that a **matrix** (plural, **matrices**) is an array of numbers or letters. A matrix is *not* the same as a determinant. You cannot find the value of a matrix. Each number (or letter) is an **element** in the matrix rather than some part of an overall value. Again we have rows and columns. For matrices we are allowed different numbers of rows and columns. An $m \times n$ (i.e. m by n) matrix has m rows and n columns. We write that the order of the matrix is $m \times n$. The order of writing rows and then columns is important. Where the numbers of rows and columns are the same we have a **square** matrix. You will remember that a determinant must be square.

The **null** or **zero** matrix has only elements of 0. For example
$$\begin{pmatrix} 0 & 0 & 0 \\ 0 & 0 & 0 \\ 0 & 0 & 0 \end{pmatrix}$$ is a 3×3 null matrix.

The **unit** matrix has elements of 1 along the leading diagonal and elements of 0 elsewhere. For example $\begin{pmatrix} 1 & 0 & 0 \\ 0 & 1 & 0 \\ 0 & 0 & 1 \end{pmatrix}$ is a 3×3 unit matrix.

When we looked at determinants we calculated their values. We can do more with matrices. First of all let us look at addition and subtraction. We can only add/subtract matrices of exactly the same order, i.e. the same number of rows and the same number of columns. In Examples 1.11 we will look at some possible additions/subtractions.

▚▚▚▚▚▚ **Examples 1.11** ▚▚▚▚▚▚▚▚▚▚▚▚▚▚▚▚▚▚▚▚▚▚▚▚

i)   We can add/subtract a 2×3 matrix and a 2×3 matrix.
ii) We *cannot* add/subtract a 2×3 matrix (2 rows) and a 3×3 matrix (3 rows). This is because of the different numbers of rows.
iii) We *cannot* add/subtract a 2×3 matrix and a 3×2 matrix. This is because neither the numbers of rows nor columns correspond.

When we add/subtract matrices we do this with corresponding elements in the matrices. In this chapter we are going to concentrate on 3×3 matrices, but the principles apply generally. Adding/subtracting 3×3 matrices means the answer will be another 3×3 matrix.

▚▚▚▚▚▚ **Examples 1.12** ▚▚▚▚▚▚▚▚▚▚▚▚▚▚▚▚▚▚▚▚▚▚▚▚

If $A = \begin{pmatrix} 1 & -2 & 12 \\ 8 & 9 & 7 \\ 3 & -4 & -6 \end{pmatrix}$, $B = \begin{pmatrix} -3 & 4 & 6 \\ 15 & 2 & 0 \\ 18 & 5 & 1 \end{pmatrix}$ and $C = \begin{pmatrix} -8 & 4 & 5 \\ 10 & 11 & -1 \\ -9 & 8 & 0 \end{pmatrix}$

write the following as one simplified 3×3 matrix

i) $A + B$,   ii) $A - C$,   iii) $A + B + C$,   iv) $A - B - C$.

Notice how the matrices are represented by capital letters in bold type. This is due to their connection with vectors.

i)   For $A + B = \begin{pmatrix} 1 & -2 & 12 \\ 8 & 9 & 7 \\ 3 & -4 & -6 \end{pmatrix} + \begin{pmatrix} -3 & 4 & 6 \\ 15 & 2 & 0 \\ 18 & 5 & 1 \end{pmatrix}$

we add $\begin{pmatrix} \text{row 1, col 1} & \text{row 1, col 2} & \text{row 1, col 3} \\ \text{row 2, col 1} & \text{row 2, col 2} & \text{row 2, col 3} \\ \text{row 3, col 1} & \text{row 3, col 2} & \text{row 3, col 3} \end{pmatrix}$

to get $\begin{pmatrix} 1 + (-3) & -2 + 4 & 12 + 6 \\ 8 + 15 & 9 + 2 & 7 + 0 \\ 3 + 18 & -4 + 5 & -6 + 1 \end{pmatrix}$

$= \begin{pmatrix} -2 & 2 & 18 \\ 23 & 11 & 7 \\ 21 & 1 & -5 \end{pmatrix}$.

ii)   For $A - C = \begin{pmatrix} 1 & -2 & 12 \\ 8 & 9 & 7 \\ 3 & -4 & -6 \end{pmatrix} - \begin{pmatrix} -8 & 4 & 5 \\ 10 & 11 & -1 \\ -9 & 8 & 0 \end{pmatrix}$

we subtract $\begin{pmatrix} \text{row 1, col 1} & \text{row 1, col 2} & \text{row 1, col 3} \\ \text{row 2, col 1} & \text{row 2, col 2} & \text{row 2, col 3} \\ \text{row 3, col 1} & \text{row 3, col 2} & \text{row 3, col 3} \end{pmatrix}$

to get $\begin{pmatrix} 1 - (-8) & -2 - 4 & 12 - 5 \\ 8 - 10 & 9 - 11 & 7 - (-1) \\ 3 - (-9) & -4 - 8 & -6 - 0 \end{pmatrix}$

$\boxed{(-)(-) = +.}$

$= \begin{pmatrix} 9 & -6 & 7 \\ -2 & -2 & 8 \\ 12 & -12 & -6 \end{pmatrix}$.

iii)   For $A + B + C$, already we have part of this result for $A + B$.

$A + B + C = \begin{pmatrix} -2 & 2 & 18 \\ 23 & 11 & 7 \\ 21 & 1 & -5 \end{pmatrix} + \begin{pmatrix} -8 & 4 & 5 \\ 10 & 11 & -1 \\ -9 & 8 & 0 \end{pmatrix}$

$= \begin{pmatrix} -2 + (-8) & 2 + 4 & 18 + 5 \\ 23 + 10 & 11 + 11 & 7 + (-1) \\ 21 + (-9) & 1 + 8 & -5 + 0 \end{pmatrix}$

$= \begin{pmatrix} -10 & 6 & 23 \\ 33 & 22 & 6 \\ 12 & 9 & -5 \end{pmatrix}$.

Alternatively we could have completed the calculation in one move as

$A + B + C$

$$= \begin{pmatrix} 1 & -2 & 12 \\ 8 & 9 & 7 \\ 3 & -4 & -6 \end{pmatrix} + \begin{pmatrix} -3 & 4 & 6 \\ 15 & 2 & 0 \\ 18 & 5 & 1 \end{pmatrix} + \begin{pmatrix} -8 & 4 & 5 \\ 10 & 11 & -1 \\ -9 & 8 & 0 \end{pmatrix}$$

$$= \begin{pmatrix} 1 + (-3) + (-8) & -2 + 4 + 4 & 12 + 6 + 5 \\ 8 + \quad 15 + 10 & 9 + 2 + 11 & 7 + 0 + (-1) \\ 3 + \quad 18 + (-9) & -4 + 5 + 8 & -6 + 1 + 0 \end{pmatrix}$$

$$= \begin{pmatrix} -10 & 6 & 23 \\ 33 & 22 & 6 \\ 12 & 9 & -5 \end{pmatrix}.$$

iv) We may write $A - B - C$ as $A - C - B$ or $(A - C) - B$ and use our result for $A - C$.

Then $(A - C) - B = \begin{pmatrix} 9 & -6 & 7 \\ -2 & -2 & 8 \\ 12 & -12 & -6 \end{pmatrix} - \begin{pmatrix} -3 & 4 & 6 \\ 15 & 2 & 0 \\ 18 & 5 & 1 \end{pmatrix}.$

$$= \begin{pmatrix} 9 - (-3) & (-6) - 4 & 7 - 6 \\ -2 - 15 & -2 - 2 & 8 - 0 \\ 12 - 18 & -12 - 5 & -6 - 1 \end{pmatrix}$$

$$= \begin{pmatrix} 12 & -10 & 1 \\ -17 & -4 & 8 \\ -6 & -17 & -7 \end{pmatrix}$$

Alternatively we could have completed the calculation in one move, as $A - B - C$. This is left as an exercise for you to complete.

There are two other simple arithmetic operations we can perform with matrices. Each one is the reverse process of the other. For the first one we can multiply a matrix by a scalar, i.e. **scalar multiplication**. For this process each element in the matrix is multiplied by the scalar.

Example 1.13

If $\boldsymbol{B} = \begin{pmatrix} -3 & 4 & 6 \\ 15 & 2 & 0 \\ 18 & 5 & 1 \end{pmatrix}$ then we can write down the matrix for $4\boldsymbol{B}$.

4 is the scalar multiplier.

Then $\quad 4\boldsymbol{B} = 4 \begin{pmatrix} -3 & 4 & 6 \\ 15 & 2 & 0 \\ 18 & 5 & 1 \end{pmatrix}$

$$= \begin{pmatrix} 4 \times -3 & 4 \times 4 & 4 \times 6 \\ 4 \times 15 & 4 \times 2 & 4 \times 0 \\ 4 \times 18 & 4 \times 5 & 4 \times 1 \end{pmatrix}$$

$$= \begin{pmatrix} -12 & 16 & 24 \\ 60 & 8 & 0 \\ 72 & 20 & 4 \end{pmatrix}.$$

The reverse process to scalar multiplication is removing a common factor. We must remove that factor from *all* the elements.

If possible, remove a factor from the matrices

i) $\begin{pmatrix} -3 & 6 & 39 \\ 9 & 27 & 3 \\ 15 & 21 & 3 \end{pmatrix}$, ii) $\begin{pmatrix} -3 & 6 & 39 \\ 9 & 27 & 3 \\ 4 & 21 & 0 \end{pmatrix}$.

i) In $\begin{pmatrix} -3 & 6 & 39 \\ 9 & 27 & 3 \\ 15 & 21 & 3 \end{pmatrix}$ each term has a common factor of 3.

We may write the matrix as the alternative $3 \begin{pmatrix} -1 & 2 & 13 \\ 3 & 9 & 1 \\ 5 & 7 & 1 \end{pmatrix}$.

ii) In this matrix we could remove a common factor of 3 from most of the elements. The element 4 is not an obvious multiple of 3. We would not benefit from removing that factor of 3. If it was necessary we might do so, though this would introduce a fraction as

$3 \begin{pmatrix} -1 & 2 & 13 \\ 3 & 9 & 1 \\ \frac{4}{3} & 7 & 0 \end{pmatrix}$.

■■■■ EXERCISE 1.4 ■■■■■■■■

$$\text{Let } A = \begin{pmatrix} 2 & 6 & 12 \\ 5 & 9 & -7 \\ 3 & 1 & -8 \end{pmatrix}, \; B = \begin{pmatrix} 7 & -3 & -7 \\ 0 & 15 & -5 \\ 0 & -4 & 16 \end{pmatrix}, \; C = \begin{pmatrix} 4 & -6 & 20 \\ 11 & 2 & \frac{1}{2} \\ -2 & 5 & 14 \end{pmatrix},$$

$$D = \begin{pmatrix} -10 & 13 & 17 \\ -7 & 8 & -1 \\ 25 & 30 & 10 \end{pmatrix} \quad \text{and} \quad Z = \begin{pmatrix} 0 & 0 & 0 \\ 0 & 0 & 0 \\ 0 & 0 & 0 \end{pmatrix}.$$

For the following matrix additions and subtractions give each answer as a single 3 × 3 matrix.

1 $A + B$	9 $2A + B + Z$
2 $A - B$	10 $\frac{1}{2}C + D - Z$
3 $A + B + C$	11 $3Z + B - D$
4 $A - B + D$	12 $A + 2B - 3C$
5 $A - B - D$	13 $C - \frac{1}{2}A$
6 $B - A$	14 $5D - 3C + B$
7 $B + A$	15 $B - A + 2C - D$
8 $2D - 3C$	

In Exercise 1.4 we have demonstrated two simple relationships. As might be expected, the answers to questions 1 and 7 are the same. The order of addition does not affect the answer. Generally we may write

$$A + B = B + A.$$

This says that matrix addition is **commutative**, or that matrix addition obeys the **commutative law**.

Answers to questions 2 and 6 differ, but only by a minus sign. The order of subtraction is important. Generally we may write

$$A - B \neq B - A, \quad \text{or} \quad A - B = -(B - A).$$

Matrix subtraction is *not* commutative, i.e. matrix subtraction does *not* obey the commutative law.

Matrix multiplication

The order in which we multiply matrices is important. Generally matrix multiplication is *not* commutative, though there are a few exceptions, i.e. $AB \neq BA$.

The size (i.e. order) of the matrices is also vital. Suppose we have 2 matrices, A and B, and wish to multiply them together as AB. This is possible only if the number of columns of *A* is the same as the number of rows of *B*, i.e. multiplying *AB* we need matrices $m \times n$ and $n \times p$ noticing the same *ns*. Our answer will be an $m \times p$ matrix. The values for *m* and *p* do *not* matter. If they are the same we can also find *BA*. If they are different we will be *unable* to find *BA*.

In this next set of examples we look at which matrices we may multiply together.

Examples 1.15

$A = \begin{pmatrix} 2 & 3 & 1 \\ 0 & -5 & 6 \end{pmatrix}$ is a 2 × 3 matrix,

$B = \begin{pmatrix} 4 & 0 & 1 \\ 0 & 0 & -4 \\ 2 & 0 & 7.5 \\ 1 & -7 & 6 \end{pmatrix}$ is a 4 × 3 matrix,

$C = \begin{pmatrix} 5 \\ -2 \\ 0 \end{pmatrix}$ is a 3 × 1 matrix.

Remember the column and row rule. The number of columns of the first matrix must be the same as the number of rows of the second matrix. Now we can decide which matrices we might be able to multiply together. Matrices *A* and *B* each have 3 columns, while matrix *C* has 3 rows. We may multiply *AC* and *BC*. You might like to check for yourself that there are no other possible multiplications.

For the rest of this chapter we are going to use 3 × 3 and 3 × 1 matrices. Then we will use them to solve simultaneous linear equations.

The general idea when multiplying matrices is to multiply particular elements and add those results. Remember from Volume 1, we multiply row elements from the first matrix with column elements from the second matrix. Let us have a look at some general cases before the numerical examples.

Suppose we have $A = \begin{pmatrix} a_1 & b_1 & c_1 \\ e_1 & f_1 & g_1 \\ k_1 & l_1 & m_1 \end{pmatrix}$ and $B = \begin{pmatrix} a_2 \\ e_2 \\ k_2 \end{pmatrix}$.

A is a 3 × 3 matrix and

> Number of columns is 3.

B is a 3 × 1 matrix.

> Number of rows is 3.

The column and row rule allows us to multiply *AB*. The matrix *AB* will be of order 3 × 1. These are the number of rows from *A* and the number of columns form *B*. We will look at *AB* in stages:

First row
$$\begin{pmatrix} a_1 & b_1 & c_1 \end{pmatrix} \begin{pmatrix} a_2 \\ e_2 \\ k_2 \end{pmatrix} = \begin{pmatrix} a_1a_2 + b_1e_2 + c_1k_2 \end{pmatrix}$$

Second row
$$\begin{pmatrix} e_1 & f_1 & g_1 \end{pmatrix} \begin{pmatrix} a_2 \\ e_2 \\ k_2 \end{pmatrix} = \begin{pmatrix} e_1a_2 + f_1e_2 + g_1k_2 \end{pmatrix}$$

Third row
$$\begin{pmatrix} & & \\ k_1 & l_1 & m_1 \end{pmatrix} \begin{pmatrix} a_2 \\ e_2 \\ k_2 \end{pmatrix} = \begin{pmatrix} & \\ k_1a_2 + l_1e_2 + m_1k_2 \end{pmatrix}.$$

We link together these stages to give a complete matrix solution of

$$AB = \begin{pmatrix} a_1 & b_1 & c_1 \\ e_1 & f_1 & g_1 \\ k_1 & l_1 & m_1 \end{pmatrix} \begin{pmatrix} a_2 \\ e_2 \\ k_2 \end{pmatrix} = \begin{pmatrix} a_1a_2 + b_1e_2 + c_1k_2 \\ e_1a_2 + f_1e_2 + g_1k_2 \\ k_1a_2 + l_1e_2 + m_1k_2 \end{pmatrix}.$$

Look closely at the final matrix. You should notice the pattern of letters and subscripts. It is consistent with the original matrices.

Examples 1.16

Multiply together the matrices

i) $\begin{pmatrix} 1 & 2 & 5 \\ 3 & 4 & 6 \\ 9 & 7 & 8 \end{pmatrix} \begin{pmatrix} 0 \\ -1 \\ -3 \end{pmatrix}$, ii) $\begin{pmatrix} -3 & 14 & 8 \\ -7 & -2 & 1 \\ 0 & -1 & 4 \end{pmatrix} \begin{pmatrix} -4 \\ 5 \\ 6 \end{pmatrix}.$

We use the rules involving multiplication and addition that we have just looked at generally.

i)
$$\begin{pmatrix} 1 & 2 & 5 \\ 3 & 4 & 6 \\ 9 & 7 & 8 \end{pmatrix} \begin{pmatrix} 0 \\ -1 \\ -3 \end{pmatrix} = \begin{pmatrix} 1 \times 0 + 2 \times (-1) + 5 \times (-3) \\ 3 \times 0 + 4 \times (-1) + 6 \times (-3) \\ 9 \times 0 + 7 \times (-1) + 8 \times (-3) \end{pmatrix}$$

$$= \begin{pmatrix} -17 \\ -22 \\ -31 \end{pmatrix}.$$

ii)
$$\begin{pmatrix} -3 & 14 & 8 \\ -7 & -2 & 1 \\ 0 & -1 & 4 \end{pmatrix} \begin{pmatrix} -4 \\ 5 \\ 6 \end{pmatrix} = \begin{pmatrix} -3 \times (-4) + 14 \times 5 + 8 \times 6 \\ -7 \times (-4) + -2 \times 5 + 1 \times 6 \\ 0 \times (-4) + -1 \times 5 + 4 \times 6 \end{pmatrix}$$

$$= \begin{pmatrix} 130 \\ 24 \\ 19 \end{pmatrix}.$$

You need to look back at these examples. Notice how the original column and row rule allows us to multiply these matrices. Look at the pattern of numbers in the working of the calculation. Finally check that the first matrix had 3 rows and the second matrix had 1 column. This means our answer must have, and does have, 3 rows and 1 column.

Let us look at two more general matrices,

$$A = \begin{pmatrix} a_1 & b_1 & c_1 \\ e_1 & f_1 & g_1 \\ k_1 & l_1 & m_1 \end{pmatrix} \quad \text{and} \quad C = \begin{pmatrix} a_2 & b_2 & c_2 \\ e_2 & f_2 & g_2 \\ k_2 & l_2 & m_2 \end{pmatrix}.$$

We can check out the column and row rule to allow us to attempt the multiplication. Also, we know that AC will be a 3×3 matrix. In stages, we have:

First row, first column

$$\begin{pmatrix} a_1 & b_1 & c_1 \end{pmatrix} \begin{pmatrix} a_2 \\ e_2 \\ k_2 \end{pmatrix} = \begin{pmatrix} a_1 a_2 + b_1 e_2 + c_1 k_2 \end{pmatrix}$$

First row, second column

$$\begin{pmatrix} a_1 & b_1 & c_1 \end{pmatrix} \begin{pmatrix} b_2 \\ f_2 \\ l_2 \end{pmatrix} = \begin{pmatrix} a_1 b_2 + b_1 f_2 + c_1 l_2 \end{pmatrix}$$

First row, third column

$$\begin{pmatrix} a_1 & b_1 & c_1 \end{pmatrix} \begin{pmatrix} c_2 \\ g_2 \\ m_2 \end{pmatrix} = \begin{pmatrix} a_1 c_2 + b_1 g_2 + c_1 m_2 \end{pmatrix}$$

Second row, first column

$$\begin{pmatrix} e_1 & f_1 & g_1 \end{pmatrix} \begin{pmatrix} a_2 \\ e_2 \\ k_2 \end{pmatrix} = \begin{pmatrix} e_1 a_2 + f_1 e_2 + g_1 k_2 \end{pmatrix}$$

Second row, second column

$$\begin{pmatrix} e_1 & f_1 & g_1 \end{pmatrix} \begin{pmatrix} b_2 \\ f_2 \\ l_2 \end{pmatrix} = \begin{pmatrix} e_1 b_2 + f_1 f_2 + g_1 l_2 \end{pmatrix}$$

Second row, third column

$$\begin{pmatrix} e_1 & f_1 & g_1 \end{pmatrix} \begin{pmatrix} c_2 \\ g_2 \\ m_2 \end{pmatrix} = \begin{pmatrix} e_1 c_2 + f_1 g_2 + g_1 m_2 \end{pmatrix}$$

Third row, first column

$$\begin{pmatrix} & & \\ k_1 & l_1 & m_1 \end{pmatrix} \begin{pmatrix} a_2 \\ e_2 \\ k_2 \end{pmatrix} = \begin{pmatrix} \\ k_1a_2 + l_1e_2 + m_1k_2 \end{pmatrix}$$

Third row, second column

$$\begin{pmatrix} & & \\ k_1 & l_1 & m_1 \end{pmatrix} \begin{pmatrix} b_2 \\ f_2 \\ l_2 \end{pmatrix} = \begin{pmatrix} \\ k_1b_2 + l_1f_2 + m_1l_2 \end{pmatrix}$$

Third row, third column

$$\begin{pmatrix} & & \\ k_1 & l_1 & m_1 \end{pmatrix} \begin{pmatrix} c_2 \\ g_2 \\ m_2 \end{pmatrix} = \begin{pmatrix} \\ k_1c_2 + l_1g_2 + m_1m_2 \end{pmatrix}$$

Again, we can bring together these stages for one complete solution. In the next set of examples we work in stages and then gather together our complete solutions.

Examples 1.17

Multiply together the matrices

i) $\begin{pmatrix} 1 & 2 & -3 \\ 3 & 4 & -7 \\ 5 & -8 & 11 \end{pmatrix} \begin{pmatrix} 14 & 0 & -1 \\ 9 & 6 & -2 \\ -4 & 7 & 10 \end{pmatrix}$,

ii) $\begin{pmatrix} 14 & 0 & -1 \\ 9 & 6 & -2 \\ -4 & 7 & 10 \end{pmatrix} \begin{pmatrix} 1 & 2 & -3 \\ 3 & 4 & -7 \\ 5 & -8 & 11 \end{pmatrix}$

We look at each example in stages, bringing those stages together as a complete solution.

i) In stages, we have:

First row, first column

$$\begin{pmatrix} 1 & 2 & -3 \end{pmatrix} \begin{pmatrix} 14 \\ 9 \\ -4 \end{pmatrix} = \begin{pmatrix} 14 + 18 + 12 \end{pmatrix}$$

First row, second column

$$\begin{pmatrix} 1 & 2 & -3 \end{pmatrix} \begin{pmatrix} 0 \\ 6 \\ 7 \end{pmatrix} = \begin{pmatrix} 0 + 12 - 21 \end{pmatrix}$$

First row, third column

$$
\begin{pmatrix} 1 & 2 & -3 \\ & & \\ & & \end{pmatrix}
\begin{pmatrix} & & -1 \\ & & -2 \\ & & 10 \end{pmatrix}
=
\begin{pmatrix} & & -1-4-30 \\ & & \\ & & \end{pmatrix}
$$

Second row, first column

$$
\begin{pmatrix} & & \\ 3 & 4 & -7 \\ & & \end{pmatrix}
\begin{pmatrix} 14 \\ 9 \\ -4 \end{pmatrix}
=
\begin{pmatrix} \\ 42+36+28 \\ \end{pmatrix}
$$

Second row, second column

$$
\begin{pmatrix} & & \\ 3 & 4 & -7 \\ & & \end{pmatrix}
\begin{pmatrix} 0 \\ 6 \\ 7 \end{pmatrix}
=
\begin{pmatrix} \\ 0+24-49 \\ \end{pmatrix}
$$

Second row, third column

$$
\begin{pmatrix} & & \\ 3 & 4 & -7 \\ & & \end{pmatrix}
\begin{pmatrix} -1 \\ -2 \\ 10 \end{pmatrix}
=
\begin{pmatrix} \\ -3-8-70 \\ \end{pmatrix}
$$

Third row, first column

$$
\begin{pmatrix} & & \\ & & \\ 5 & -8 & 11 \end{pmatrix}
\begin{pmatrix} 14 \\ 9 \\ -4 \end{pmatrix}
=
\begin{pmatrix} \\ \\ 70-72-44 \end{pmatrix}
$$

Third row, second column

$$
\begin{pmatrix} & & \\ & & \\ 5 & -8 & 11 \end{pmatrix}
\begin{pmatrix} 0 \\ 6 \\ 7 \end{pmatrix}
=
\begin{pmatrix} \\ \\ 0-48+77 \end{pmatrix}
$$

Third row, third column

$$
\begin{pmatrix} & & \\ & & \\ 5 & -8 & 11 \end{pmatrix}
\begin{pmatrix} -1 \\ -2 \\ 10 \end{pmatrix}
=
\begin{pmatrix} \\ \\ -5+16+110 \end{pmatrix}
$$

Bringing together all these stages gives us

$$
\begin{pmatrix} 1 & 2 & -3 \\ 3 & 4 & -7 \\ 5 & -8 & 11 \end{pmatrix}
\begin{pmatrix} 14 & 0 & -1 \\ 9 & 6 & -2 \\ -4 & 7 & 10 \end{pmatrix}
=
\begin{pmatrix} 44 & -9 & -35 \\ 106 & -25 & -81 \\ -46 & 29 & 121 \end{pmatrix}.
$$

ii) In stages we have:

First row, first column

$$\begin{pmatrix} 14 & 0 & -1 \\ & & \end{pmatrix} \begin{pmatrix} 1 \\ 3 \\ 5 \end{pmatrix} = \begin{pmatrix} 14 + 0 - 5 \\ \\ \end{pmatrix}$$

First row, second column

$$\begin{pmatrix} 14 & 0 & -1 \\ & & \end{pmatrix} \begin{pmatrix} 2 \\ 4 \\ -8 \end{pmatrix} = \begin{pmatrix} 28 + 0 + 8 \\ \\ \end{pmatrix}$$

First row, third column

$$\begin{pmatrix} 14 & 0 & -1 \\ & & \end{pmatrix} \begin{pmatrix} -3 \\ -7 \\ 11 \end{pmatrix} = \begin{pmatrix} -42 + 0 - 11 \\ \\ \end{pmatrix}$$

Second row, first column

$$\begin{pmatrix} 9 & 6 & -2 \\ & & \end{pmatrix} \begin{pmatrix} 1 \\ 3 \\ 5 \end{pmatrix} = \begin{pmatrix} 9 + 18 - 10 \\ \\ \end{pmatrix}$$

Second row, second column

$$\begin{pmatrix} 9 & 6 & -2 \\ & & \end{pmatrix} \begin{pmatrix} 2 \\ 4 \\ -8 \end{pmatrix} = \begin{pmatrix} 18 + 24 + 16 \\ \\ \end{pmatrix}$$

Second row, third column

$$\begin{pmatrix} 9 & 6 & -2 \\ & & \end{pmatrix} \begin{pmatrix} -3 \\ -7 \\ 11 \end{pmatrix} = \begin{pmatrix} -27 - 42 - 22 \\ \\ \end{pmatrix}$$

Third row, first column

$$\begin{pmatrix} & & \\ -4 & 7 & 10 \end{pmatrix} \begin{pmatrix} 1 \\ 3 \\ 5 \end{pmatrix} = \begin{pmatrix} \\ -4 + 21 + 50 \end{pmatrix}$$

Third row, second column

$$\begin{pmatrix} & & \\ -4 & 7 & 10 \end{pmatrix} \begin{pmatrix} 2 \\ 4 \\ -8 \end{pmatrix} = \begin{pmatrix} \\ -8 + 28 - 80 \end{pmatrix}$$

Third row, third column

$$\begin{pmatrix} & & \\ & & \\ -4 & 7 & 10 \end{pmatrix}\begin{pmatrix} & -3 & \\ & -7 & \\ & 11 & \end{pmatrix} = \begin{pmatrix} & & \\ & & \\ & & 12-49+110 \end{pmatrix}$$

Bringing together all these stages gives us

$$\begin{pmatrix} 14 & 0 & -1 \\ 9 & 6 & -2 \\ -4 & 7 & 10 \end{pmatrix}\begin{pmatrix} 1 & 2 & -3 \\ 3 & 4 & -7 \\ 5 & -8 & 11 \end{pmatrix} = \begin{pmatrix} 9 & 36 & -53 \\ 17 & 58 & -91 \\ 67 & -60 & 73 \end{pmatrix}.$$

Now look back over these examples, noticing how the original matrices have been written. Because the answers are different they demonstrate the commutative law failing, $AB \neq BA$, which we mentioned earlier.

Most examples of matrix multiplication do *not* obey the commutative law. However, there is one exception. It involves the **unit** matrix.

Examples 1.18

Multiply together the matrices

i) $\begin{pmatrix} 5 & -7 & 9 \\ 2 & 10 & 4 \\ -3 & 6 & 3 \end{pmatrix}\begin{pmatrix} 1 & 0 & 0 \\ 0 & 1 & 0 \\ 0 & 0 & 1 \end{pmatrix}$, ii) $\begin{pmatrix} 1 & 0 & 0 \\ 0 & 1 & 0 \\ 0 & 0 & 1 \end{pmatrix}\begin{pmatrix} 5 & -7 & 9 \\ 2 & 10 & 4 \\ -3 & 6 & 3 \end{pmatrix}.$

Remember the pattern of rows and columns as we multiply together these 3×3 matrices.

i) $\begin{pmatrix} 5 & -7 & 9 \\ 2 & 10 & 4 \\ -3 & 6 & 3 \end{pmatrix}\begin{pmatrix} 1 & 0 & 0 \\ 0 & 1 & 0 \\ 0 & 0 & 1 \end{pmatrix} =$

$$\begin{pmatrix} 5\times1- 7\times0+9\times0 & 5\times0- 7\times1+9\times0 & 5\times0- 7\times0+9\times1 \\ 2\times1+ 10\times0+4\times0 & 2\times0+ 10\times1+4\times0 & 2\times0+ 10\times0+4\times1 \\ -3\times1+ 6\times0+3\times0 & -3\times0+ 6\times1+3\times0 & -3\times0+ 6\times0+3\times1 \end{pmatrix}$$

$$= \begin{pmatrix} 5 & -7 & 9 \\ 2 & 10 & 4 \\ -3 & 6 & 3 \end{pmatrix}.$$

ii) $\begin{pmatrix} 1 & 0 & 0 \\ 0 & 1 & 0 \\ 0 & 0 & 1 \end{pmatrix}\begin{pmatrix} 5 & -7 & 9 \\ 2 & 10 & 4 \\ -3 & 6 & 3 \end{pmatrix} =$

$$\begin{pmatrix} 1\times5+0\times2+0\times(-3) & 1\times(-7)+0\times10+0\times6 & 1\times9+0\times4+0\times3 \\ 0\times5+1\times2+0\times(-3) & 0\times(-7)+1\times10+0\times6 & 0\times9+1\times4+0\times3 \\ 0\times5+0\times2+1\times(-3) & 0\times(-7)+0\times10+1\times6 & 0\times9+0\times4+1\times3 \end{pmatrix}$$

$$= \begin{pmatrix} 5 & -7 & 9 \\ 2 & 10 & 4 \\ -3 & 6 & 3 \end{pmatrix}.$$

The pattern of matrix multiplication and the answers apply to any order of matrices. Suppose I is the unit matrix and A is any other matrix of the correct order. Then,

$$AI = IA = A.$$

We can extend our multiplication beyond just two matrices. Our next example looks at three matrices using an earlier example.

Example 1.19

Multiply together the three matrices

$$\begin{pmatrix} 0 & 1 & -3 \\ 11 & -2 & 3 \\ 5 & -6 & 2 \end{pmatrix} \begin{pmatrix} 1 & 2 & 5 \\ 3 & 4 & 6 \\ 9 & 7 & 8 \end{pmatrix} \begin{pmatrix} 0 \\ -1 \\ -3 \end{pmatrix}.$$

We multiplied together the second and third matrices in Examples 1.16. Using that result we simplify our matrices to

$$\begin{pmatrix} 0 & 1 & -3 \\ 11 & -2 & 3 \\ 5 & -6 & 2 \end{pmatrix} \begin{pmatrix} 1 & 2 & 5 \\ 3 & 4 & 6 \\ 9 & 7 & 8 \end{pmatrix} \begin{pmatrix} 0 \\ -1 \\ -3 \end{pmatrix} = \begin{pmatrix} 0 & 1 & -3 \\ 11 & -2 & 3 \\ 5 & -6 & 2 \end{pmatrix} \begin{pmatrix} -17 \\ -22 \\ -31 \end{pmatrix}$$

$$= \begin{pmatrix} 0 - 22 + 93 \\ -187 + 44 - 93 \\ -85 + 132 - 62 \end{pmatrix}$$

$$= \begin{pmatrix} 71 \\ -236 \\ -15 \end{pmatrix}.$$

EXERCISE 1.5

Multiply together the following matrices.

1 $\begin{pmatrix} 2 & 6 & -1 \\ 4 & 3 & 2 \\ -5 & 0 & 11 \end{pmatrix} \begin{pmatrix} 1 \\ 0 \\ 7 \end{pmatrix}$

3 $\begin{pmatrix} 0 & -5 & 0 \\ 6 & 0 & -1 \\ \frac{1}{2} & -2 & 4 \end{pmatrix} \begin{pmatrix} 10 \\ -3 \\ 12 \end{pmatrix}$

2 $\begin{pmatrix} 3 & 0 & -2 \\ -6 & 5 & 7 \\ 8 & 1 & -4 \end{pmatrix} \begin{pmatrix} 1 \\ 2 \\ 3 \end{pmatrix}$

4 $\begin{pmatrix} -2 & 5 & -3 \\ 6 & 8 & \frac{1}{2} \\ 10 & 9 & 15 \end{pmatrix} \begin{pmatrix} 4 \\ \frac{1}{2} \\ 1 \end{pmatrix}$

5 $\begin{pmatrix} 5 & 2 & -3 \\ 1 & 6 & 4 \\ -1 & -8 & 10 \end{pmatrix} \begin{pmatrix} 2 & 3 & -6 \\ 12 & 1 & 11 \\ 0 & 0 & 7 \end{pmatrix}$

6 $\begin{pmatrix} 2 & -1 & 4 \\ 0 & -2 & 3 \\ 0 & 15 & 2 \end{pmatrix} \begin{pmatrix} 5 & 0 & 7 \\ 7 & 11 & 6 \\ 1 & 5 & 0 \end{pmatrix}$

7 $\begin{pmatrix} 1 & 2 & 11 \\ 0 & 2 & -1 \\ -5 & 0 & 3 \end{pmatrix} \begin{pmatrix} 5 & 3 & -2 \\ -2 & 4 & -2 \\ -4 & 3 & 10 \end{pmatrix}$

8 $\begin{pmatrix} 10 & 8 & 0 \\ 9 & 0 & 8 \\ -1 & 0 & 15 \end{pmatrix} \begin{pmatrix} -3 \\ 1 \\ 2 \end{pmatrix}$

9 $\begin{pmatrix} 1 & 6 & -5 \\ 0 & 2 & -1 \\ 3 & 0 & 4 \end{pmatrix} \begin{pmatrix} 10 & 8 & 0 \\ 9 & 0 & 8 \\ -1 & 0 & 15 \end{pmatrix} \begin{pmatrix} -3 \\ 1 \\ 2 \end{pmatrix}$

10 $\begin{pmatrix} 3 & -6 & 8 \\ 2 & 4 & -5 \\ 1 & 0 & 1 \end{pmatrix} \begin{pmatrix} -2 & 6 & 10 \\ 0 & 6 & \frac{1}{2} \\ 3 & 8 & 15 \end{pmatrix} \begin{pmatrix} 2 \\ 5 \\ -1 \end{pmatrix}$

Determinant of a matrix

In the introduction we mentioned it is possible to find the determinant of a matrix. Remember a determinant has the same number of rows as columns. This means we can only find the determinant of a square matrix. Simply, if $A = \begin{pmatrix} a_1 & b_1 & c_1 \\ e_1 & f_1 & g_1 \\ k_1 & l_1 & m_1 \end{pmatrix}$ then $\det A = \begin{vmatrix} a_1 & b_1 & c_1 \\ e_1 & f_1 & g_1 \\ k_1 & l_1 & m_1 \end{vmatrix}$. The usual calculation rules apply in the calculation of $\det A$.

▓▓▓▓▓▓ **Example 1.20** ▓▓▓▓▓▓

Calculate the determinant of the matrix $\begin{pmatrix} 1 & -2 & 12 \\ 8 & 9 & 7 \\ 3 & -4 & -6 \end{pmatrix}$.

$$\Delta = \begin{vmatrix} 1 & -2 & 12 \\ 8 & 9 & 7 \\ 3 & -4 & -6 \end{vmatrix} = 1\begin{vmatrix} 9 & 7 \\ -4 & -6 \end{vmatrix} - -2\begin{vmatrix} 8 & 7 \\ 3 & -6 \end{vmatrix} + 12\begin{vmatrix} 8 & 9 \\ 3 & -4 \end{vmatrix}$$

$$= 1(-54 - -28) + 2(-48 - 21) + 12(-32 - 27)$$
$$= 1(-26) + 2(-69) + 12(-59)$$
$$= -26 - 138 - 708$$
$$= -872.$$

It is possible that the determinant of a matrix may have a value of 0. We need to note this case. It refers to a singular matrix.

A singular matrix has a determinant of zero value.

━━━━━ **Examples 1.21** ━━━━━━━━━━━━━━━━━━━━━━━━━

i) If $A = \begin{pmatrix} 2 & 3 & -1 \\ 1 & -1 & 1 \\ 2 & 3 & -1 \end{pmatrix}$ then $\det A = \begin{vmatrix} 2 & 3 & -1 \\ 1 & -1 & 1 \\ 2 & 3 & -1 \end{vmatrix} = 0.$

ii) If $B = \begin{pmatrix} 3 & -1 & 11 \\ 9 & -3 & 33 \\ -1 & 1 & 4 \end{pmatrix}$ then $\det B = \begin{vmatrix} 3 & -1 & 11 \\ 9 & -3 & 33 \\ -1 & 1 & 4 \end{vmatrix} = 0.$

We saw both these determinants in Examples 1.10. They come from sets of simultaneous equations that have no solution. We will meet this situation again when we look at the matrix solution of simultaneous equations.

Transpose of a matrix

When we interchange the rows and columns of a matrix we get the **transpose** of the original matrix.

If $A = \begin{pmatrix} a_1 & b_1 & c_1 \\ e_1 & f_1 & g_1 \\ k_1 & l_1 & m_1 \end{pmatrix}$ then we write the transpose of A as

$A^T = \begin{pmatrix} a_1 & e_1 & k_1 \\ b_1 & f_1 & l_1 \\ c_1 & g_1 & m_1 \end{pmatrix}.$

You can transpose any size of matrix; it does *not* have to be square. We concentrate on square matrices because we are working towards finding inverse matrices. In the next example we look at transposing a matrix and calculating its determinant.

━━━━━ **Example 1.22** ━━━━━━━━━━━━━━━━━━━━━━━━━

Write down the transpose of the matrix $\begin{pmatrix} 5 & 7 & 0 \\ 2 & 2 & 4 \\ 3 & -4 & 1 \end{pmatrix}.$

Also calculate the determinant of the transposed matrix.

If $A = \begin{pmatrix} 5 & 7 & 0 \\ 2 & 2 & 4 \\ 3 & -4 & 1 \end{pmatrix}$ then $A^T = \begin{pmatrix} 5 & 2 & 3 \\ 7 & 2 & -4 \\ 0 & 4 & 1 \end{pmatrix}.$

Also $\det A^T = \begin{vmatrix} 5 & 2 & 3 \\ 7 & 2 & -4 \\ 0 & 4 & 1 \end{vmatrix} = 5 \begin{vmatrix} 2 & -4 \\ 4 & 1 \end{vmatrix} - 2 \begin{vmatrix} 7 & -4 \\ 0 & 1 \end{vmatrix} + 3 \begin{vmatrix} 7 & 2 \\ 0 & 4 \end{vmatrix}$

$= 5(2 - -16) - 2(7 - 0) + 3(28 - 0)$

$= 5(18) - 2(7) + 3(28)$

$= 160.$

Remember that **writing the rows as columns and the columns as rows leaves the value of the determinant unchanged**. This means that the determinants of a matrix and its transpose are the same.

■■■■ EXERCISE 1.6 ■■■■

For the following matrices
i) write down their transposes, and
ii) calculate the determinants of the transposed matrices.

1 $\begin{pmatrix} 2 & -1 & 2 \\ 5 & 4 & 2 \\ 6 & 0 & 8 \end{pmatrix}$ 4 $\begin{pmatrix} 0 & 3 & -17 \\ 1 & 2 & 3 \\ 7 & 31 & -1 \end{pmatrix}$

2 $\begin{pmatrix} 0 & 11 & 5 \\ 1 & -2 & -6 \\ 2 & 4 & 7 \end{pmatrix}$ 5 $\begin{pmatrix} 5 & 2 & 6 \\ 11 & -6 & 5 \\ 1 & 6 & 4 \end{pmatrix}$

3 $\begin{pmatrix} 3 & 10 & 0 \\ 8 & -6 & 6 \\ 1 & 5 & 7 \end{pmatrix}$

Minors and cofactors of a matrix

We are working towards finding the inverse of a matrix. Again we concentrate on the 3×3 matrix.

Firstly, let us introduce the double suffix notation for the elements of the matrix, e.g. a_{13}. a_{13} means the element in row 1, column 3.

For our general 3×3 matrix suppose $A = \begin{pmatrix} a_{11} & a_{12} & a_{13} \\ a_{21} & a_{22} & a_{23} \\ a_{31} & a_{32} & a_{33} \end{pmatrix}$.

Let us find the minor of a_{11}. We strike out the row and column containing a_{11}, i.e. strike out row 1 and column 1. Whatever determinant remains is the minor, i.e. $\begin{vmatrix} a_{22} & a_{23} \\ a_{32} & a_{33} \end{vmatrix}$.

We can repeat this process for each element in turn. For the minor of a_{21} we strike out row 2 and column 1. What remains is $\begin{vmatrix} a_{12} & a_{13} \\ a_{32} & a_{33} \end{vmatrix}$.

For all our elements we can calculate the minors. This means that for our 3×3 matrix we have 9 minors.

The cofactor involves the minor together with a sign adjustment. The adjustment uses the addition of the subscripts. -1 is raised to the power of that addition.

The cofactor of a_{11} is $\quad (-1)^{(1+1)} \begin{vmatrix} a_{22} & a_{23} \\ a_{32} & a_{33} \end{vmatrix} = \begin{vmatrix} a_{22} & a_{23} \\ a_{32} & a_{33} \end{vmatrix},$

the cofactor of a_{12} is $\quad (-1)^{(1+2)} \begin{vmatrix} a_{21} & a_{23} \\ a_{31} & a_{33} \end{vmatrix} = - \begin{vmatrix} a_{21} & a_{23} \\ a_{31} & a_{33} \end{vmatrix},$

the cofactor of a_{13} is $\quad (-1)^{(1+3)} \begin{vmatrix} a_{21} & a_{22} \\ a_{31} & a_{32} \end{vmatrix} = \begin{vmatrix} a_{21} & a_{22} \\ a_{31} & a_{32} \end{vmatrix},$

the cofactor of a_{21} is $\quad (-1)^{(2+1)} \begin{vmatrix} a_{12} & a_{13} \\ a_{32} & a_{33} \end{vmatrix} = - \begin{vmatrix} a_{12} & a_{13} \\ a_{32} & a_{33} \end{vmatrix}.$

This pattern continues for all the elements of the matrix, ending with the element a_{33}.

The cofactor of a_{33} is $\quad (-1)^{(3+3)} \begin{vmatrix} a_{11} & a_{12} \\ a_{21} & a_{22} \end{vmatrix} = \begin{vmatrix} a_{11} & a_{12} \\ a_{21} & a_{22} \end{vmatrix}.$

You will notice here, and in later examples, that the different powers of -1 cause the $+/-$ signs to alternate.

Generally the cofactor of a_{ij} uses $(-1)^{(i+j)}$.

Cofactor of $a_{ij} = (-1)^{(i+j)} \times$ minor of a_{ij}.

Inverse of a matrix

Only a square matrix has an inverse. If the original matrix is A then the inverse matrix is A^{-1}. A^{-1} is just a symbol; do not write it in any other form. If we multiply together the original and inverse matrices in any order we get the unit matrix, I, i.e.

$$AA^{-1} = I = A^{-1}A.$$

This is a rare exception to the order of matrix multiplication being important.

Only a non-singular matrix has an inverse, i.e. the determinant of the matrix is non-zero. Hence, first find the value of the determinant when attempting to find the inverse of a matrix.

Before we attempt an example we can go through the stages in finding an inverse matrix.

Suppose we have a matrix, A.

Calculate det A. If det $A \neq 0$ then we can proceed.

For each element find each minor.

Now find each cofactor.

Write down the matrix of cofactors, say B.

Transpose B to get B^T. B^T is the adjoint matrix of A. $\boxed{B^T = \text{adj } A.}$

The inverse matrix A^{-1} is $A^{-1} = \dfrac{1}{\det A} \times \text{adj} A$.

Examples 1.23

Find the inverses of the matrices

i) $\begin{pmatrix} 1 & 2 & 0 \\ 9 & 12 & 6 \\ 4 & -3 & 3 \end{pmatrix}$, ii) $\begin{pmatrix} 2 & -7 & 5 \\ 3 & 4 & 8 \\ 0 & -1 & 1 \end{pmatrix}$.

i) Let $A = \begin{pmatrix} 1 & 2 & 0 \\ 9 & 12 & 6 \\ 4 & -3 & 3 \end{pmatrix}$.

We calculate the determinant of A,

i.e. $\det A = \begin{vmatrix} 1 & 2 & 0 \\ 9 & 12 & 6 \\ 4 & -3 & 3 \end{vmatrix} = 1\begin{vmatrix} 12 & 6 \\ -3 & 3 \end{vmatrix} - 2\begin{vmatrix} 9 & 6 \\ 4 & 3 \end{vmatrix} + 0\begin{vmatrix} 9 & 12 \\ 4 & -3 \end{vmatrix}$

$$= 48.$$

Now we find the minors and cofactors. Notice how we calculate them in an orderly way.

Element	Minor		Cofactor	
a_{11} is 1,	$\begin{vmatrix} 12 & 6 \\ -3 & 3 \end{vmatrix}$	$= 54,$	$(-1)^{(1+1)}54$	$= 54.$
a_{12} is 2,	$\begin{vmatrix} 9 & 6 \\ 4 & 3 \end{vmatrix}$	$= 3,$	$(-1)^{(1+2)}3$	$= -3.$
a_{13} is 0,	$\begin{vmatrix} 9 & 12 \\ 4 & -3 \end{vmatrix}$	$= -75,$	$(-1)^{(1+3)}(-75)$	$= -75.$
a_{21} is 9,	$\begin{vmatrix} 2 & 0 \\ -3 & 3 \end{vmatrix}$	$= 6,$	$(-1)^{(2+1)}6$	$= -6.$
a_{22} is 12,	$\begin{vmatrix} 1 & 0 \\ 4 & 3 \end{vmatrix}$	$= 3,$	$(-1)^{(2+2)}3$	$= 3.$
a_{23} is 6,	$\begin{vmatrix} 1 & 2 \\ 4 & -3 \end{vmatrix}$	$= -11,$	$(-1)^{(2+3)}(-11)$	$= 11.$
a_{31} is 4,	$\begin{vmatrix} 2 & 0 \\ 12 & 6 \end{vmatrix}$	$= 12,$	$(-1)^{(3+1)}12$	$= 12.$
a_{32} is -3,	$\begin{vmatrix} 1 & 0 \\ 9 & 6 \end{vmatrix}$	$= 6,$	$(-1)^{(3+2)}6$	$= -6.$
a_{33} is 3,	$\begin{vmatrix} 1 & 2 \\ 9 & 12 \end{vmatrix}$	$= -6,$	$(-1)^{(3+3)}(-6)$	$= -6.$

The matrix of cofactors is $\begin{pmatrix} 54 & -3 & -75 \\ -6 & 3 & 11 \\ 12 & -6 & -6 \end{pmatrix}$.

Transposing the matrix of cofactors we get

$$\text{Adj } A \text{ is } \begin{pmatrix} 54 & -6 & 12 \\ -3 & 3 & -6 \\ -75 & 11 & -6 \end{pmatrix}.$$

Hence $\quad A^{-1} = \dfrac{1}{48} \times \begin{pmatrix} 54 & -6 & 12 \\ -3 & 3 & -6 \\ -75 & 11 & -6 \end{pmatrix}.$

We could write this answer in an alternative way, applying the fraction to each element.

You ought to test this answer. Remember that $AA^{-1} = I$. When you multiply these matrices you should get

$$\frac{1}{48} \times \begin{pmatrix} 48 & 0 & 0 \\ 0 & 48 & 0 \\ 0 & 0 & 48 \end{pmatrix} = \begin{pmatrix} 1 & 0 & 0 \\ 0 & 1 & 0 \\ 0 & 0 & 1 \end{pmatrix}, \text{ removing a common factor to}$$

confirm your answer is correct.

ii) Let $A = \begin{pmatrix} 2 & -7 & 5 \\ 3 & 4 & 8 \\ 0 & -1 & 1 \end{pmatrix}.$

We calculate the determinant of A,

i.e. $\det A = \begin{vmatrix} 2 & -7 & 5 \\ 3 & 4 & 8 \\ 0 & -1 & 1 \end{vmatrix} = 2\begin{vmatrix} 4 & 8 \\ -1 & 1 \end{vmatrix} - -7\begin{vmatrix} 3 & 8 \\ 0 & 1 \end{vmatrix} + 5\begin{vmatrix} 3 & 4 \\ 0 & -1 \end{vmatrix}$

$$= 30.$$

Now we find the minors and cofactors. Notice how we calculate them in an orderly way.

Element	Minor		Cofactor	
a_{11} is 2,	$\begin{vmatrix} 4 & 8 \\ -1 & 1 \end{vmatrix}$	$= 12,$	$(-1)^{(1+1)}12$	$= 12.$
a_{12} is -7,	$\begin{vmatrix} 3 & 8 \\ 0 & 1 \end{vmatrix}$	$= 3,$	$(-1)^{(1+2)}3$	$= -3.$
a_{13} is 5,	$\begin{vmatrix} 3 & 4 \\ 0 & -1 \end{vmatrix}$	$= -3,$	$(-1)^{(1+3)}(-3)$	$= -3.$
a_{21} is 3,	$\begin{vmatrix} -7 & 5 \\ -1 & 1 \end{vmatrix}$	$= -2,$	$(-1)^{(2+1)}(-2)$	$= 2.$

a_{22} is 4, $\quad \begin{vmatrix} 2 & 5 \\ 0 & 1 \end{vmatrix} = 2, \quad (-1)^{(2+2)}3 = 2.$

a_{23} is 8, $\quad \begin{vmatrix} 2 & -7 \\ 0 & -1 \end{vmatrix} = -2, \quad (-1)^{(2+3)}(-2) = 2.$

a_{31} is 0, $\quad \begin{vmatrix} -7 & 5 \\ 4 & 8 \end{vmatrix} = -76, \quad (-1)^{(3+1)}(-76) = -76.$

a_{32} is -1, $\quad \begin{vmatrix} 2 & 5 \\ 3 & 8 \end{vmatrix} = 1, \quad (-1)^{(3+2)}1 = -1.$

a_{33} is 1, $\quad \begin{vmatrix} 2 & -7 \\ 3 & 4 \end{vmatrix} = 29, \quad (-1)^{(3+3)}29 = 29.$

The matrix of cofactors is $\begin{pmatrix} 12 & -3 & -3 \\ 2 & 2 & 2 \\ -76 & -1 & 29 \end{pmatrix}$

Transposing the matrix of cofactors we get

$$\text{Adj } A \text{ is } \begin{pmatrix} 12 & 2 & -76 \\ -3 & 2 & -1 \\ -3 & 2 & 29 \end{pmatrix}.$$

Hence $\qquad A^{-1} = \dfrac{1}{30} \times \begin{pmatrix} 12 & 2 & -76 \\ -3 & 2 & -1 \\ -3 & 2 & 29 \end{pmatrix}.$

You ought to test this answer. Remember that $AA^{-1} = I$. When you multiply these matrices you should get

$$\frac{1}{30} \times \begin{pmatrix} 30 & 0 & 0 \\ 0 & 30 & 0 \\ 0 & 0 & 30 \end{pmatrix} = \begin{pmatrix} 1 & 0 & 0 \\ 0 & 1 & 0 \\ 0 & 0 & 1 \end{pmatrix}, \text{ removing a common factor to}$$

confirm your answer is correct.

EXERCISE 1.7

Find the inverse of each matrix.

1 $\begin{pmatrix} 4 & 3 & 0 \\ 5 & -4 & 1 \\ 8 & -1 & 2 \end{pmatrix}$

2 $\begin{pmatrix} -6 & 0 & 2 \\ 17 & 1 & 6 \\ 0 & 3 & 0 \end{pmatrix}$

3 $\begin{pmatrix} 4 & 5 & 0 \\ -3 & \frac{1}{2} & 1 \\ 6 & 4 & 2 \end{pmatrix}$

4 $\begin{pmatrix} 3 & -2 & 2 \\ 4 & 10 & 1 \\ 6 & 20 & -4 \end{pmatrix}$

5 $\begin{pmatrix} 5 & 3 & 5 \\ 3 & 7 & 1 \\ 0 & 8 & 8 \end{pmatrix}$

Simultaneous equations 2

Already in this chapter we have looked at sets of simultaneous linear equations. In Example 1.9 we had:

$$4x - 2y + z = 3,$$
$$3x - 8y - 4z = 10$$

and $7x \quad + 5z = -1.$

On the left of these equations we may separate the coefficients from the variables x, y and z. We write this in matrix form as

$$\begin{pmatrix} 4 & -2 & 1 \\ 3 & -8 & -4 \\ 7 & 0 & 5 \end{pmatrix} \begin{pmatrix} x \\ y \\ z \end{pmatrix} = \begin{pmatrix} 3 \\ 10 \\ -1 \end{pmatrix}.$$

You can check for yourself that the left-hand side multiplies out to give $4x - 2y + z$, $3x - 8y - 4z$ and $7x + 5z$.

Generally we may write $AX = B$, where $A = \begin{pmatrix} 4 & -2 & 1 \\ 3 & -8 & -4 \\ 7 & 0 & 5 \end{pmatrix}$, $X = \begin{pmatrix} x \\ y \\ z \end{pmatrix}$

and $B = \begin{pmatrix} 3 \\ 10 \\ -1 \end{pmatrix}.$

Now $AX = B$ is a general equation. Remember from Volume 1 that whatever we do to one side we must also do to the other side, i.e. our operation must be consistent throughout the equation. We will multiply both sides by the inverse matrix, A^{-1},

i.e. $A^{-1}AX = A^{-1}B$

i.e. $IX = A^{-1}B$

i.e. $X = A^{-1}B$

$A^{-1}A = I.$

$IX = X.$

Let us look at this in practice.

Example 1.24

Solve the simultaneous equations $4x - 2y + z = 3,$
$$3x - 8y - 4z = 10$$
and $7x \quad + 5z = -1.$

In matrix form these equations are $\begin{pmatrix} 4 & -2 & 1 \\ 3 & -8 & -4 \\ 7 & 0 & 5 \end{pmatrix} \begin{pmatrix} x \\ y \\ z \end{pmatrix} = \begin{pmatrix} 3 \\ 10 \\ -1 \end{pmatrix}.$

Our general solution is $X = A^{-1}B$. This means we must find the inverse

matrix of $\begin{pmatrix} 4 & -2 & 1 \\ 3 & -8 & -4 \\ 7 & 0 & 5 \end{pmatrix}.$

First, we calculate the determinant of A,

i.e. $\det A = \begin{vmatrix} 4 & -2 & 1 \\ 3 & -8 & -4 \\ 7 & 0 & 5 \end{vmatrix} = 4\begin{vmatrix} -8 & -4 \\ 0 & 5 \end{vmatrix} - -2\begin{vmatrix} 3 & -4 \\ 7 & 5 \end{vmatrix} + 1\begin{vmatrix} 3 & -8 \\ 7 & 0 \end{vmatrix}$

$$= -18.$$

Now we find the minors and cofactors. Again we calculate them in an orderly way.

Element	Minor		Cofactor	
a_{11} is 4,	$\begin{vmatrix} -8 & -4 \\ 0 & 5 \end{vmatrix}$	$= -40,$	$(-1)^{(1+1)}(-40)$	$= -40.$
a_{12} is -2,	$\begin{vmatrix} 3 & -4 \\ 7 & 5 \end{vmatrix}$	$= 43,$	$(-1)^{(1+2)}43$	$= -43.$
a_{13} is 1,	$\begin{vmatrix} 3 & -8 \\ 7 & 0 \end{vmatrix}$	$= 56,$	$(-1)^{(1+3)}56$	$= 56.$
a_{21} is 3,	$\begin{vmatrix} -2 & 1 \\ 0 & 5 \end{vmatrix}$	$= -10,$	$(-1)^{(2+1)}(-10)$	$= 10.$
a_{22} is -8,	$\begin{vmatrix} 4 & 1 \\ 7 & 5 \end{vmatrix}$	$= 13,$	$(-1)^{(2+2)}13$	$= 13.$
a_{23} is -4,	$\begin{vmatrix} 4 & -2 \\ 7 & 0 \end{vmatrix}$	$= 14,$	$(-1)^{(2+3)}14$	$= -14.$
a_{31} is 7,	$\begin{vmatrix} -2 & 1 \\ -8 & -4 \end{vmatrix}$	$= 16,$	$(-1)^{(3+1)}16$	$= 16.$
a_{32} is 0,	$\begin{vmatrix} 4 & 1 \\ 3 & -4 \end{vmatrix}$	$= -19,$	$(-1)^{(3+2)}(-19)$	$= 19.$
a_{33} is 5,	$\begin{vmatrix} 4 & -2 \\ 3 & -8 \end{vmatrix}$	$= -26,$	$(-1)^{(3+3)}(-26)$	$= -26.$

The matrix of cofactors is $\begin{pmatrix} -40 & -43 & 56 \\ 10 & 13 & -14 \\ 16 & 19 & -26 \end{pmatrix}$.

Transposing the matrix of cofactors we get

$$\text{Adj } A \text{ is } \begin{pmatrix} -40 & 10 & 16 \\ -43 & 13 & 19 \\ 56 & -14 & -26 \end{pmatrix}.$$

Hence $$A^{-1} = \frac{1}{-18} \times \begin{pmatrix} -40 & 10 & 16 \\ -43 & 13 & 19 \\ 56 & -14 & -26 \end{pmatrix}.$$

Using $\qquad X = A^{-1}B$

we have
$$\begin{pmatrix} x \\ y \\ z \end{pmatrix} = \frac{1}{-18} \times \begin{pmatrix} -40 & 10 & 16 \\ -43 & 13 & 19 \\ 56 & -14 & -26 \end{pmatrix} \begin{pmatrix} 3 \\ 10 \\ -1 \end{pmatrix}$$

$$= \frac{1}{-18} \begin{pmatrix} -120 + 100 - 16 \\ -129 + 130 - 19 \\ 168 - 140 + 26 \end{pmatrix}$$

$$= \begin{pmatrix} \dfrac{-36}{-18} \\ \dfrac{-18}{-18} \\ \dfrac{54}{-18} \end{pmatrix}$$

i.e. $\qquad \begin{pmatrix} x \\ y \\ z \end{pmatrix} = \begin{pmatrix} 2 \\ 1 \\ -3 \end{pmatrix}$ or $x = 2$, $y = 1$, $z = -3$.

As usual, we can check these solutions by substituting into either of the original equations.

ASSIGNMENT

We return to our original simultaneous linear equations from the assembly teams; $3x + 3.5y + 5.5z = 28$,
$$4x + 3y + 6z = 30$$
and $5x + 2.5y + 5.5z = 29$.

In matrix form we have $\begin{pmatrix} 3 & 3.5 & 5.5 \\ 4 & 3 & 6 \\ 5 & 2.5 & 5.5 \end{pmatrix} \begin{pmatrix} x \\ y \\ z \end{pmatrix} = \begin{pmatrix} 28 \\ 30 \\ 29 \end{pmatrix}$.

We aim to find the inverse matrix of $\begin{pmatrix} 3 & 3.5 & 5.5 \\ 4 & 3 & 6 \\ 5 & 2.5 & 5.5 \end{pmatrix}$.

Let this be A. We calculate the determinant of A,

i.e. $\det A = \begin{vmatrix} 3 & 3.5 & 5.5 \\ 4 & 3 & 6 \\ 5 & 2.5 & 5.5 \end{vmatrix} = 5$.

Now we find the minors and cofactors, leaving you to check the working.

Element	*Minor*	*Cofactor*		
a_{11} is 3,	1.5,	$(-1)^{(1+1)}1.5$	$=$	1.5.
a_{12} is 3.5,	-8,	$(-1)^{(1+2)}(-8)$	$=$	8.
a_{13} is 5.5,	-5,	$(-1)^{(1+3)}(-5)$	$=$	-5.
a_{21} is 4,	5.5,	$(-1)^{(2+1)}5.5$	$=$	-5.5.
a_{22} is 3,	-11,	$(-1)^{(2+2)}(-11)$	$=$	-11.
a_{23} is 6,	-10,	$(-1)^{(2+3)}(-10)$	$=$	10.
a_{31} is 5,	4.5,	$(-1)^{(3+1)}4.5$	$=$	4.5.
a_{32} is 2.5,	-4,	$(-1)^{(3+2)}(-4)$	$=$	4.
a_{33} is 5.5,	-5,	$(-1)^{(3+3)}(-5)$	$=$	-5.

The matrix of cofactors is $\begin{pmatrix} 1.5 & 8 & -5 \\ -5.5 & -11 & 10 \\ 4.5 & 4 & -5 \end{pmatrix}$.

Transposing the matrix of cofactors we get

$$\text{Adj}\,A \text{ is } \begin{pmatrix} 1.5 & -5.5 & 4.5 \\ 8 & -11 & 4 \\ -5 & 10 & -5 \end{pmatrix}.$$

Hence $$A^{-1} = \frac{1}{5} \times \begin{pmatrix} 1.5 & -5.5 & 4.5 \\ 8 & -11 & 4 \\ -5 & 10 & -5 \end{pmatrix}$$

Using $$X = A^{-1}B$$

we have $$\begin{pmatrix} x \\ y \\ z \end{pmatrix} = \frac{1}{5} \times \begin{pmatrix} 1.5 & -5.5 & 4.5 \\ 8 & -11 & 4 \\ -5 & 10 & -5 \end{pmatrix} \begin{pmatrix} 28 \\ 30 \\ 29 \end{pmatrix}$$

$$= \frac{1}{5} \times \begin{pmatrix} 42 - 165 + 130.5 \\ 224 - 330 + 116 \\ -140 + 300 - 145 \end{pmatrix}$$

$$= \begin{pmatrix} \dfrac{7.5}{5} \\ \dfrac{10}{5} \\ \dfrac{15}{5} \end{pmatrix}$$

i.e. $$\begin{pmatrix} x \\ y \\ z \end{pmatrix} = \begin{pmatrix} 1.5 \\ 2 \\ 3 \end{pmatrix} \qquad \text{or } x = 1.5,\ y = 2,\ z = 3.$$

You will notice that these values agree with the answers we worked out using the determinant method.

▬▬ EXERCISE 1.8 ▬▬▬▬▬▬▬▬▬▬▬▬▬▬▬▬▬▬

In each question write the simultaneous equations in matrix form. Hence solve them for x, y and z.

1　$x + y + z = 7,$
　　$x + y - z = 15$ and
　　$x - y - z = 5$

2　$x + 3y + z = 13,$
　　$x + 4y - 2z = -1$ and
　　$x - 2y + 3z = 18$

3　$3y = x + 2z - 1,$
　　$5y = 20 - x - 5z$ and
　　$z = x + y - 6$

4　$4x + y - 2z = 0,$
　　$4x - 5y + 4z = 3$ and
　　$x + 2y - z = 1$

5　$3y - 5x - 2z = 1\frac{1}{2},$
　　$y + \quad 3z = 7$ and
　　$y + 2x - z = -2$

6　$x + 2y - 4z = -2,$
　　$2x - y + z = 4$ and
　　$x - y + z = 1$

7　$2x - 3y + 3z = -1,$
　　$x \quad + z = 0$ and
　　$3x + 2y - 4z = -12$

8　$7x - 6y + 4z = 18,$
　　$6y = 7x + 4z - 12$ and
　　$2x + 3y + 4z = 6$

9　$3x + y + z - 6 = 0,$
　　$x + 3y \quad - 11 = 0$ and
　　$2x - y + z \quad = 2$

10　$2x - 5y - z + 23 = 0,$
　　$15y = 69 + 6x - 3z$ and
　　$x + y - z = 33$

The final set of exercises involves the solution of practical simultaneous linear equations. For extra practice you can attempt the solutions using i) determinants and ii) matrices.

▬▬ EXERCISE 1.9 ▬▬▬▬▬▬▬▬▬▬▬▬▬▬▬▬▬▬

1 A local company assembles timing mechanisms for the heating industry. The process for each mechanism can be split into three basic sections. These are: assembly which takes x minutes, inspection which takes y minutes and packaging which takes z minutes. x, y and z are related by the simultaneous equations $\quad 2x + 4y + 3z = 53,$

$$x + 5y + 2z = 40$$

and $\quad 4x + y + 4z = 72.$

Solve these equations to find the time for each part of the process.

2 The electrical circuit shows 2 emfs together with various resistors. The sums of the relevant voltages produce the simultaneous equations.

They are in terms of the currents, I_1, I_2 and I_3 amps.

$$6I_1 + 2(I_1 - I_2) = 20,$$
$$2(I_2 - I_1) + 4I_2 + 1.5(I_2 - I_3) = 0$$

and $\quad 1.5I_3 + 1.5(I_3 - I_2) = 15.$

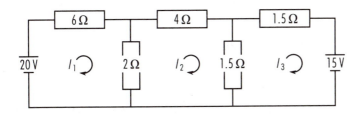

These equations will simplify. Show that they may be reduced to

$$8I_1 - 2I_2 = 20,$$
$$-2I_1 + 7.5I_2 - 1.5I_3 = 0$$

and
$$-1.5I_2 + 3I_3 = 15.$$

Now solve these simultaneous equations for I_1, I_2 and I_3.

3   Two aircraft will collide if they attempt to occupy the same coordinates (x, y, z) in space at the same time. These coordinates are measured in kilometres from a navigation beacon.

i)   They are connected by the simultaneous equations

$$-5x - 10y + 24z = 13,$$
$$3x + 4y - z = 13$$

and $6x + y - 3z = 3.$

Find the coordinates for a potential collision to occur.

ii)   Suppose the equations are

$$-5x - 10y + 24z = 13,$$
$$3x + 4y - z = 13$$

and $6x + 8y - 2z = 3.$

Why can there be no collision?

4   An engineering company part-produces and assembles two models of paint spraying equipment. The table below shows the number of minutes per item allocated to part-manufacture, assembly and packaging/warehousing.

	Manufacturing	*Assembly*	*Pack/Warehouse*
Model One	8	12	10
Model Two	10	7	10

For these products the production manager generally allocates daily timings. These are 395 minutes for manufacturing and 380 minutes for assembly, each including slack time of z minutes. The 400 minutes for packaging/warehousing is exact. Within these times the company produces x number of Model One and y number of Model Two. x, y and z are connected by the simultaneous equations

$$8x + 10y + z = 395,$$
$$12x + 7y + z = 380$$
and $10x + 10y \quad = 400.$

By solving these equations decide how many of each model are made and the slack time involved.

5   The electrical circuit below shows an emf together with various resistors. The sums of the relevant voltages produce the simultaneous equations.

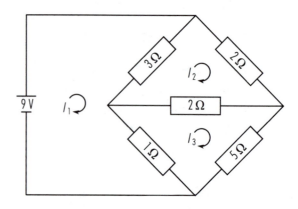

They are in terms of the currents, I_1, I_2 and I_3 amps and form the simultaneous equations

$$4I_1 - 3I_2 - I_3 = 9,$$
$$-3I_1 + 7I_2 - 2I_3 = 0$$
and $-I_1 - 2I_2 + 8I_3 = 0.$

Solve these simultaneous equations for I_1, I_2 and I_3.

2 Complex Numbers

The objectives of this chapter are to:

1   Understand the limit of the real number system.
2   Define j to be $\sqrt{-1}$.
3   Define a complex number in Cartesian form.
4   Solve a quadratic equation with complex roots.
5   Represent a complex number using an Argand diagram.
6   Add and subtract complex numbers.
7   Define the complex conjugate of a complex number.
8   Multiply and divide complex numbers in Cartesian form.
9   Define a complex number in polar form using modulus and argument.
10 Multiply and divide complex numbers in polar form.
11 Convert Cartesian form to polar form and vice versa.
12 Find the square roots of a complex number.
13 Apply complex numbers to alternating current (a.c.) circuits.

Introduction

Firstly the 'complex' of **complex numbers** does not mean complicated. The number system so far has been based on **real numbers**. These are numbers that actually exist. Real numbers include whole numbers (integers), negative numbers and decimal (or fractional) numbers. In practical applications these numbers can have real meanings. The numbers have rank order, i.e. we know which ones are large and which ones are small. For example, in statistics we can place a list of numbers in rank order and find the median.

Very quickly in this chapter you will see that real numbers let us down. This means we need to extend our system of numbers to include **complex numbers**. In this chapter we will look at two versions of complex numbers, i) the Cartesian form, and ii) the polar form.

■■■■■ ASSIGNMENT ■■■■■

The Assignment for this chapter looks at how complex numbers are applied in electrical circuits. We look at the elements of resistors, capacitors and inductors in an a.c. (alternating current) circuit. Ohm's law of $V = IR$ applies to d.c. (direct current) circuits. V is the voltage, I the current and R the resistance. For a.c. circuits this becomes $V = IZ$ where Z is the impedance. The representation and value of Z depends on the inclusion of resistors and/or capacitors and/or inductors. It also depends on them being in series or in parallel. As we learn about complex numbers we will look at their application to a.c. theory.

Quadratic equations

We have seen quadratic equations before, in Volume 1. We have solved them by various methods:

  i)   graphically,
 ii)   by factorisation,
iii)   by completing the square and
 iv)   by the formula.

The general quadratic equation $ax^2 + bx + c = 0$ may be solved using the formula

$$x = \frac{-b \pm \sqrt{b^2 - 4ac}}{2a}.$$

Example 2.1

Use the formula to solve the quadratic equation $x^2 - 6x + 25 = 0$.

In our quadratic equation we have $a = 1$, $b = -6$ and $c = 25$. We substitute these values into our formula

$$x = \frac{-b \pm \sqrt{b^2 - 4ac}}{2a}$$

to get

$$x = \frac{-(-6) \pm \sqrt{(-6)^2 - 4(1)(25)}}{2(1)}$$

$$= \frac{6 \pm \sqrt{36 - 100}}{2}$$

$$= \frac{6 \pm \sqrt{-64}}{2}.$$

We know the square root of 64 is ± 8. However we do *not* know about the square root of -64. Try this on your calculator by inputting $64\rfloor$ $^{+}\!/\!_{-}\rfloor$ $\sqrt{\,}\rfloor$. You will see the error message -E- displayed. We *cannot* find the real square roots of any negative values. This is where the real number system lets us down. We need to extend the system to deal with square roots of negative numbers.

Cartesian complex numbers

We define $\sqrt{-1}$ **to be** j. Pure Mathematics textbooks use i instead of j. To avoid confusion with electrical current, i, we choose j.

$x + jy$ **is a general Cartesian** (or **rectangular** or **algebraic**) **complex number** where x and y may be any numbers. x **is the real part**. y **is the imaginary part** because it is attached to the j. We often use z for the complex number written as $z = x + jy$.

Later in this chapter we will see that we can plot complex numbers. We will see the real and imaginary axes are at 90° to each other. These axes are similar to the axes we use during graph plotting.

▨▨▨ **Examples 2.2** ▨▨▨▨▨▨▨▨▨▨▨▨▨▨▨▨▨▨▨▨▨▨

  i) $3 + j8$;   real part is 3, imaginary part is 8.

 ii) $3 - j8$;   real part is 3, imaginary part is -8.

iii) 5 or $5 + j0$;   real part is 5, imaginary part is 0.

 iv) $j7$ or $0 + j7$;   real part is 0, imaginary part is 7.

  v) $-1.62 + j4.3$;   real part is -1.62, imaginary part is 4.3.

With this new knowledge we can try our original example again.

▨▨▨ **Example 2.3** ▨▨▨▨▨▨▨▨▨▨▨▨▨▨▨▨▨▨▨▨▨▨▨▨▨

Remember we are trying to solve the quadratic equation $x^2 - 6x + 25 = 0$ and have reached

$$x = \frac{6 \pm \sqrt{-64}}{2}.$$

We can split -64 into factors of -1 and 64 so that

$$x = \frac{6 \pm \sqrt{-1 \times 64}}{2}$$

$$= \frac{6 \pm \sqrt{-1} \times \sqrt{64}}{2} \qquad \boxed{j = \sqrt{-1} \text{ and } 8 = \sqrt{64}.}$$

$$= \frac{6 \pm j8}{2}$$

$$= \frac{2(3 \pm j4)}{2}$$

i.e. $x = 3 + j4,\ 3 - j4$ are the complex solutions (roots) of our quadratic equation.

EXERCISE 2.1

Solve the following quadratic equations

1   $x^2 + 4x + 7 = 0$.

2   $2x^2 - 3x + 11 = 0$.

3   $x^2 + x + 1 = 0$.

4   $3x^2 - x = -5$.

5   $7x^2 - 4x + 1 = 0$.

Let us return to our definition of j.

Using $j = \sqrt{-1}$

we can square both sides to get

$$j^2 = (\sqrt{-1})^2 = -1.$$

This is useful when we multiply complex numbers.

We can go a little further and look at larger powers of j.

Now $j^3 = j \times j^2 = j \times -1 = -j$.

Also $j^4 = j^2 \times j^2 = (-1) \times (-1) = 1$

and $j^5 = j \times j^4 = j \times 1 = j$.

If we continue with more powers of j this pattern repeats itself. You should check for yourself that $j^6 = -1$, $j^7 = -j$, $j^8 = 1$, etc.

The Argand diagram

The **Argand diagram** is a pair of **Cartesian** (or **rectangular**) axes used when representing complex numbers. The **horizontal axis** is the **real axis** and the **vertical axis** is the **imaginary axis**. Points on the diagram represent the complex numbers in a similar way to graphical coordinates. The straight line joining the origin to the coordinate point is called the **phasor**.

In Figs. 2.1 we see the complex number $2.5 + j4$ represented as a coordinate point and by a phasor.

If we label the origin O and the coordinate point Z then the phasor is $\overrightarrow{OZ}$, similar to a vector.

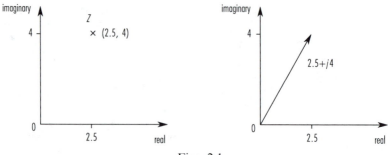

Figs. 2.1

i) Represent the complex numbers $-2 - j3$ and $-j1.5$ on an Argand diagram by points,

ii) Represent the complex numbers $4 - j$ and -4 on an Argand diagram by phasors.

We simply follow the ideas shown in Figs. 2.1 to get

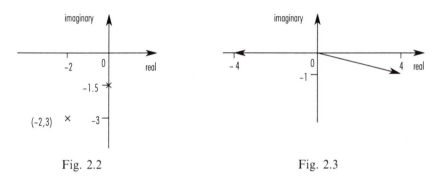

Fig. 2.2   Fig. 2.3

EXERCISE 2.2

Represent the following complex numbers by a point on an Argand diagram.

1	$3 + j4$	5	$-1.5 - j7$
2	$5 - j1.2$	6	6
3	$-8 + j6$	7	$-j4$
4	0		

Represent the following complex numbers by a phasor on an Argand diagram.

8	-3	12	$2 + j3$
9	$j2$	13	$3 + j2$
10	$-3 + j2$	14	$6 - j4$
11	$-3 - j2$	15	$-j5$

Addition and subtraction

The usual rules of algebra apply here. We add and subtract like terms, i.e. separately we add and subtract real parts and imaginary parts.

██████ **Examples 2.5** ████████████████████████████████████

Let $z_1 = 3 + j4$, $z_2 = 2.5 - j$ and $z_3 = -1 + j5$. We demonstrate addition and subtraction using these complex numbers.

i) $z_1 + z_2$ $= (3 + j4) + (2.5 - j)$

$= 3 + j4 + 2.5 - j$

$= (3 + 2.5) + j(4 - 1)$

| Separating real and imaginary parts. |

$= 5.5 + j3.$

ii) $z_1 + z_2 + z_3$ $= (3 + j4) + (2.5 - j) + (-1 + j5)$

$= 3 + j4 + 2.5 - j - 1 + j5$

$= (3 + 2.5 - 1) + j(4 - 1 + 5)$

| Separating parts as before. |

$= 4.5 + j8.$

Alternatively we could have used our earlier result to give

$z_1 + z_2 + z_3$ $= (5.5 + j3) + (-1 + j5)$

$= 5.5 + j3 - 1 + j5$

$= (5.5 - 1) + j(3 + 5)$

$= 4.5 + j8.$

iii) $z_1 - z_3$ $= (3 + j4) - (-1 + j5)$

$= 3 + j4 + 1 - j5$

| Minus sign between the brackets affects -1 and $+j5$. |

$= (3 + 1) + j(4 - 5)$

$= 4 - j.$

iv) $z_3 + z_1 - z_2$ $= (-1 + j5) + (3 + j4) - (2.5 - j)$

$= -1 + j5 + 3 + j4 - 2.5 + j$

$= (-1 + 3 - 2.5) + j(5 + 4 + 1)$

$= -0.5 + j10.$

You will quickly become more familiar with complex numbers. Then you will be able to write down the answers with fewer lines of working.

Addition and subtraction of complex numbers using Argand diagrams

We can repeat this set of examples using phasors on an Argand diagram. You will see that the ideas are very similar to those of vectors from Volume 1.

██████ **Examples 2.6** ████████████████████████████████████

Again we use $z_1 = 3 + j4$, $z_2 = 2.5 - j$ and $z_3 = -1 + j5$. This time we demonstrate addition and subtraction of these complex numbers using an Argand diagram. We also show the parallelogram law.

i) For $z_1 + z_2$ we start in Fig. 2.4 with the phasor for $3 + j4$. We draw a line from the origin to the point with coordinates $(3, 4)$. Next we add $2.5 - j$ by moving 2.5 horizontally to the right and 1 vertically downwards. We move down rather than up because of the minus sign.

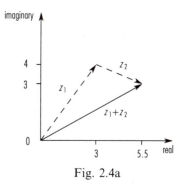

Fig. 2.4a

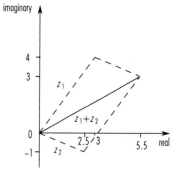

Fig. 2.4b

You can see that we have reached the point $(5.5, 3)$. We represent the result by the phasor joining the origin to this point, i.e. $z_1 + z_2 = 5.5 + j3$.

ii) For $z_1 + z_2 + z_3$ we use our previous result for $z_1 + z_2$ so that $z_1 + z_2 + z_3 = (5.5 + j3) + (-1 + j5)$. In Fig. 2.5 we start with the phasor $5.5 + j3$. We add $-1 + j5$ by moving 1 horizontally to the left (because of the minus sign) and 5 vertically upwards. You can see that we have reached the point $(4.5, 8)$.

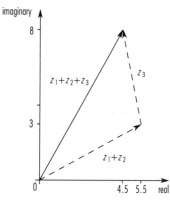

Fig. 2.5a

We represent the result by the phasor joining the origin to this point, i.e. $z_1 + z_2 + z_3 = 4.5 + j8$.

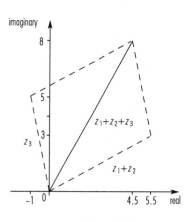

Fig. 2.5b

iii) For $z_1 - z_3$ we remember that subtraction is the addition of a negative value, i.e. $z_1 + (-z_3)$. We start in Fig. 2.6 with the phasor for $3 + j4$, drawing a line from the origin to the point with coordinates $(3, 4)$. Because $z_3 = -1 + j5$ we can write $-z_3 = -(-1 + j5) = 1 - j5$.

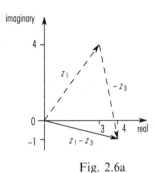

Fig. 2.6a

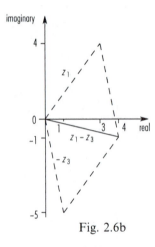

Fig. 2.6b

This means we add $1 - j5$ by moving 1 horizontally to the right and 5 vertically downwards. We move down rather than up because of the minus sign. You can see that we have reached the point $(4, -1)$. We represent the result by the phasor joining the origin to this point, i.e. $z_1 - z_3 = 4 - j$.

iv) The last example, $z_3 + z_1 - z_2$, is left as an exercise for you to complete. Here are some hints. Start with a phasor for z_3 and then add a phasor for z_1. Re-write z_2 as $-z_2$. Now add $-z_2$ to your previous result. You should reach the point with coordinates $(-0.5, 10)$, allowing you to draw the resultant phasor.

▬▬▬ ASSIGNMENT ▬▬▬

Let us look at part of an a.c. circuit. We need to know how to represent elements of the circuit using complex numbers.

A **resistor**, ⎯⊏▭▭⊐⎯ , has an **impedance of R**.

It is a real value.

A **capacitor**, ⎯⊣⊢⎯ , has an **impedance of $\dfrac{1}{j\omega C}$**.

It is an imaginary value indicated by the j. It can be re-written with the j in the numerator. We multiply both the numerator and denominator by j,

i.e. $\dfrac{1}{j\omega C} \times \dfrac{j}{j} = \dfrac{j}{j^2\omega C} = -\dfrac{j}{\omega C} = -jX_C$

where $X_C = \dfrac{1}{\omega C}$

$\dfrac{j}{j}$ cancels to 1.

$j^2 = -1$.

An **inductor**, 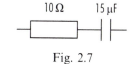 , has an **impedance of** $j\omega L = jX_L$ where $X_L = \omega L$. Again, this is an imaginary value indicated by the j. As usual, $\omega = 2\pi f$ where f is the frequency.

If the elements are in series we simply add the impedances together.

Suppose we have a resistor of $10\,\Omega$ and a capacitor of $15\,\mu F$ in series, with a frequency of $1\,kHz$ (i.e. $f = 1 \times 10^3\,Hz$), shown in Fig. 2.7.

For the resistor the impedance is $10\,\Omega$.

Fig. 2.7

The capacitor of $15\,\mu F$ must be used with consistent units of farads (F), i.e. $15\,\mu F = 15 \times 10^{-6}\,F$. Also $\omega = 2\pi f$ is $\omega = 2\pi \times 10^3$.

This means the impedance is $\dfrac{-j}{2\pi \times 10^3 \times 15 \times 10^{-6}} = -j10.61\,\Omega$.

Adding together these impedances because the elements are in series we have: total impedance $= 10 - j10.61\,\Omega$.

$$Z = R - jX_C$$

EXERCISE 2.3

In this set of exercises let $z_1 = 2 + j3$, $z_2 = 0.5 - j1.5$, $z_3 = -4 - j5$ and $z_4 = 6 - j$.

Add and subtract the following combinations of complex numbers. For each question check your answers by drawing the phasors on an Argand diagram.

1 $z_2 + z_4$

2 $z_1 + z_2 + z_4$

3 $z_3 + z_4 + z_1$

4 $z_2 - z_4$

5 $z_4 - z_2$

6 $z_4 - z_2 + z_3$

7 $z_4 - z_2 - z_3$

8 $-z_3 - z_4 - z_1$

9 $z_1 - z_2 + z_3 - z_4$

10 $z_4 + z_3 + z_2 - z_1$

Scalar multiplication

We can think of a real number as a scalar. This section is similar to the scalar multiplication section you met in the vectors chapter (Chapter 13) of Volume 1. If $z = 3 + j4$ then the 2 in $2z$ is the scalar multiplier. Multiplying a complex number by a scalar affects both parts in the same way. This means that the 2 affects both the **real** and **imaginary parts**.

i) If $z = 3 + j4$ then $2z = 2(3 + j4)$
$$= (2 \times 3) + (2 \times j4)$$
$$= 6 + j8.$$

ii) If $z = 3 - j5$ then $-1.5z = -1.5(3 - j5)$
$$= (-1.5 \times 3) + (-1.5 \times -j5)$$
$$= -4.5 + j7.5.$$

Multiplication in Cartesian form

The usual rules of algebra apply when multiplying together two complex numbers. We start with a simple example where one of the complex numbers is just j.

Example 2.8

Multiply $3 + j4$ by j. Display both $3 + j4$ and the answer on an Argand diagram.

We write $(3 + j4)j = j3 + j^2 4$
$$= j3 - 4$$

$$\boxed{j^2 = -1.}$$

or $-4 + j3$.

In Fig. 2.8 we see the original complex number, $3 + j4$, and the answer, $-4 + j3$. Notice how multiplication by j rotates the original phasor $90°$ anti-clockwise.

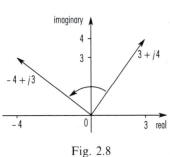

Fig. 2.8

Now we can combine the ideas of scalar multiplication and multiplication by j. Each part of the first complex number is multiplied by each part of the second complex number.

Examples 2.9

Multiply together the complex numbers and simplify the answers

i) $(3 + j4)(2 + j)$, ii) $(5 - j2)(6 + j7)$, iii) $(5 - j2)(6 + j7)(1 - j3)$.

i)   This example is a combination of two earlier examples. We multiply $3 + j4$ by 2 and by j. The original addition of 2 and j (as $2 + j$) means our result is the addition of our two previous answers, i.e. we have $6 + j8 + -4 + j3 = 2 + j11$.

However, we are more likely to attempt the multiplication from the beginning, as

$$(3 + j4)(2 + j) = 6 + j^2 4 + j3 + j8$$
$$= 6 - 4 + j3 + j8 \qquad \boxed{j^2 = -1.}$$
$$= 2 + j11.$$

ii) Also $\quad (5 - j2)(6 + j7) = 30 - j^2 14 + j35 - j12$
$$= 30 + 14 + j35 - j12 \qquad \boxed{j^2 = -1.}$$
$$= 44 + j23.$$

iii) In our final example we use the previous result for $(5 - j2)(6 + j7)$ so that

$$(5 - j2)(6 + j7)(1 - j3) = (44 + j23)(1 - j3)$$
$$= 44 - j^2 69 - j132 + j23$$
$$= 44 + 69 - j132 + j23 \qquad \boxed{j^2 = -1.}$$
$$= 113 - j109.$$

■ EXERCISE 2.4 ■

Multiply out the following brackets fully simplifying your complex number answer.

1 $\quad 3(5 + j4)$	11 $\quad (1 + j)(5 - j10)$
2 $\quad -2(3 + j)$	12 $\quad (6 - j2)(3 - j5)$
3 $\quad (3 - j7)2$	13 $\quad (3 + j4)(3 - j4)$
4 $\quad j(6 + j3)$	14 $\quad (5 - j12)(5 + j12)j$
5 $\quad j(9 - j1.5)$	15 $\quad 3(3 - j)(2 + j7)$
6 $\quad j4(9 - j1.5)$	16 $\quad j(5 + j4)(3 + j2)$
7 $\quad (9 - j1.5)j4$	17 $\quad (3 + j2)(3 - j)(2 + j7)$
8 $\quad (5 + j4)j2$	18 $\quad (10 - j)(5 + j4)(3 + j2)$
9 $\quad (5 + j4)(3 + j2)$	19 $\quad (3 + j1.5)(2 + j8)(1 - j)$
10 $\quad (3 - j)(2 + j7)$	20 $\quad (2.5 - j)(1 + j)(1 - j)$

The complex conjugate

$x + jy$ and $x - jy$ are **complex conjugate numbers of each other**. Their real parts are identical whilst their imaginary parts differ by $+/-$. On an Argand diagram this means they are equally inclined to the real axis. One of them is above and one below the real axis.

▄▄▄▄▄ **Examples 2.10** ▄▄▄▄▄▄▄▄▄▄▄▄▄▄▄▄▄▄▄▄▄▄▄▄▄▄▄▄▄▄▄▄▄▄

i)　The complex conjugate of $5 + j2$ is $5 - j2$.

ii)　The complex conjugate of $-3 + j4$ is $-3 - j4$.

In Fig. 2.9 we have drawn both original and both conjugate complex numbers on an Argand diagram.

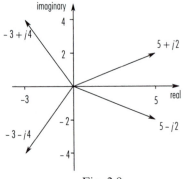

Fig. 2.9

▄▄▄▄▄ **Examples 2.11** ▄▄▄▄▄▄▄▄▄▄▄▄▄▄▄▄▄▄▄▄▄▄▄▄▄▄▄▄▄▄▄▄▄▄

i)　The complex conjugate of $2 - j6$ is $2 + j6$.

ii)　The complex conjugate of $-j$ is j.

iii)　The complex conjugate of 4 (i.e $4 + j0$) is still 4. This is because addition and subtraction of $j0$ have the same effect, leaving 4 unaltered.

For yourself, you should plot these complex numbers and their conjugates on an Argand diagram.

▄▄▄▄▄ **Example 2.12** ▄▄▄▄▄▄▄▄▄▄▄▄▄▄▄▄▄▄▄▄▄▄▄▄▄▄▄▄▄▄▄▄▄▄

Multiply the complex number $5 + j2$ by its complex conjugate.

From Examples 2.10 (i) we know that the conjugate is $5 - j2$. Now we can write

$$(5 + j2)(5 - j2) = 25 - j^2 4 - j10 + j10$$
$$= 25 + 4 - j10 + j10$$
$$= 29.$$

Notice how the imaginary parts cancel out, one being positive and the other being negative. This will always happen when you multiply together a complex number and its conjugate.

We can demonstrate our numerical result more generally using a general complex number and its complex conjugate, i.e.

$$(x + jy)(x - jy) = x^2 - j^2 y^2 - jxy + jxy$$
$$= x^2 + y^2 - jxy + jxy$$
$$= x^2 + y^2$$

which is a real result.

Division in Cartesian form

When we multiply together complex numbers we always get an answer. Generally it has both real and imaginary parts. Sometimes we do get $j0$ as the imaginary part. In every case we can draw our answer on an Argand diagram. For example we found that $(3+j4)(2+j)=2+j11$. We can represent this by the phasor from the origin to the point $(2, 11)$.

Now $\dfrac{2+j}{3+j4}$ has no obvious answer and no obvious position on an Argand diagram.

Any number divided by itself is 1, e.g. $\dfrac{5}{5}$. We can apply this idea to complex numbers too, e.g. $\dfrac{3-j4}{3-j4}=1$.

The complex conjugate of the denominator, $3+j4$, is $3-j4$.

Also a complex number multiplied by its complex conjugate gives a real value. This is useful in the denominator. In Example 2.13 we link together these ideas.

Example 2.13

Divide the complex number $2+j$ by $3+j4$.

We write
$$\frac{2+j}{3+j4} = \frac{(2+j)}{(3+j4)} \times 1$$

$$= \frac{(2+j)}{(3+j4)} \times \frac{(3-j4)}{(3-j4)}.$$

We multiply both the numerator and denominator by the complex conjugate $(3-j4)$ of the denominator $(3+j4)$. This means we create a real number in the denominator. Because we originally multiply by 1 we do *not* change the overall value.

This gives
$$\frac{6-j^24-j8+j3}{9-j^216-j12+j12}$$

$$= \frac{6+4-j8+j3}{9+16-j12+j12}$$

$\boxed{j^2 = -1.}$

$$= \frac{10-j5}{25}$$

or $\dfrac{10}{25} - \dfrac{j5}{25}$ or $0.4 - j0.2$.

We always find the complex conjugate of the denominator. Then we multiply both the numerator and the denominator by this complex conjugate. This always makes the denominator into a real number.

███████ **Examples 2.14** ████████████████████████████████████

Express as complex numbers in the form $a + jb$

i) $\dfrac{(3 - j2)(1 + j7)}{-1 + j}$, ii) $\dfrac{2 + j5}{(3 + j)(-4 - j)}$.

In both cases we attempt the multiplication of the brackets before thinking about the division.

i) $\dfrac{(3 - j2)(1 + j7)}{-1 + j} = \dfrac{3 - j^2 14 + j21 - j2}{-1 + j}$

$= \dfrac{3 - (-14) + j19}{-1 + j}$ $\boxed{-(-14) = 14.}$

$= \dfrac{17 + j19}{-1 + j}$ $\boxed{3 + 14 = 17.}$

$= \dfrac{(17 + j19)(-1 - j)}{(-1 + j)(-1 - j)}$ $\boxed{\begin{array}{l}\text{Multiplication by}\\ \text{complex conjugate.}\end{array}}$

$= \dfrac{-17 - j^2 19 - j17 - j19}{1 + j - j - j^2}$

$= \dfrac{-17 - (-1)19 - j36}{1 - (-1)}$ $\boxed{j^2 = -1.}$

$= \dfrac{2 - j36}{2}$

$= \dfrac{2(1 - j18)}{2}$ $\boxed{\begin{array}{l}\text{2 is a common factor}\\ \text{to cancel.}\end{array}}$

$= 1 - j18.$

ii) $\dfrac{2 + j5}{(3 + j)(-4 - j)} = \dfrac{2 + j5}{-12 - j^2 - j3 - j4}$

$= \dfrac{2 + j5}{-12 - (-1) - j7}$ $\boxed{j^2 = -1.}$

$= \dfrac{2 + j5}{-11 - j7}$

$= \dfrac{(2 + j5)(-11 + j7)}{(-11 - j7)(-11 + j7)}$ $\boxed{\begin{array}{l}\text{Multiplication by}\\ \text{compex conjugate.}\end{array}}$

$= \dfrac{-22 + j^2 35 + j14 - j55}{121 - j^2 49 - j77 + j77}$

$= \dfrac{-22 + (-1)35 - j41}{121 - (-1)49}$ $\boxed{j^2 = -1.}$

$= \dfrac{-57 - j41}{170}$

or $-0.335 - j0.241.$

▮▮▮▮ ASSIGNMENT ▮▮▮▮

Let us return to our a.c. circuit, extending Fig. 2.7 to include an inductor of 5 mH (5×10^{-3} H) in parallel with the other elements. Our circuit is given in Fig. 2.10.

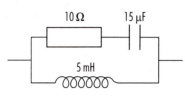

Fig. 2.10

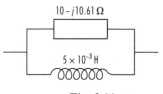

Fig. 2.11

We know that for the resistor and capacitor in series the impedance is $10 - j10.61\ \Omega$. This means we can simplify our circuit, as shown in Fig. 2.11.

For impedances in parallel we add their reciprocals when finding the resultant,

i.e. $\dfrac{1}{Z} = \dfrac{1}{Z_1} + \dfrac{1}{Z_2} + \ldots$ where Z is the total impedance.

As before $f = 10^3$ Hz in $\omega = 2\pi f$.

For the inductor the impedance, $j\omega L$, is $j2\pi \times 10^3 \times 5 \times 10^{-3}$

$$j10\pi \quad \text{or} \quad j31.42.$$

Bringing together these impedances in parallel we have

$$\frac{1}{Z} = \frac{1}{10 - j10.61} + \frac{1}{j31.42}$$

$Z_1 = 10 - j10.61.$
$Z_2 = j31.42.$

$$= \frac{j31.42 + 10 - j10.61}{(10 - j10.61)j31.42}$$

i.e. $\dfrac{1}{Z} = \dfrac{10 + j20.81}{333.37 + j314.2}.$

When we re-arrange this equation to make Z the subject we get

$$Z = \frac{333.37 + j314.2}{10 + j20.81}$$

$$= \frac{333.37 + j314.2}{10 + j20.81} \times \frac{10 - j20.81}{10 - j20.81}$$

Multiplication by complex conjugate.

$$= \frac{9872.20 - j3795.43}{533.06}$$

$$= 18.52 - j7.12\ \Omega \quad \text{is the total impedance.}$$

▮ EXERCISE 2.5 ▮

Divide and multiply as necessary the following complex numbers. Leave your answers in the form $a + jb$.

1 $\dfrac{2 + j}{3 - j4}$

2 $\dfrac{6 - j2}{3 + j5}$

3 $\dfrac{2 + j}{6 + j8}$

4 $\dfrac{3 + j1.5}{-5 + j12}$

5 $\dfrac{(5 + j4)(3 + j)}{3 - j4}$

6 $\dfrac{(2 + j)(3 + j2)}{6 + j8}$

7 $\dfrac{(3 - j7)(2 + j7)}{j}$

8 $\dfrac{(5 + j8)(10 + j)}{j(1 + j)}$

9 $\dfrac{(3 + j2)(1 - j)}{1 + j}$

10 $\dfrac{2 - j3}{(2.5 - j2)(4 - j3)}$

Polar form

The **polar form** is an alternative form for a complex number, z. In shortened form we write it as $z = r\underline{/\theta}$.

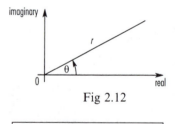

Fig 2.12

r is the **modulus** (or **size** or **length** or **magnitude**).

> Plural of modulus is moduli.

We write this as $r = |z|$, knowing that r is always positive. θ is the **argument**. We write this as $\theta = \arg z$. θ is the inclination of the complex number to the positive real axis. We label the positive real axis as $\theta = 0°$. Because θ is an angle we may use either degrees or radians. To account for all possibilities on the Argand diagram, θ has a range of values. This range may be either

(a)   or   (b)

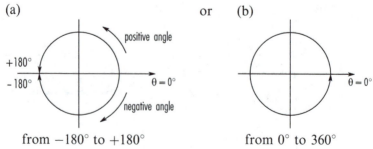

from $-180°$ to $+180°$ from $0°$ to $360°$

Fig. 2.13

We have to be careful here. Now $-180°$ and $+180°$ give the same angular position. Also $0°$ and $360°$ give another similar angular position. To avoid any confusion we do not quite include one of these extreme values in each case.

We use $\quad -180° < \theta \leqslant 180°,$ | Preferring 180° to −180°. |

i.e. $\quad\quad -\pi < \theta \leqslant \pi.$

or $\quad\quad 0° \leqslant \theta < 360°,$ | Preferring 0° to 360°. |

i.e. $\quad\quad 0 \leqslant \theta < 2\pi.$

Our examples concentrate on the range $-180° < \theta \leqslant 180°$. This range is the **range of principal values** of θ.

Examples 2.15

Represent the following polar form complex numbers on Argand diagrams   i) $5\underline{/35°}$, $6.7\underline{/0°}$ and $2.4\underline{/-90°}$;   ii) $2.8\underline{/160°}$, $2.8\underline{/-20°}$ and $3.1\underline{/-142°}$.

i)   ii)

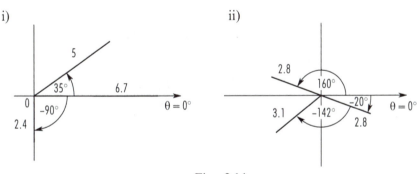

Figs. 2.14

EXERCISE 2.6

Represent the following polar form complex numbers on an Argand diagram.

1	$10\underline{/45°}$	6	$1.8\underline{/150°}$
2	$7.5\underline{/90°}$	7	$4.5\underline{/-117°}$
3	$3.6\underline{/20°}$	8	$9\underline{/117°}$
4	$4\underline{/-16°}$	9	$11\underline{/-161°}$
5	$10\underline{/-45°}$	10	$6.5\underline{/-75°}$

Multiplication and division in polar form

For two complex numbers, $z_1 = r_1\underline{/\theta_1}$ and $z_2 = r_2\underline{/\theta_2}$ we have rules for their multiplication and division.

We **multiply** according to $z_1 z_2 = r_1 r_2 \underline{/\theta_1 + \theta_2}$,

i.e. we **multiply the moduli** and **add the arguments**.

We **divide** according to $\dfrac{z_1}{z_2} = \dfrac{r_1}{r_2}\underline{/\theta_1 - \theta_2}$,

i.e. we **divide the moduli** and **subtract the arguments**.

We link together these rules in the following table.

Complex Numbers	Moduli	Arguments
Multiplication	×	+
Division	÷	−

In the next examples we demonstrate these ideas and extend them to more than two complex numbers.

Examples 2.16

In each case express the following as one complex number

i) $5\underline{/60°} \times 2.4\underline{/45°}$, ii) $5\underline{/60°} \times 2.4\underline{/-45°}$,

iii) $7.1\underline{/157°} \times 3.3\underline{/45°}$, iv) $4.9\underline{/23°} \times 1.2\underline{/4°} \times 6\underline{/-10°}$.

In each case we obey the rules, multiplying the moduli and adding the arguments.

i) $5\underline{/60°} \times 2.4\underline{/45°} \quad = 5 \times 2.4\underline{/60° + 45°}$

$\qquad\qquad\qquad\qquad = 12\underline{/105°}$.

ii) $5\underline{/60°} \times 2.4\underline{/-45°} \quad = 5 \times 2.4\underline{/60° + -45°}$

$\qquad\qquad\qquad\qquad\quad = 12\underline{/15°}$.

iii) $7.1\underline{/157°} \times 3.3\underline{/45°} = 7.1 \times 3.3\underline{/157° + 45°}$

$\qquad\qquad\qquad\qquad\quad = 23.43\underline{/202°}$.

Now 202° is outside our range of principal values, $-180° < \theta \leqslant 180°$. The Argand diagram shows we need to trace in a negative direction from the positive real axis (i.e. $\theta = 0°$) through 158° as an alternative to reach this position, i.e. $23.43\underline{/-158°}$.

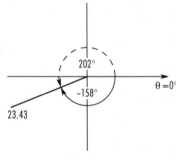

Fig. 2.15

iv) $4.9\underline{/23^\circ} \times 1.2\underline{/4^\circ} \times 6\underline{/-10^\circ} = 4.9 \times 1.2 \times 6\underline{/23^\circ + 4^\circ - 10^\circ}$

$$= 35.28\underline{/17^\circ}.$$

Examples 2.17

If $z = 5\underline{/60^\circ}$ write down the values of   i) $3z$,   ii) $\dfrac{z}{2}$ or $\dfrac{1}{2}z$,   iii) $\dfrac{4}{5}z$.

i) In $3z$ we may think of 3 as a scalar multiplier or a real number or a polar form complex number $3\underline{/0^\circ}$. Because we are working with polar forms we will use this last option to give

$$3z = 3\underline{/0^\circ} \times 5\underline{/60^\circ}$$
$$= 3 \times 5\underline{/0^\circ + 60^\circ}$$
$$= 15\underline{/60^\circ}.$$

ii) In a similar way we can write $\dfrac{1}{2}$ as $\dfrac{1}{2}\underline{/0^\circ}$ to give

$$\frac{1}{2}z = \frac{1}{2}\underline{/0^\circ} \times 5\underline{/60^\circ}$$
$$= \frac{1}{2} \times 5\underline{/0^\circ + 60^\circ}$$
$$= 2.5\underline{/60^\circ}.$$

iii) Applying the same idea again we have

$$\frac{4}{5}z = \frac{4}{5}\underline{/0^\circ} \times 5\underline{/60^\circ}$$
$$= \frac{4}{5} \times 5\underline{/0^\circ + 60^\circ}$$
$$= 4\underline{/60^\circ}.$$

Notice that the argument remains unchanged. Only the modulus is affected by the scalar multiplier.

EXERCISE 2.7

In each case simplify to one complex number in polar form.

1 $10\underline{/45^\circ} \times 1.8\underline{/13^\circ}$

2 $7.5\underline{/90^\circ} \times 10\underline{/-45^\circ}$

3 $3.6\underline{/20^\circ} \times 5\underline{/45^\circ}$

4 $2 \times 4.5\underline{/-117^\circ}$

5 $\dfrac{1}{2} \times 11\underline{/-161^\circ}$

6 $4.5\underline{/117^\circ} \times 4\underline{/83^\circ}$

7 $6\underline{/-136^\circ} \times 2\underline{/-75^\circ}$

8 $\dfrac{5}{13} \times 6.5\underline{/-75^\circ}$

9 $7\underline{/14^\circ} \times 2.5\underline{/132^\circ} \times 3\underline{/-64^\circ}$

10 $1.6\underline{/169^\circ} \times 2.4\underline{/43^\circ} \times 10\underline{/81^\circ}$

Examples 2.18

In each case express the following as one complex number

i) $12\angle 115^\circ \div 2.5\angle 73^\circ$, ii) $12\angle 115^\circ \div 2.5\angle -73^\circ$,

iii) $\dfrac{2.9\angle 12^\circ \times 4.8\angle 150^\circ}{6\angle 34^\circ}$, iv) $\dfrac{36\angle 145^\circ}{2.4\angle 30^\circ \times 1.8\angle -14^\circ}$.

In each case we obey the rules, dividing the moduli and subtracting the arguments.

i) $$12\angle 115^\circ \div 2.5\angle 73^\circ = \frac{12}{2.5}\angle 115^\circ - 73^\circ$$
$$= 4.8\angle 42^\circ.$$

ii) $$12\angle 115^\circ \div 2.5\angle -73^\circ = \frac{12}{2.5}\angle 115^\circ - -73^\circ$$
$$= 4.8\angle 188^\circ$$

Now 188° is outside our range of principal values, $-180^\circ < \theta \leqslant 180^\circ$. The Argand diagram shows we need to trace in a negative direction from the positive real axis (i.e. $\theta = 0^\circ$) through 172° as an alternative to reach this position, i.e. $4.8\angle -172^\circ$.

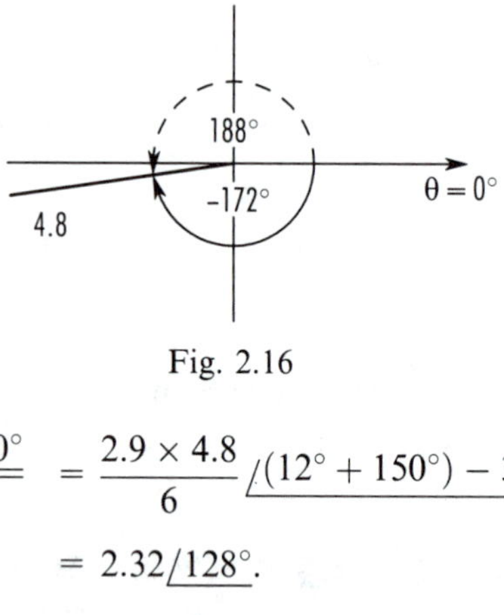

Fig. 2.16

iii) $$\frac{2.9\angle 12^\circ \times 4.8\angle 150^\circ}{6\angle 34^\circ} = \frac{2.9 \times 4.8}{6}\angle (12^\circ + 150^\circ) - 34^\circ$$
$$= 2.32\angle 128^\circ.$$

iv) $$\frac{36\angle 145^\circ}{2.4\angle 30^\circ \times 1.8\angle -14^\circ} = \frac{36}{2.4 \times 1.8}\angle 145^\circ - (30^\circ + -14^\circ)$$
$$= 8.\bar{3}\angle 129^\circ.$$

Notice that multiplication and division are easier using polar forms. There is no need to use the complex conjugate during division. However, we must remember that we *cannot* add or subtract using polar forms. For these two arithmetic operations we must use the Cartesian form. This indicates we need to be able to convert between the Cartesian and polar forms. After the next exercise we will look at the conversions.

■■■ **EXERCISE 2.8** ■■■■■■■■■■■■■

In each case simplify to one complex number in polar form.

1 $10\underline{/45^\circ} \div 5\underline{/13^\circ}$

2 $7.5\underline{/110^\circ} \div 4.5\underline{/-36^\circ}$

3 $\dfrac{12\underline{/-114^\circ}}{25\underline{/32^\circ}}$

4 $\dfrac{1.6\underline{/-173^\circ}}{0.2\underline{/-45^\circ}}$

5 $\dfrac{4.8\underline{/-155^\circ}}{1.5\underline{/72^\circ}}$

6 $\dfrac{7\underline{/67^\circ}}{5\underline{/132^\circ}}$

7 $\dfrac{4\underline{/56^\circ} \times 3.5\underline{/67^\circ}}{2.4\underline{/78^\circ}}$

8 $\dfrac{5.6\underline{/135^\circ}}{2\underline{/10^\circ} \times 6\underline{/81^\circ}}$

9 $\dfrac{20\underline{/-165^\circ} \times 2}{4.5\underline{/-35^\circ} \times 12\underline{/40^\circ}}$

10 $\dfrac{7\underline{/14^\circ} \times 2.4\underline{/43^\circ} \times 10\underline{/75^\circ}}{5\underline{/-30^\circ} \times 1.8\underline{/70^\circ}}$

Conversion between Cartesian and polar forms

In this chapter we have seen the two general forms for a complex number, z. The **Cartesian** form is $z = x + jy$ and the **polar** form is $z = r\underline{/\theta}$. In Fig. 2.17 we give the forms together in one diagram. Notice that we have completed a right-angled triangle in the first (positive) quadrant.

We have some general rules for the conversions. However we must emphasise that the techniques are easier with the aid of a diagram.

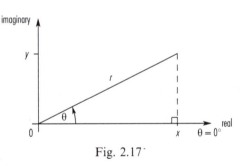

Fig. 2.17˙

When **converting from polar to Cartesian** form

$x = r\cos\theta$ gives the **real** part

and $y = r\sin\theta$ gives the **imaginary** part.

> Using trigonometry in the right-angled triangle.

When **converting from Cartesian to polar** form

$r = \sqrt{x^2 + y^2}$ gives the **modulus**

> Pythagoras' theorem. Positive root as we need only size.

and $\tan\theta = \dfrac{y}{x}$ allows us to find the **argument**.

We apply these conversion rules in the following examples. Afterwards we look at how simple it is to do the conversions with a calculator.

▓▓▓▓▓▓ **Examples 2.19** ▓▓▓▓▓▓

With the aid of a diagram convert the following complex numbers from polar to Cartesian form   i) $z_1 = 6\underline{/30^\circ}$,   ii) $z_2 = 9.5\underline{/125^\circ}$, iii) $z_3 = 2.75\underline{/-50^\circ}$,   iv) $z_4 = 11\underline{/-149^\circ}$.

Fig. 2.18 shows the position of the four complex numbers, one in each quadrant. For the conversions we apply the general rules. Remember that $x = r\cos\theta$ gives the real part and $y = r\sin\theta$ gives the imaginary part.

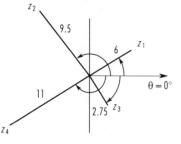

Fig. 2.18

i) For $z_1 = 6\underline{/30^\circ}$, $x = r\cos\theta$

                is $x = 6\cos 30^\circ$ $= 5.20.$

                    $y = r\sin\theta$

                is $y = 6\sin 30^\circ$ $= 3.00.$

    $\therefore$ $z_1 = 5.20 + j3.$

ii) For $z_2 = 9.5\underline{/125^\circ}$, $x = 9.5\cos 125^\circ$ $= -5.45.$

                and $y = 9.5\sin 125^\circ$ $= 7.78.$

    $\therefore$ $z_2 = -5.45 + j7.78.$

iii) For $z_3 = 2.75\underline{/-50^\circ}$, $x = 2.75\cos(-50^\circ) = 1.77.$

              and $y = 2.75\sin(-50^\circ) = -2.11.$

    $\therefore$ $z_3 = 1.77 - j2.11.$

iv) For $z_4 = 11\underline{/-149^\circ}$, $x = 11\cos(-149^\circ) = -9.43.$

              and $y = 11\sin(-149^\circ) = -5.67.$

    $\therefore$ $z_4 = -9.43 - j5.67.$

▓▓▓▓▓▓ **Examples 2.20** ▓▓▓▓▓▓

With the aid of diagrams convert the following complex numbers from Cartesian to polar form i) $z_1 = 3 + j4$, ii) $z_2 = -5 + j12$, iii) $z_3 = 2 - j$, iv) $z_4 = -8 - j6$.

i) For $z_1 = 3 + j4$,

$$r = \sqrt{x^2 + y^2} \qquad \tan\theta = \frac{y}{x}$$

    is $r = \sqrt{3^2 + 4^2}$ is $\tan\theta = \dfrac{4}{3}$

        $= \sqrt{9 + 16}$ i.e. $\theta = 53.13^\circ$

        $= \sqrt{25}$

  i.e. $r = 5.$

    $\therefore$ $z_1 = 5\underline{/53.13^\circ}.$

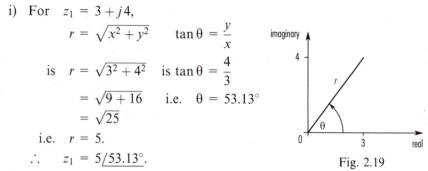

Fig. 2.19

ii) For $z_2 = -5 + j12,$

$$r = \sqrt{(-5)^2 + 12^2}$$

i.e. $r = 13.$

When we attempt to find the argument we need to be careful. The calculator does *not* give an immediate answer. We use an acute angle, α, with the *sizes* of the opposite and adjacent sides,

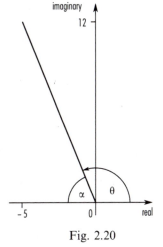

i.e. $$\tan \alpha = \frac{12}{5}$$

gives $\alpha = 67.38°.$

This means that $\theta = 180° - \alpha$

$$= 180° - 67.38°$$

$$= 112.62°.$$

$\therefore \quad z_2 = 13\underline{/112.62°}.$

Fig. 2.20

iii) For $z_3 = 2 - j,$

$$r = \sqrt{2^2 + (-1)^2}$$

i.e. $r = 2.24.$

Again we use the *sizes* of the triangles' sides so that

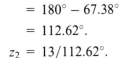

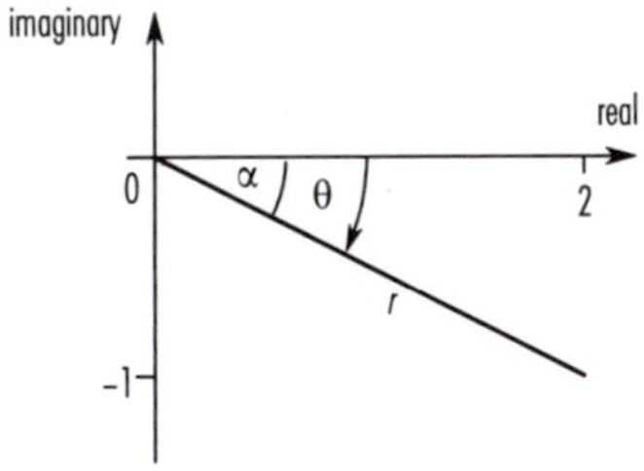

$$\tan \alpha = \frac{1}{2}$$

gives $\alpha = 26.57°.$

We know this clockwise direction is negative, i.e. $\theta = -\alpha = -26.57°.$

$\therefore \quad z_3 = 2.24\underline{/-26.57°}.$

Fig. 2.21

iv) For $z_4 = -8 - j6,$

$$r = \sqrt{(-8)^2 + (-6)^2}$$

i.e. $r = 10.$

Once again we use the *sizes* of the triangles' sides so that

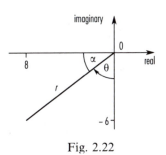

$$\tan \alpha = \frac{6}{8}$$

gives $\alpha = 36.87°.$

Fig. 2.22

We label θ in our diagram, remembering it is in a negative direction,

i.e. $\quad \theta = -(180° - \alpha)$

$\qquad = -(180° - 36.87°)$

$\qquad = -143.13°.$

$\therefore \quad z_4 = 10\underline{/-143.13°}.$

Now we know the principles behind the conversions. In this next section we attempt conversions using a calculator. The function keys we need are $\boxed{P \rightarrow R}$ and $\boxed{R \rightarrow P}$. The R stands for **rectangular** (i.e. **Cartesian**) and the P stands for **polar**. $\boxed{P \rightarrow R}$ converts polar to Cartesian form and $\boxed{R \rightarrow P}$ converts Cartesian to polar form.

▓▓▓▓ Examples 2.21 ▓▓▓▓▓▓▓▓▓▓▓▓▓▓▓▓▓▓▓▓▓▓▓▓

Using a calculator, convert the following complex numbers from polar to Cartesian form   i) $5\underline{/60°}$,   ii) $7.4\underline{\Big/\dfrac{3\pi}{4}}$ or $7.4\underline{/2.356\ldots}$

i)   First we must check that the calculator is in the degree mode. For $5\underline{/60°}$ our order of key operations is;

$\boxed{5}\ \boxed{P \rightarrow R}\ \boxed{60}\ \boxed{=}$ $\qquad$ to display the real part 2.5

followed by $\boxed{X \leftrightarrow Y}$ $\qquad$ to display the imaginary part 4.33.

$\therefore\ 5\underline{/60°} = 2.5 + j4.33.$

ii) This time we need to use radian mode. For $7.4\underline{/2.356}$ our order of key operations is

$\boxed{7.4}\ \boxed{P \rightarrow R}\ \boxed{2.356}\ \boxed{=}$ $\qquad$ to display the real part -5.23

followed by $\boxed{X \leftrightarrow Y}$ $\qquad$ to display the imaginary part 5.23.

$\therefore\ 7.4\underline{/2.356} = -5.23 + j5.23.$

▓▓▓▓ Examples 2.22 ▓▓▓▓▓▓▓▓▓▓▓▓▓▓▓▓▓▓▓▓▓▓▓▓

Using a calculator, convert the following complex numbers from Cartesian to polar form   i) $3 + j4$,   ii) $-5 - j12$.

In these examples we may use either degree or radian mode.

i)   For $3 + j4$ our order of key operations is

$\boxed{3}\ \boxed{R \rightarrow P}\ \boxed{4}\ \boxed{=}$ $\qquad$ to display the modulus 5

followed by $\boxed{X \leftrightarrow Y}$ $\qquad$ to display the argument 53.13°.

In radian mode the argument is 0.927.

$\therefore \quad 3 + j4 = 5\underline{/53.13°}.$

ii) For $-5 - j12$ our order of key operations is

$\underline{5|}$ $\underline{^{+/-}|}$ $\underline{R{\to}P|}$ $\underline{12|}$ $\underline{^{+/-}|}$ $\underline{=|}$ to display the modulus 13

followed by $\underline{X{\leftrightarrow}Y|}$ to display the argument $-112.62°$.

In radian mode the argument is -1.966 .

$\therefore$ $-5 - j12 = 13\underline{/-112.62°}$.

■ ASSIGNMENT ■

We take another look at our a.c. circuit and its total impedance of $18.52 - j7.12 \, \Omega$. Using a calculator we can convert this to a polar form complex number. The order of key operations is

$\underline{18.52|}$ $\underline{R{\to}P|}$ $\underline{7.12|}$ $\underline{^{+/-}|}$ $\underline{=|}$ to display the modulus 19.84

followed by $\underline{X{\leftrightarrow}Y|}$ to display the argument $-21.03°$.

This means we can write alternatively that the total impedance is $19.84\underline{/-21.03°} \, \Omega$.

■ EXERCISE 2.9 ■

Convert the following complex numbers from polar to Cartesian form.

1 $1.8\underline{/13°}$

2 $4.5\underline{/117°}$

3 $2.5\underline{/-76°}$

4 $11.4\underline{/-135°}$

5 $2.7\underline{/-90°}$

Convert the following complex numbers from Cartesian to polar form.

6 $5 + j2$

7 $6 - j7$

8 $-1 + j3$

9 $-3 - j4$

10 $-15 - j15$

Square root of a complex number

We need our complex number to be in polar form rather than Cartesian form. Remember that a square root of a number multiplied by itself gives that original number. Let us write this mathematically for some original complex number, $z = r\underline{/\theta}$.

Now $\sqrt{z} \times \sqrt{z} = z$.

When multiplying in polar form we know to multiply the moduli and to add the arguments. For the moduli this means we need to choose $\sqrt{r}$ and for the arguments $\frac{\theta}{2}$. Then the multiplication of $\sqrt{r}$ and itself gives us r.

Also the addition of $\frac{\theta}{2}$ and itself gives us θ. Mathematically this is

$$\sqrt{r}\underline{/\frac{\theta}{2}} \times \sqrt{r}\underline{/\frac{\theta}{2}} = \sqrt{r} \times \sqrt{r}\underline{/\frac{\theta}{2} + \frac{\theta}{2}}$$
$$= r\underline{/\theta}.$$

Also, each number has two square roots. Now we know that θ and $(360° + \theta)$ occupy the same angular position. In our build up to a square root we used $\frac{\theta}{2}$ and now suggest we also use $\frac{(360° + \theta)}{2}$, i.e. $\left(180° + \frac{\theta}{2}\right)$ as the other argument.

We bring together all these features. For a complex number $z = r\underline{/\theta}$ the square roots are given by

$$\sqrt{z} = \sqrt{r}\underline{/\frac{\theta}{2}} \quad \text{and} \quad \sqrt{r}\underline{/180° + \frac{\theta}{2}}.$$

Examples 2.23

Find the square roots of the complex numbers

i) $z_1 = 17\underline{/25°}$, ii) $z_2 = 3.69\underline{/-154°}$, iii) $z_3 = -5 + j8$.

i) For $z_1 = 17\underline{/25°}$ we see that $r = 17$ and $\theta = 25°$.

Then $\sqrt{r} = \sqrt{17} = 4.12$.

Also $\theta = \frac{25°}{2} = 12.5°$ and $180° + \frac{\theta}{2} = 192.5°$.

The second value is outside the range of principal values. In the usual way we bring it back into the range as $-167.5°$.

The square roots are $4.12\underline{/12.5°}$ and $4.12\underline{/-167.5°}$. It is useful to see these square roots displayed on an Argand diagram. You can see in Fig. 2.23 that they both lie on the same straight line. They are similarly inclined to the horizontal axis, one above and one below. This is an expected result.

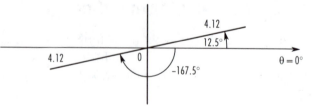

Fig. 2.23

ii) For $z_2 = 3.69 \underline{/-154°}$ we see that $r = 3.69$ and $\theta = -154°$.

Then $\sqrt{r} = \sqrt{3.69} = 1.92$.

Also $\dfrac{\theta}{2} = \dfrac{-154°}{2} = -77°$ and $180° + \dfrac{\theta}{2} = 103°$.

The square roots are $1.92\underline{/-77°}$ and $1.92\underline{/103°}$.

As an exercise, draw these roots on an Argand diagram. Check that they are correctly inclined to the horizontal axis.

iii) For $z_3 = -5 + j8$ we need to convert z_3 to polar form. The order of key operations is

$\boxed{5} \quad \boxed{^{+}/_{-}} \quad \boxed{R{\to}P} \quad \boxed{8} \quad \boxed{=}$ to display the modulus 9.43

followed by $\boxed{X{\leftrightarrow}Y}$ to display the argument 122°.

Then $\sqrt{r} = \sqrt{9.43} = 3.07$.

Also $\dfrac{\theta}{2} = \dfrac{122°}{2} = 61°$ and $180° + \dfrac{\theta}{2} = 241°$.

The second value is outside the range of principal values. In the usual way we bring it back into the range as $-119°$.

The square roots are $3.07\underline{/61°}$ and $3.07\underline{/-119°}$.

◼◼◼◼ EXERCISE 2.10 ◼◼◼◼◼◼◼◼◼

Find the square roots of the following complex numbers.

1	$49\underline{/60°}$	6	$3 + j4$
2	$121\underline{/135°}$	7	$-12 + j5$
3	$13.6\underline{/-24°}$	8	$24 - j7$
4	$6.2\underline{/-118°}$	9	$-13 - j10$
5	$25\underline{/180°}$	10	$j9$

The final exercise applies complex numbers to a.c. theory. You should recall the series and parallel impedances from the assignment. All the other important formulae are given in each question.

◼◼◼◼ EXERCISE 2.11 ◼◼◼◼◼◼◼◼◼

1 A circuit with resistance and inductive reactance has an impedance, Z, given by $Z = 35 + j20\,\Omega$. If the admittance, Y, is given by $\dfrac{1}{Z}$ calculate its value giving your answers in both Cartesian and polar form.

2 The total impedance, Z, of two impedances, z_1 and z_2, in parallel is given by $Z = \dfrac{z_1 z_2}{z_1 + z_2}$.

$z_1 = 40 + j50\,\Omega$ and $z_2 = 20\,\Omega$. Find the value of $z_1 + z_2$ in Cartesian form and then convert it to polar form.

Also convert z_1 and z_2 to polar form before calculating Z.

3 Impedances $z_1 = 2 - j3\,\Omega$ and $z_2 = 6 + j5\,\Omega$ are connected. Find the total impedance, Z, when in

i)   series $Z = z_1 + z_2$,

ii) parallel $\dfrac{1}{Z} = \dfrac{1}{z_1} + \dfrac{1}{z_2}$.

4 A voltage $V = 50 + j45\,$V is applied across an impedance $Z = 4.5 + j1.5\,\Omega$. Using $V = IZ$ find the current I amps in Cartesian form. What is the size of this current?

5 In a circuit the power, P, is related to the current, I, and the impedance, Z, by $P = I^2 Z$. For $P = 40\underline{/110^\circ}$ and $Z = 30\underline{/45^\circ}$ find I^2 in polar form. Hence find I.

6 Convert the impedances $z_1 = 3 + j5\,\Omega$ and $z_2 = 5 + j250\,\Omega$ into polar form.

For a transmission line the propagation constant, P, is given by the formula $P = 10^{-4} \times \sqrt{z_1 z_2}$. Find P in both polar and Cartesian forms. The correct value for P must lie in the positive quadrant so discard any other answers.

7 A resistor of $20\,\Omega$ is connected in series with an inductor of $0.25\,$H. The terminal voltage, V, is $230\,$V and the frequency, f, is $50\,$Hz. Calculate ω where $\omega = 2\pi f$. Calculate the total impedance Z in polar form. Write down the voltage in polar form, and using $V = IZ$ find the current I also in polar form.

8 The total impedance, Z, across the elements in parallel is given by

$Z = \dfrac{z_1 z_2}{z_1 + z_2}$.

Given that $z_1 = 2.5 + j3.75\,\Omega$ and $z_2 = 4.5 + j2.5\,\Omega$ find the value of Z in Cartesian form.

9 The characteristic impedance, Z, of a transmission line is given by $Z^2 = \dfrac{R + j\omega L}{G + j\omega C}$. Calculate Z for $R = 2.5\,\Omega$, $L = 0.2 \times 10^{-3}\,$H, $G = 4 \times 10^{-6}\,\Omega$, $C = 10^{-9}\,$F and $\omega = 10^4\,$Hz.

10 The circuit diagram relates to a simple low pass filter. The open circuit input impedance, z_{oc}, is given by $z_{oc} = \dfrac{z_2(z_1 + z_2)}{2z_2 + z_1}$. The short

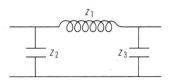

circuit input impedance, z_{sc}, is given by $z_{sc} = \dfrac{z_1 z_2}{z_1 + z_2}$. The characteristic impedance, z_c, is given by $z_c = \sqrt{z_{oc} z_{sc}}$.

For $z_1 = j\,800\,\Omega$ and $z_2 = -j\,90\,\Omega$ find z_{oc}, z_{sc} and z_c in polar form. Convert your answers to Cartesian form.

3 Trigonometrical Graphs

The objectives of this chapter are to:

1 Revise the degrees and radians relationship.

2 Define amplitude, period and frequency.

3 Sketch the graphs of $\sin x$, $2\sin x$, $\frac{1}{2}\sin x$, $\sin 2x$, $\sin\frac{1}{2}x$ and $\cos x$, $2\cos x$, $\frac{1}{2}\cos x$, $\cos 2x$, $\cos\frac{1}{2}x$ for values of x between $0°$ and $360°$.

4 Sketch general graphs of the form $A\sin ax$ and $A\cos ax$ where A and a are constants.

5 Define angular velocity.

6 In **4** above replace ax with ωt.

7 Sketch the graphs of $\sin^2 x$ and $\cos^2 x$ for values of x between $0°$ and $360°$.

8 State the approximations for $\sin x$, $\cos x$ and $\tan x$ where x is small.

9 Define phase angle and measure it where appropriate.

10 Determine a single wave resulting from a combination of two waves of the same frequency.

11 Determine graphically a single wave resulting from a combination of two waves of different frequencies.

12 Show that the wave resulting from a combination of two sine waves of different frequencies is non-sinusoidal.

Introduction

The majority of this chapter looks at sine and cosine **graphs (waveforms)**. We start with the standard waves for $\sin x$ and $\cos x$ and develop them with different multipliers. We can use either degrees or radians and so remind you of the conversion techniques.

■■■■■ ASSIGNMENT ■■■■■

Our Assignment for this chapter looks at a pair of parallel a.c. voltages. $v_1 = 15\sin 5t$ and $v_2 = 8\cos 5t$. We will look at the waveforms of these voltages separately. Later we will add them to find a total source voltage.

Degrees and radians

The definition of a radian is based on a circle. Draw a circle with centre O and radius r. Mark a point A on the circumference and measure an arc equal in length to the radius. If this arc length ends at B, the angle AOB ($\angle AOB$), is defined to be of size 1 radian. This is shown in Fig. 3.1.

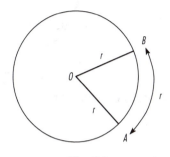

A formal definition is that one radian is the angle **subtended** at the centre of a circle by an arc equal in length to the radius of the circle. We use the short form *rad* instead of radian.

Fig. 3.1

You will remember from Volume 1 that there is a connection between degrees and radians. The circumference of the circle in Fig. 3.1 is given as $2\pi r$. To draw a circle we need to turn through 1 revolution (360°). We see that the greater the angle we turn for a given radius the greater the arc formed until we have completed a full circle.

An arc of length r is linked with an angle of 1 radian;
an arc of length $2r$ is linked with an angle of 2 radians; and so
an arc of length $2\pi r$ is linked with an angle of 2π radians, which is the circumference. This gives us

$$1 \text{ rev} = 360° = 2\pi \text{ rad.}$$

Now we can remind ourselves of the conversions between degrees and radians.

Using $\qquad 360° = 2\pi$

we have $\qquad \dfrac{360°}{360} = \dfrac{2\pi}{360}$
> Dividing by 360.

i.e. $\qquad 1° = \dfrac{\pi}{180} \text{ rad.}$

This is the conversion factor to move from degrees to radians.

Examples 3.1

Convert i) 210° and ii) $\theta°$ into radians.

i) 210° is $210 \times 1° = 210 \times \dfrac{\pi}{180}$

$$= \dfrac{7\pi}{6} \text{ or } 3.665 \text{ rad.}$$

ii) $\theta°$ is $\theta \times 1°$ $= \theta \times \dfrac{\pi}{180}$ or 0.017θ rad.

Using $2\pi = 360°$

we have $\dfrac{2\pi}{2\pi} = \dfrac{360°}{2\pi}$

$\boxed{\text{Dividing by } 2\pi.}$

i.e.   **1 rad** $= \dfrac{\mathbf{180°}}{\pi}.$

This is the conversion factor to move from radians to degrees. A calculator check finds that 1 radian is slightly less than 60°, approximately 57.3° (3 sig. fig.).

Examples 3.2

Convert   i)   0.35 rad and   ii) θ rad into degrees.

i)   0.35 rad is $0.35 \times 1 = 0.35 \times \dfrac{180°}{\pi}$

$= 20.05°$ (4 sf).

ii) θ is $\theta \times 1$ $= \theta \times \dfrac{180°}{\pi}$

$= 57.3\theta°$ (3 sf).

Most of our discussion in this chapter will use the range 0° to 360°. There is no reason to stop there. We can keep going round and round by further revolutions using our earlier relationship that $1 \text{ rev} = 360° = 2\pi$ rad. All revolutions will be multiples of this relationship,

e.g. $2 \text{ rev} = 2 \times 2\pi$ $= 4\pi$ rad,

$5\frac{1}{2} \text{ rev} = 5\frac{1}{2} \times 2\pi$ $= 11\pi$ rad.

EXERCISE 3.1

Convert the following angles from degrees into radians.

1	25°	**4**	109°
2	277.5°	**5**	380°
3	−60°		

Convert the following radian measures into degrees, giving your answer correct to 2 decimal places.

6	1.47	**9**	$\dfrac{5\pi}{12}$
7	$\dfrac{7\pi}{4}$	**10**	1.35
8	2.15		

Graphs of sin *x* and cos *x*

In Fig. 3.2 we refresh our memories with the graphs of $y = \sin x$ and $y = \cos x$ on a pair of axes. You need to look at them carefully and understand their shapes. Look for any similarities and any differences between them. Pay attention to where they cross each axis. They are important because much of our later work is based on these waves. We are going to discuss their features in the next few sections.

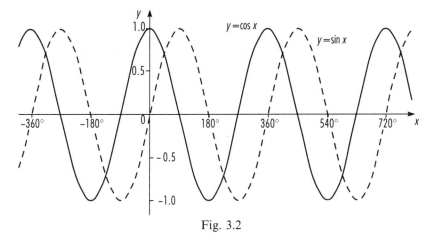

Fig. 3.2

Amplitude

The **amplitude (peak value)** is the maximum value from some mean (or equilibrium) position. For many, but by no means all, trigonometric **graphs (curves** or **waveforms)** the mean position is the horizontal axis. We understand the amplitude to be a positive value.

The amplitudes of the sine and cosine waves are both 1. On the vertical axis the waveforms lie between -1 and 1. Their mean positions are at the horizontal axis. We may think of

$$y = \sin x \qquad \text{and} \qquad y = \cos x$$

as $\quad y = 1 \sin x \qquad$ and $\qquad y = 1 \cos x.$ | 1 is the amplitude. |

More generally $y = A \sin x$ and $y = A \cos x$, where each have an amplitude of A.

▓▓▓▓ **Examples 3.3** ▓▓▓▓

In Fig. 3.3 we sketch the graphs of

i) $y = \sin x,$

ii) $y = 2 \sin x,$

iii) $y = \dfrac{1}{2} \sin x.$

| Amplitude = 1.
| Amplitude = 2.
| Amplitude = $\dfrac{1}{2}$.

Sometimes you may see $\frac{1}{2}\sin x$ written as $\frac{\sin x}{2}$

We omit the complete table but include two specimen calculations.

For $y = 2\sin x$ when $x = 60°$ we use the calculator as

$\underline{60°|}$ $\underline{\sin|}$ to display 0.8660...

and $\underline{\times|}$ $\underline{2|}$ to display an answer of 1.732 (4 sf).

For $y = \frac{1}{2}\sin x$ when $x = 135°$ we use the calculator as

$\underline{135°|}$ $\underline{\sin|}$ to display 0.7071...

and $\underline{\times|}$ $\underline{0.5|}$ to display an answer of 0.354 (3 dp).

As an exercise for yourself you should complete a table and plot the graphs on one set of axes.

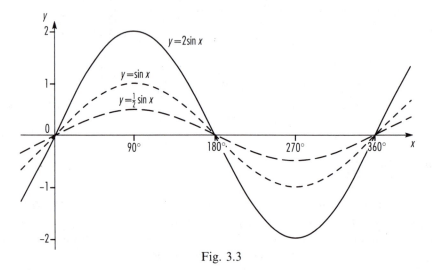

Fig. 3.3

Examples 3.4

In Fig. 3.4 we sketch the graphs of

i) $y = \cos x$, | Amplitude = 1.

ii) $y = 2\cos x$, | Amplitude = 2.

iii) $y = \frac{1}{2}\cos x$. | Amplitude = $\frac{1}{2}$.

Sometimes you may see $\frac{1}{2}\cos x$ written as $\frac{\cos x}{2}$.

We omit the complete table. As an exercise for yourself you should complete a table and plot the graphs on one set of axes. Fig. 3.4 shows the graphs.

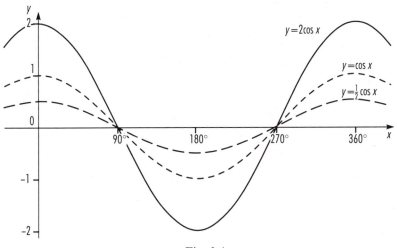

Fig. 3.4

Period

The sine and cosine graphs are continuous waveforms that repeat themselves. This means they are periodic. $y = \sin x$ and $y = \cos x$ repeat themselves every 360° (2π radians). We may think of

$$y = \sin x \qquad \text{and} \qquad y = \cos x$$

as $\qquad y = \sin 1x \qquad$ and $\qquad y = \cos 1x$

$$\boxed{\dfrac{360°}{1} = 360°.}$$

More generally $y = \sin ax$ and $y = \cos ax$ repeat themselves every $\dfrac{360}{a}$°.

This means the graphs of $y = \sin x$ and $y = \cos x$ are periodic, with period 360° (2π radians).

Examples 3.5

In Fig. 3.5 we sketch the graphs of

i) $y = \sin x$, $\qquad a = 1$, $\qquad$ period $= \dfrac{360°}{1} = 360°$.

ii) $y = \sin 2x$, $\qquad a = 2$, $\qquad$ period $= \dfrac{360°}{2} = 180°$.

iii) $y = \sin \frac{1}{2}x$, $\qquad a = \frac{1}{2}$, $\qquad$ period $= \dfrac{360°}{\frac{1}{2}} = 720°$.

The cycle for $y = \sin 2x$ is compressed into 180°. In contrast, the cycle for $y = \sin \frac{1}{2}x$ is extended over 720°. We omit the complete table but include two specimen calculations.

For $\quad y = \sin 2x$ when $x = 45°$ we use the calculator with

$\underline{45°} \quad \underline{\times} \quad \underline{2} \quad \underline{=} \quad$ to display 90°

and $\qquad\qquad\qquad \underline{\sin} \quad$ to display an answer of 1.

For $y = \sin \frac{1}{2} x$ when $x = 150°$ we use the calculator as

$\boxed{150°}$ $\boxed{\times}$ $\boxed{0.5}$ $\boxed{=}$ to display 75°

and $\underline{\sin}$ to display an answer of 0.966.

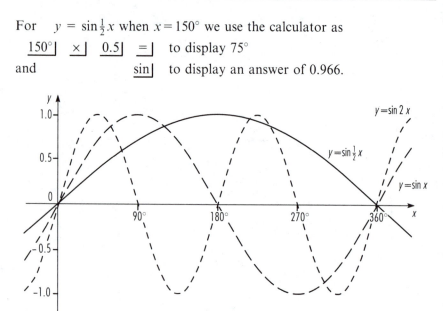

Fig. 3.5

Examples 3.6

In Fig. 3.6 we sketch the graphs of

i) $y = \cos x$, $a = 1$,   period $= \dfrac{360°}{1} = 360°$.

ii) $y = \cos 2x$, $a = 2$,   period $= \dfrac{360°}{2} = 180°$.

iii) $y = \cos \frac{1}{2} x$, $a = \frac{1}{2}$,   period $= \dfrac{360°}{\frac{1}{2}} = 720°$

The cycle for $y = \cos 2x$ is compressed into 180°. In contrast, the cycle for $y = \cos \frac{1}{2} x$ is extended over 720°.

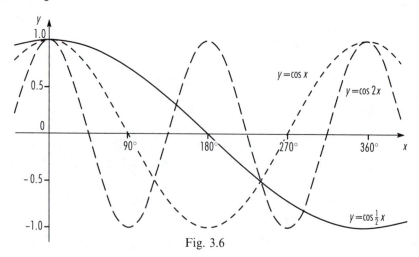

Fig. 3.6

We often refer to **period** as **time period**. The units for time are *not* degrees. Instead of $y = \sin ax$ and $y = \cos ax$ we use time, t, $y = \sin \omega t$ and $y = \cos \omega t$ with radians.

$$\text{Time period, } T = \frac{2\pi}{\omega} \text{ seconds.}$$

Examples 3.7

In terms of time, t, we look at time period using radians.

i) $y = \cos t$, $\omega = 1$,   time period $= \dfrac{2\pi}{1} = 2\pi$.

ii) $y = \cos 2t$, $\omega = 2$,   time period $= \dfrac{2\pi}{2} = \pi$.

iii) $y = \cos \frac{1}{2} t$ $\omega = \frac{1}{2}$,   time period $= \dfrac{2\pi}{\frac{1}{2}} = 4\pi$.

We have looked at the basic waves of $y = A\sin ax$ and $y = A\cos ax$ separately, altering A and a. Just as easily, we can alter them both together. In Examples 3.8 we look at the necessary order of calculator operations. A complete table and plot is left for you to do in Exercise 3.2.

Examples 3.8

i) For $y = 3 \sin 4x$, let $x = 75°$. We find y by using

    $\boxed{4}$ $\boxed{\times}$ $\boxed{75°}$ $\boxed{=}$ to display $300°$,

                              $\boxed{\sin}$ to display $-0.8660\ldots$

and $\boxed{\times}$ $\boxed{3}$ $\boxed{=}$ to display an answer of -2.598.

ii) We can think of $y = 2.5 \cos \dfrac{1}{3} x$ as $y = 2.5 \cos \dfrac{x}{3}$. Let $x = 150°$ and find y by using

    $\boxed{150°}$ $\boxed{\div}$ $\boxed{3}$ $\boxed{=}$ to display $50°$,

                              $\boxed{\cos}$ to display $0.642\ldots$

and $\boxed{\times}$ $\boxed{2.5}$ $\boxed{=}$ to display an answer of 1.607.

For our next variation we see what happens when we add a number rather than simply use multiplication.

Example 3.9

For $y = 2 + \sin x$ our original sine wave is shifted vertically upwards by 2. This vertical shift has no effect on the size of the amplitude, angular velocity, period or frequency. For this example involving $\sin x$ these values remain as 1, 1, 2π and $\dfrac{1}{2\pi}$ respectively. Again we omit the complete table but include a specimen calculation. We let $x = 35°$ and find y by using

35°⎤ sin⎤   to display 0.5735...

and   +⎤ 2⎤ =⎤   to display 2.574.

We show the wave in Fig. 3.7.

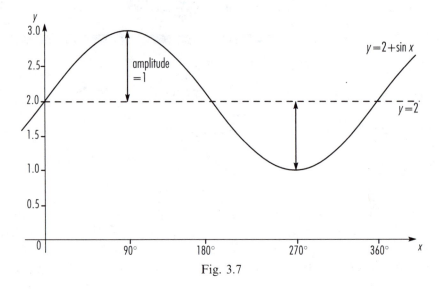

Fig. 3.7

Frequency

We define **frequency, f,** to be the number of cycles every second. The units of frequency are Hertz (Hz).

$$\textbf{Frequency} = \frac{\textbf{Number of cycles}}{\textbf{1 second}}$$

$$= \frac{1 \text{ cycle}}{\text{Time period}}$$

$$= \frac{1}{T} = \frac{1}{\frac{2\pi}{\omega}}$$

i.e. $f = \dfrac{\omega}{2\pi}$ **Hz.**

Also $f = \dfrac{1}{T}$.

$\omega = 2\pi f$ is the **angular velocity** with units of radians per second (rads^{-1}).

▨▨▨▨▨ **Example 3.10** ▨▨▨▨▨▨▨▨▨▨▨▨▨▨▨▨▨▨▨▨

$$y = 4\sin\frac{3}{5}t \qquad\qquad y = A\sin\omega t.$$

has $\omega = \dfrac{3}{5}$ or 0.6.

This means we have an angular velocity of 0.6 rads^{-1}.

We have a time period of $\dfrac{2\pi}{0.6} = 10.5$ seconds. $T = \dfrac{2\pi}{\omega}$

Also we have a frequency of $\dfrac{0.6}{2\pi} = 0.095$ Hz. $f = \dfrac{\omega}{2\pi}.$

ASSIGNMENT

Let us look at our a.c. voltages, $v_1 = 15\sin 5t$ and $v_2 = 8\cos 5t$.

For v_1 the amplitude is 15 and for v_2 the amplitude is 8. For both voltages $\omega = 5$ and so the time period is $\dfrac{2\pi}{5} = 0.4\pi = 1.256\,\text{s}$. Using $f = \dfrac{1}{T}$ the frequency is 0.80 Hz.

In Fig. 3.8 we plot these waves together on one set of axes. Again we omit the table of values, but you will have plenty of practice in the next exercise.

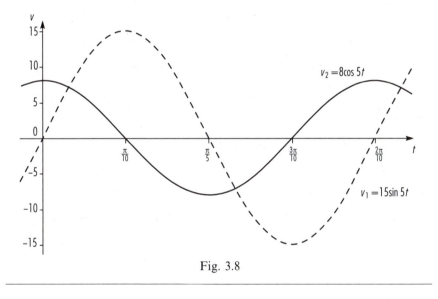

Fig. 3.8

EXERCISE 3.2

For the trigonometrical waveforms in Questions 1 to 5 write down the
  i) amplitude,
 ii) angular velocity,
iii) period (degrees and radians), and
 iv) frequency.

1 $y = \sin 3x.$

2 $y = 5\cos 2t.$

3 $y = \cos 3x.$

4 $y = 2\sin 3x.$

5 $y = -1 + \cos t.$

6   Over 1 cycle sketch the waveforms in Questions 1 to 5.

7   Write down the equation of the sinusoidal wave with an amplitude of 4 and a frequency of 2.5.

8   You are given the equation $y = 2\cos 3t$. Write down the sine wave with the same amplitude and twice the frequency.

9   On one pair of labelled axes over 1 cycle sketch the graphs of i) $y_1 = \sin 3x$, ii) $y_2 = 2\sin 3x$, iii) $y_3 = 4 + 2\sin 3x$. Clearly label where each curve cuts the axes.

10   For the wave $y = 3\sin 4x$ construct a table of values and plot the graph on a fully labelled pair of axes. Use values of x from $0°$ to $100°$ at intervals of $5°$.

11   In the range 0 to 2π radians on one set of labelled axes sketch i) $y_1 = \cos x$, ii) $y_2 = \cos 2x$, iii) $y_3 = 1 + \cos 2x$. Clearly label where each curve cuts the axes.

12   Over 1 cycle on the same axes sketch the graphs of $y = 2\sin x$ and $y = -2\sin x$. Clearly label each curve.

13   For the wave $y = 4 + 2\sin 3t$ construct a table of values and plot the graph on a fully labelled pair of axes. Use values of t from 0 to $\dfrac{2\pi}{3}$ radians at intervals of $\dfrac{\pi}{18}$.

14   What is the period of the wave $y = \dfrac{1}{2}(1 - \cos 2x)$? Construct a table of values from $0°$ over this period at intervals of $15°$. Hence plot a fully labelled graph of y against x.

15   Plot the graph of $y = \dfrac{1}{2}(1 + \cos 2x)$ over 1 cycle. Your table of values should be from $x = 0°$ at intervals of $15°$. Fully label your graph.

Graphs of $\sin^2 x$ and $\cos^2 x$

One of our original waves was $y = \sin x$. We base the graph of $y = \sin^2 x$ on this wave, understanding $\sin^2 x$ to be $(\sin x)^2$, i.e. we square the values of our original sine wave. This makes all the negative values positive and so the wave does *not* lie below the horizontal axis. We omit the complete table but include a specimen calculation. Let $x = 240°$ so our order of calulator operations is

$$\underline{240°}\mid\ \underline{\sin}\mid\quad \text{to display } -0.8660\ldots$$

and $\qquad\qquad\qquad \underline{x^2}\mid\quad \text{to display } 0.75.$

Fig. 3.9 shows both waves on one set of axes.

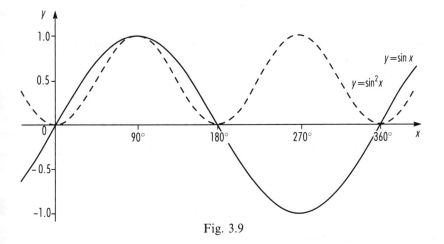

Fig. 3.9

Exactly the same ideas apply to $y = \cos^2 x$, being the squared values of our original cosine wave. Again, we include a specimen calculation with $x = 135°$ so that

$$135°| \quad \cos| \quad \text{to display} -0.7071\ldots$$
and $\qquad\qquad x^2| \quad$ to display 0.5.

Fig. 3.10 shows both waves on one set of axes

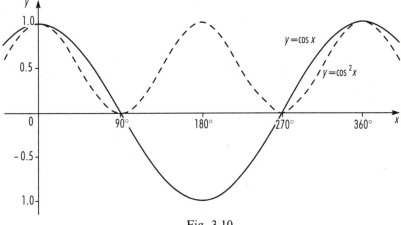

Fig. 3.10

In Fig. 3.11 we bring both $y = \sin^2 x$ and $y = \cos^2 x$ together on one set of axes. Notice how they oscillate about the line $y = 0.5$, i.e. this is their mean position. Then the amplitude (maximum displacement from this mean position) is 0.5. Also, the waves repeat themselves every $180°$ (π radians).

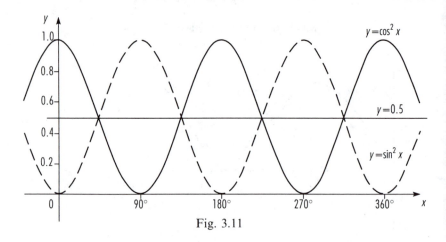

Fig. 3.11

Approximations

We have **approximations in radians** for $\sin x$, $\cos x$ and $\tan x$. These are, provided x is small, generally less than 0.10 rad ($5°$ or $6°$). Remember that $1° = \dfrac{\pi}{180}$ radians.

$\sin x \approx x$.

$\cos x \approx 1 - \dfrac{x^2}{2}$.

$\tan x \approx x$.

> $\approx$ is "equals approximately".

> x is in radians.

To avoid early approximation we use the calculator memory and quite a few decimal places.

▰▰▰ Examples 3.11 ▰▰▰

i) Let $x = 2.5° = 0.0436\ldots$ rad.

Then $\sin x = \sin 0.0436\ldots$ $= 0.0436\ldots$

only differs in the 5th decimal place.

$\tan x = \tan 0.0436\ldots$ $= 0.0436\ldots$

only differs in the 5th decimal place.

$\cos x = \cos 0.0436\ldots$ $= 0.999048\ldots$

and $1 - \dfrac{x^2}{2} = 1 - \dfrac{(0.0436\ldots)^2}{2} = 0.999048\ldots$

is correct to the first 6 decimal places.

ii) Let $x = 8° = 0.1396$ rad.

Then $\sin x = \sin 0.1396$ $= 0.13917\ldots$

only differs in the 4th decimal place.

$$\tan x = \tan 0.1396 \qquad = 0.1405\ldots$$

differs in the 2nd decimal place.

$$\cos x = \cos 0.1396\ldots \qquad = 0.990268\ldots$$

and $\quad 1 - \dfrac{x^2}{2} = 1 - \dfrac{(0.13966\ldots)^2}{2} = 0.990252\ldots$

only differs in the 5th decimal place.

In Figs. 3.12(a), (b) and (c) we look at the graphs of $\sin x$, $\cos x$ and $\tan x$ for small values of x. You can see how good the approximations are close to the vertical axis (i.e. $x=0$). Only as we move away with the size of x increasing do the graphs and their approximations diverge.

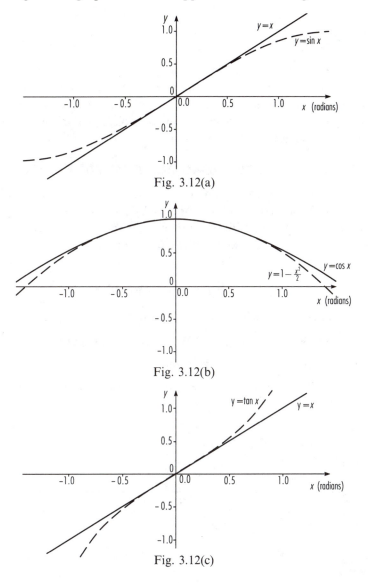

Fig. 3.12(a)

Fig. 3.12(b)

Fig. 3.12(c)

▌▬▬▬ EXERCISE 3.3 ▬▬▬▬▬▬▬▬▬▬▬

For $\sin x$, $\cos x$ and $\tan x$ test the accuracy of the approximations. In each case state at which decimal place the trigonometric function and its approximation start to differ.

1	0.100 rad		**6**	8.5°
2	0.125 rad		**7**	−10°
3	−0.150 rad		**8**	−7.5°
4	0.200 rad		**9**	12.5°
5	−0.185 rad		**10**	11°

Phase angle

So far we have seen sine and cosine waves crossing the horizontal axis at 90°, 180°, 270° and 360° etc. When we alter the frequency we get multiples and sub-multiples of these values. Also we have seen a vertical shift of our graphs, e.g. $y = 2 + \sin x$. In this section we look at a horizontal shift. A horizontal shift creates a **phase difference**.

▬▬▬▬ Example 3.12 ▬▬▬▬▬▬▬▬▬▬▬▬▬▬▬▬▬▬▬▬▬▬▬▬▬

In Fig. 3.13 we sketch the graphs of $y = \sin x$ and $y = \sin(x + 30°)$. You can see we shift the first wave horizontally by 30° ($\pi/6$ radians) to get the second wave. Again, we omit the complete table but include two specimen calculations. When $x = 20°$ we use the calculator with

$\boxed{20°}$ $\boxed{+}$ $\boxed{30°}$ $\boxed{=}$ to display 50°

and $\boxed{\sin}$ to display an answer of 0.766. . .

In radians we can rewrite 30° and our equation as either $y = \sin\left(x + \dfrac{\pi}{6}\right)$ or $y = \sin(x + 0.523 . . .)$. When $x = 1.375$ we use the calculator with

$\boxed{1.375}$ $\boxed{+}$ $\boxed{0.523. . .}$ $\boxed{=}$ to display 1.898 . . .

and $\boxed{\sin}$ to display an answer of 0.946 . . .

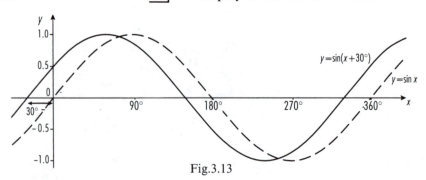

Fig.3.13

We have marked the **phase angle** of 30° on our diagram. Notice that we have shifted the graph of $y = \sin x$ horizontally to the left to get $y = \sin(x + 30°)$. $y = \sin(x + 30°)$ **leads** $y = \sin x$. We can think of this as reaching the peak value first. Alternatively, $y = \sin x$ **lags** $y = \sin(x + 30°)$.

▰▰▰▰▰▰ **Example 3.13** ▰▰▰▰▰▰▰▰▰▰▰▰▰▰▰▰▰▰▰▰▰▰▰▰▰▰▰▰

In Fig. 3.14 we sketch the graphs of $y = \sin t$ and $y = \sin\left(t - \dfrac{\pi}{4}\right)$. You can see we shift the first wave horizontally $\pi/4$ radians to get the second wave. Again, we omit the complete table. The calculations follow like those in Example 3.12 using subtraction rather than an addition. We have marked the **phase angle** of $\pi/4$ on our diagram. Notice that we have shifted the graph of $y = \sin t$ horizontally to the right to get $y = \sin\left(t - \dfrac{\pi}{4}\right)$. $y = \sin\left(t - \dfrac{\pi}{4}\right)$ **lags** $y = \sin t$. We can think of this as reaching the peak value later. Alternatively, $y = \sin t$ **leads** $y = \sin\left(t - \dfrac{\pi}{4}\right)$.

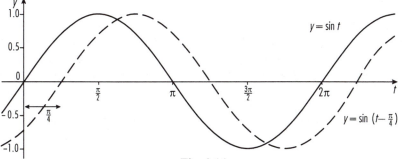

Fig. 3.14

Generally we can shift our waves horizontally and compare them against $y = R \sin \omega t$.

Then $y_1 = R\sin(\omega t + \alpha)$ **leads** $y = R \sin \omega t$

and $y_2 = R\sin(\omega t - \alpha)$ **lags** $y = R \sin \omega t$,

each with a phase angle of α.

Exactly the same ideas apply to cosine graphs:

$y_1 = R\cos(\omega t + \alpha)$ **leads** $y = R \cos \omega t$

and $y_2 = R\cos(\omega t - \alpha)$ **lags** $y = R \cos \omega t$,

each with a phase angle of α.

Let us return to our basic sine and cosine curves, shown in Fig. 3.15. You can see that the cosine curve leads the sine curve by 90° ($\pi/2$ radians). We may write that $y = \cos x$ is equivalent to $y = \sin(x + 90°)$. Alternatively, the sine curve lags the cosine curve. We may write that $y = \sin x$ is equivalent to $y = \cos(x - 90°)$.

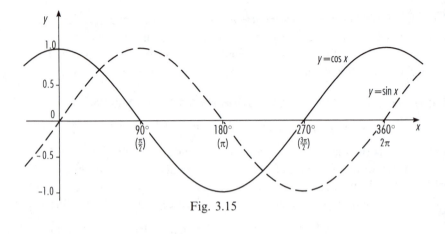

Fig. 3.15

■ EXERCISE 3.4 ■

1   Write down the cosine graph that leads $y = 4\cos t$ by $\dfrac{\pi}{3}$. Sketch both waves on one set of axes with clear labels.

2   Write down the sine graph that lags $y = 3\sin 2t$ by $\dfrac{\pi}{6}$. Sketch both waves on one set of axes with clear labels.

3   Write down the sine graph that leads $y = \sin x$ by $90°$. Sketch both these waves on one set of axes. We have seen the leading graph earlier in this chapter. Write down its alternative equation.

4   Plot the graphs of current, i, against time, t, $i_1 = \sin\left(2t + \dfrac{\pi}{6}\right)$ and $i_2 = \sin 2t$. Construct your table using values of t over 1 cycle from $t = 0$ radians at intervals of $\dfrac{\pi}{12}$ radians. Clearly label your graphs and axes together with the phase angle. Write down the lead/lag relationship between these currents.

5   Plot the graphs of voltage, v, against time, t, $v_1 = \sin\left(t + \dfrac{\pi}{3}\right)$ and $v_2 = \sin\left(t - \dfrac{\pi}{4}\right)$. Construct your table using values of t from $t = -\dfrac{\pi}{2}$ to $t = \dfrac{5\pi}{2}$. Clearly label your graphs and axes together with the phase angle. Write down the lead/lag relationship between the voltages.

Graphical wave addition – same frequencies

This is simply done. We construct a table of values and add those values where necessary. In Example 3.14 we do this in stages, noting the result is sinusoidal.

Example 3.14

Plot the graph of $y = 3\sin x + 4\cos x$ over 1 cycle.

We break this down in stages, looking separately at $3\sin x$ and $4\cos x$. Our table shows values of x in degrees but radians will work just as easily. Create your own table from $x = 0°$ to $x = 360°$ at intervals of either $10°$ or $5°$.

x	$0°$	$30°$	$45°$	$\dots$	
$\sin x$	0	0.500	0.707	$\dots$	
$3\sin x$	0	1.500	2.121	$\dots$	y_1
$\cos x$	1	0.866	0.707	$\dots$	
$4\cos x$	4	3.464	2.828	$\dots$	y_2
$3\sin x + 4\cos x$	4	4.96	4.95	$\dots$	$y_1 + y_2$

We have plotted all three graphs in Fig. 3.16. We are particularly interested in the combined wave, $y = 3\sin x + 4\cos x$. It retains the sinusoidal shape with a phase shift. Notice the amplitude is 5 and the phase angle is $53.13°$ compared with $y = 5\sin x$. In fact, our graph may be alternatively labelled as $y = 5\sin(x + 53.13°)$. $y = 3\sin x + 4\cos x$ is equivalent to $y = 5\sin(x + 53.13°)$.

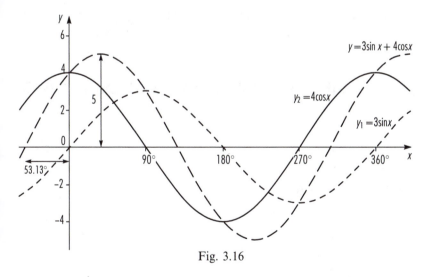

Fig. 3.16

ASSIGNMENT

We can add our a.c. voltages $v_1 = 15\sin 5t$ and $v_2 = 8\cos 5t$ graphically using the method from Example 3.14. This time we ought to use radians over 1 cycle from $t = 0$ to $t = 1.26$ s. As an exercise for yourself, construct a table and plot the combined wave of $v = 15\sin 5t + 8\cos 5t$ at intervals of 0.05 or 0.10 seconds. You can check your own attempt against our graph in Fig. 3.17. Unlike Example 3.14, we have highlighted the combined wave by plotting it alone.

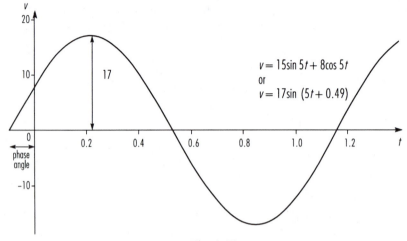

Fig. 3.17

Notice the amplitude is 17. Notice the size of our phase shift is shown as $\frac{0.49}{5} = 0.10$ (2dp) because we have plotted t rather than $5t$ horizontally. In fact, our graph may be alternatively labelled as $v = 17 \sin(5t + 0.49)$. $v = 15 \sin 5t + 8 \cos 5t$ is equivalent to $v = 17 \sin(5t + 0.49)$.

■■■■ EXERCISE 3.5 ■■■■■■■■■■■■■■■■■■■■■■■■■■■■■■■■

In the following questions, graphically add the given waves. Your table should cover 1 cycle of values. Clearly label the phase angle. State the lead/lag relationship between your equivalent sine wave and a sine wave of the same frequency.

1   $y_1 = 5 \sin x$ and $y_2 = 12 \cos x$.

2   $y_1 = \sin 2t$ and $y_2 = 2 \cos 2t$.

3   $y_1 = 12 \sin x$ and $y_2 = 12 \cos x$.

4   $y_1 = 3 \sin 2t$ and $y_2 = 6 \cos 2t$.

5   $y_1 = 9 \sin 3x$ and $y_2 = 6 \cos 3x$.

Phasor addition

We only apply this method to waves of the **same frequency**. We know from earlier graphs that the sine and cosine waves of the same frequency are 90° out of phase, i.e. the cosine leads the sine by 90°. This allows us to use them at right-angles to each other. We use the amplitudes in each case. Our method involves drawing the amplitude of the sine wave horizontally and of the cosine wave vertically. Then we complete the rectangle. The diagonal from the origin is the amplitude of the combined waves. Its inclination to the horizontal is the phase angle. Above the horizontal the phase angle is positive, and negative below. We can use either degrees or radians.

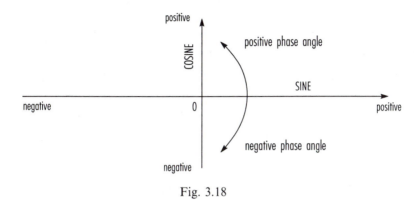

Fig. 3.18

Example 3.15

Use phasor addition to find the sine wave representing $y = 3 \sin x + 4 \cos x$.

We saw this combination of sine and cosine in Example 3.14. This time we draw a horizontal line of length 3 (amplitude of $3 \sin x$) and a vertical line of length 4 (amplitude of $4 \cos x$), as in Fig. 3.19.

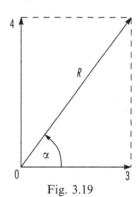

Fig. 3.19

We complete the rectangle and can measure the diagonal, R. Alternatively we can apply Pythagoras' theorem,

$$R^2 = 3^2 + 4^2$$

i.e. $R = \sqrt{9 + 16} = 5,$ using the positive root.

For the phase angle, α, we can measure the inclination of the diagonal to the horizontal. Alternatively we can use some simple trigonometry

$$\tan \alpha = \frac{4}{3} = 1.\bar{3}$$

i.e. $\alpha = 53.13°.$

Obviously we *cannot* expect such accuracy with measurement. As an alternative we could have used radians just as easily as degrees. Finally we write our combined wave as $y = 5 \sin(x + 53.13°)$, as in Example 3.14.

Example 3.16

Use phasor addition to find the sine wave representing
$y = 12 \sin x - 5 \cos x.$

We draw a horizontal line of length
12 (amplitude of $12 \sin x$) and a
vertical line of length 5 (amplitude of
$5 \cos x$) in Fig. 3.20. Notice the
vertical line is downwards, to be
consistent with a negative cosine.

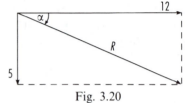

Fig. 3.20

We complete the rectangle and can
measure the diagonal, R, or apply Pythagoras' theorem,

$$R^2 = 12^2 + 5^2$$

i.e. $R = \sqrt{144 + 25} = 13,$ using the positive root.

For the phase angle, α, we can measure the inclination of the diagonal to
the horizontal or use some simple trigonometry,

$$\tan \alpha = -\frac{5}{12}$$

i.e. $\alpha = -22.62°.$

This means our combined wave is $y = 13 \sin(x - 22.62°).$

Example 3.17

Use phasor addition to find the sine wave representing
$y = 7.5 \sin x + 4 \sin(x + 60°).$

We draw a horizontal line of length 7.5 and a line of length 4 inclined at
60° above the horizontal (sine wave with leading phase angle 60°) in
Fig. 3.21. This time we complete the parallelogram. Again we can measure
for our approximate values and calculate for greater accuracy. Without
right-angled triangles we need to use the sine and cosine rules.

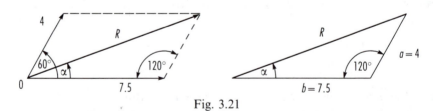

Fig. 3.21

Using the cosine rule,

$$a^2 = b^2 + c^2 - 2bc \cos A$$

we substitute to get

$$R^2 = 4^2 + 7.5^2 - 2(4)(7.5) \cos 120°$$
$$= 16 + 56.25 - 2(4)(7.5)(-0.5)$$

i.e. $R^2 = 102.25$

$\therefore$ $\qquad R = \sqrt{102.25} = 10.1$ $\qquad$ is the amplitude.

Using the sine rule,

$$\frac{a}{\sin A} = \frac{b}{\sin B}$$

we substitute to get

$$\frac{R}{\sin 120°} = \frac{4}{\sin \alpha}$$

i.e. $\qquad \sin \alpha = \dfrac{4 \times 0.866}{10.1\ldots} = 0.3426$

$\therefore$ $\qquad \alpha = 20.0°$ $\qquad$ is the phase angle.

This means our combined wave is $y = 10.1 \sin(x + 20.0°)$.

ASSIGNMENT

We can add our a.c. voltages $v_1 = 15 \sin 5t$ and $v_2 = 8 \cos 5t$ using phasor addition. We draw a horizontal line of length 15 and a vertical line of length 8 in Fig. 3.22.

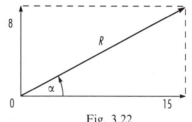

Fig. 3.22

We complete the rectangle and apply Pythagoras' theorem,

$$R^2 = 15^2 + 8^2$$

i.e. $\qquad R = \sqrt{225 + 64} = 17,$ $\qquad$ using the positive root.

For the phase angle, α, we use simple trigonometry,

$$\tan \alpha = \frac{8}{15} = 0.5\bar{3}$$

i.e. $\qquad \alpha = 0.49$ rads.

Finally we write our combined wave as $y = 17 \sin(5t + 0.49)$.

EXERCISE 3.6

Use phasor addition to add the given waves. In each case calculate the amplitude and phase angle. Write down the combined wave in the form of a sine.

1 $\quad y_1 = 12 \sin x$ $\qquad$ and $\qquad y_2 = 12 \cos x$

2 $\quad y_1 = 6 \sin 2t$ $\qquad$ and $\qquad y_2 = 4 \cos 2t$

3 $\quad y_1 = 7 \sin 3x$ $\qquad$ and $\qquad y_2 = 3.5 \cos 3x$

4 $\quad i_1 = \sin\left(2t + \dfrac{\pi}{6}\right)$ and $\qquad i_2 = \sin 2t$

5 $\quad v_1 = \sin\left(t + \dfrac{\pi}{3}\right)$ and $\qquad v_2 = \sin\left(t - \dfrac{\pi}{4}\right)$

Graphical wave addition – different frequencies

We need to add waves of different frequencies graphically. This is the only useful method. The different frequencies mean the resultant wave is no longer sinusoidal. However, we still have continuous cycles (repetition).

Example 3.18

Plot the graph of y against x where $y = 3\sin x + 4\sin 2x$.

As usual, we start with a table of values. Complete your own table from $x = 0°$ to $x = 360°$ at intervals of $5°$ or $10°$. Use some of our specimen table values as a check.

x	$0°$	$15°$	$30°$	$45°$	$60°$	...
$\sin x$	0	0.259	0.500	0.707	0.866	...
$3\sin x$	0	0.78	1.50	2.12	2.60	...
$2x$	$0°$	$30°$	$60°$	$90°$	$120°$	...
$\sin 2x$	0	0.500	0.866	1.000	0.866	...
$4\sin 2x$	0	2.00	3.46	4.00	3.46	...
$3\sin x + 4\sin 2x$	0	2.78	4.96	6.12	6.06	...

We have drawn the graph in Fig. 3.23. You can see it has a period of $360°$ but is *not* sinusoidal. We *cannot* find a simpler equation to represent it.

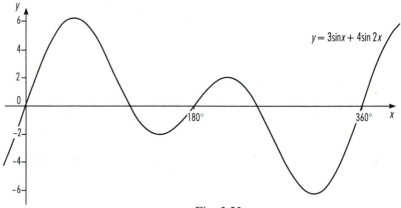

Fig. 3.23

Our final selection of questions looks at graphical addition for waves of different frequencies.

EXERCISE 3.7

1   $y = 5\sin 2x + 10\sin x$. Construct a table of values for y from $x = 0°$ to $x = 360°$ at intervals of $30°$. Plot a graph of y against x on a clearly labelled pair of axes.

2 Plot the graph of y against t where $y = \sin t + \sin 3t$. Your table of values should start with $t = 0$ and finish with $t = 2\pi$ radians at intervals of $\dfrac{\pi}{6}$ radians.

3 Using values of x from $x = -180°$ to $x = 180°$ at intervals of $30°$ plot the graph of $y = 12(\sin 2x + \cos x)$. From your graph read off the value of y when $x = 100°$. Check the accuracy of your answer by substituting for $x = 100°$ in the original equation.

4 Plot the graph of y against time, t, where $y = \sin\left(t + \dfrac{\pi}{6}\right) + \sin 2t$. Construct your table using values of t from $t = 0$ to $t = 2\pi$ radians at intervals of $\dfrac{\pi}{9}$ radians. Clearly label your graph and axes. When $y = 0.5$ what are the values of t?

5 Plot the graph of y against time, t, where
$$y = \sin\left(t + \dfrac{\pi}{3}\right) - \sin\left(t - \dfrac{\pi}{4}\right).$$
Construct your table from $t = -\pi$ to $t = \pi$ radians. Clearly label your graph and axes.

4 Trigonometrical Formulae and Equations

Introduction

The early work of this chapter concentrates on the compound angle formulae in various forms. These work for all angles. This means they are really identities and we can use $\equiv$ in place of $=$. You should use this section as a reference for these trigonometric formulae. The later work looks at solving a small selection of trigonometrical equations, sometimes applying the formulae we have deduced.

■■■■ ASSIGNMENT ■■■■

Our Assignment for this chapter looks again at the pair of parallel a.c. voltages, $v_1 = 15\sin 5t$ and $v_2 = 8\cos 5t$. We will combine these voltages using an alternative method to that of the previous chapter. Later we will solve an equation to find when a particular voltage is reached.

The reciprocal ratios

These are 3 more **trigonometric ratios** (**trigonometric operations** or **functions**) that are true for all angles. They are **cosecant** (cosec), **secant** (sec) and **cotangent** (cot). We relate them to sine, cosine and tangent by the formulae

$$\operatorname{cosec}\theta = \frac{1}{\sin\theta}, \quad \sec\theta = \frac{1}{\cos\theta}, \quad \cot\theta = \frac{1}{\tan\theta}$$

We call them **reciprocal ratios** because they are defined using "**1 over**". Most calculators have only the keys for sine, cosine and tangent. We use these keys together with the reciprocal key, $1/x$.

▨▨▨▨ **Examples 4.1** ▨▨▨▨▨▨▨▨▨▨▨▨▨▨▨▨▨▨▨▨▨▨▨▨

Using a calculator find the values of

i) cosec 40°,   ii) sec 215°,   iii) cot x where $x = 2.5$ rads.

   We can use either degrees or radians for any of our trigonometric functions.

i) Using our formula we know that $\operatorname{cosec} 40° = \dfrac{1}{\sin 40°}$.

   We use the calculator as

   $\underline{40°}$ $\underline{\sin}$ to display 0.64278...

   and $\underline{1/x}$ to display an answer of 1.556 (3 dp).

ii) Using our formula we know that $\sec 215° = \dfrac{1}{\cos 215°}$.

   We use the calculator as

   $\underline{215°}$ $\underline{\cos}$ to display −0.81915...

   and $\underline{1/x}$ to display an answer of −1.221 (3 dp).

iii) Using our formula we know that $\cot 2.5 = \dfrac{1}{\tan 2.5}$

   We use the calculator with radians as

   $\underline{2.5}$ $\underline{\tan}$ to display −0.74702...

   and $\underline{1/x}$ to display an answer of −1.339 (3 dp).

These reciprocal ratios are linked with our earlier ratios in some easy trigonometric identities. First we look at

$$\cot\theta = \frac{1}{\tan\theta} = \frac{1}{\dfrac{\sin\theta}{\cos\theta}} = \frac{\cos\theta}{\sin\theta}.$$

> Dividing by a fraction we invert and multiply.

Also, in Volume 1, Chapter 5 we deduced $\sin^2\theta + \cos^2\theta = 1$. Now we use this to deduce two more identities. In the first case we divide throughout by $\cos^2\theta$ so that

$$\sin^2 \theta + \cos^2 \theta = 1$$

becomes $\quad \dfrac{\sin^2 \theta}{\cos^2 \theta} + \dfrac{\cos^2 \theta}{\cos^2 \theta} = 1$

i.e. $\qquad \tan^2 \theta + 1 = \sec^2 \theta.$

This is usually quoted as

$$\mathbf{sec^2 \theta = 1 + tan^2 \theta.}$$

In the second case we divide throughout by $\sin^2 \theta$ so that

$$\sin^2 \theta + \cos^2 \theta = 1$$

becomes $\quad \dfrac{\sin^2 \theta}{\sin^2 \theta} + \dfrac{\cos^2 \theta}{\sin^2 \theta} = \dfrac{1}{\sin^2 \theta}$

i.e. $\qquad 1 + \cot^2 \theta = \mathrm{cosec}^2 \theta.$

This is usually quoted as

$$\mathbf{cosec^2 \theta = 1 + cot^2 \theta.}$$

Example 4.2

Using $\theta = 75°$ check that $\mathrm{cosec}^2 \theta = 1 + \cot^2 \theta.$

Now $\quad \mathrm{cosec}^2 \theta = \mathrm{cosec}^2 75°$

$$= \frac{1}{\sin^2 75°}$$

$$= \frac{1}{(0.9659\ldots)^2}$$

$$= \frac{1}{0.9330\ldots}$$

$$= 1.072.$$

Also $\qquad\qquad\qquad 1 + \cot^2 \theta = 1 + \cot^2 75°$

$$= 1 + \frac{1}{\tan^2 75°}$$

$$= 1 + \frac{1}{(3.7320\ldots)^2}$$

$$= 1 + \frac{1}{13.9282\ldots}$$

$$= 1 + 0.072$$

$$= 1.072 \text{ as expected.}$$

EXERCISE 4.1

In Questions 1 to 10 find the values of

1 $\sec 65°$

2 $\cot 45°$

3 $\mathrm{cosec}\,(-150°)$

4 $\cot 600°$

5 $\sec 390°$

6 $\mathrm{cosec}\, 250°$

7 $\cot x$ where $x = 3.5$ radians.

8 $\sec y$ where $y = -1.65$ radians.

9 $\operatorname{cosec} A$ where $A = 4.55$ radians.

10 $\sec B$ where $B = 2.75$ radians.

11 Using $\theta = 115°$ check that $\cot \theta = \dfrac{\cos \theta}{\sin \theta}$.

12 Given $\sec^2 \theta = 1 + \tan^2 \theta$ test that it is true for $\theta = 130°$.

13 Test the truth of $\operatorname{cosec}^2 \theta = 1 + \cot^2 \theta$ for $\theta = 235°$.

14 Why do we have to be careful with $\sec 90°$, $\operatorname{cosec} 0°$ and $\cot 0°$? Which other angles in the range $0°$ to $360°$ need care?

15 Using $\theta = 5.45$ radians check that $\sec^2 \theta = 1 + \tan^2 \theta$ Is the identity true for $2\theta = 10.90$ radians?

Compound angle formulae

If we have 2 angles, A and B, then the compound angles are $A + B$ and $A - B$. We may write

$$\sin (A + B) = \sin A \cos B + \cos A \sin B$$

and $\quad \sin (A - B) = \sin A \cos B - \cos A \sin B.$

They may be combined as

$$\sin (A \pm B) = \sin A \cos B \pm \cos A \sin B$$

because of their similarities, only differing by the $+/-$ signs.

Also $\quad \cos (A + B) = \cos A \cos B - \sin A \sin B$

and $\quad \cos (A - B) = \cos A \cos B + \sin A \sin B.$

They may be combined as

$$\cos (A \pm B) = \cos A \cos B \mp \sin A \sin B,$$

noticing the $+/-$ signs are reversed.

We omit the proofs of these formulae as they are *not* essential. Also we remind you that they are true for all angles.

▨▨▨▨▨ Example 4.3 ▨▨▨▨▨

In Volume 1, Chapter 5, we had a relationship between sine and cosine, $\sin(90° - \theta) = \cos \theta$. Now we use a compound angle formula to check that it is true.

Comparing $\sin(90° - \theta)$ with $\sin(A - B)$ we substitute for $A = 90°$ and $B = \theta$ to get

$$\begin{aligned}
\sin(90° - \theta) &= \sin 90° \cos \theta - \cos 90° \sin \theta \\
&= (1) \cos \theta - (0) \sin \theta \\
&= \cos \theta, \qquad \text{as expected.}
\end{aligned}$$

███████ **Example 4.4** ███████████████████████████████

Rewrite $\cos(\theta + 45°)$ in terms of $\sin\theta$ and $\cos\theta$.

Comparing $\cos(\theta + 45°)$ with $\cos(A + B)$ we substitute for $A = \theta$ and $B = 45°$ to get

$$\cos(\theta + 45°) = \cos\theta\cos 45° - \sin\theta\sin 45°$$
$$= (\cos\theta)0.7071 - (\sin\theta)0.7071$$
$$= 0.7071(\cos\theta - \sin\theta).$$

Let us apply our compound angle formulae for $\sin(A \pm B)$ and $\cos(A \pm B)$ to deduce formulae for $\tan(A \pm B)$. We remember that $\text{tangent} = \dfrac{\text{sine}}{\text{cosine}}$, i.e.

$$\tan(A + B) = \frac{\sin(A + B)}{\cos(A + B)}$$
$$= \frac{\sin A\cos B + \cos A\sin B}{\cos A\cos B - \sin A\sin B}$$

This looks less than useful, but we can convert it into terms involving $\tan A$ and $\tan B$. We treat this expression as four terms: two separated by a $+$ sign in the numerator and two separated by a $-$ sign in the denominator. We create $\tan A$ and $\tan B$ by dividing throughout by $\cos A\cos B$. Because we do this to all the terms we maintain the correct ratio of numerator to denominator.

Then $\tan(A + B) = \dfrac{\dfrac{\sin A\cos B}{\cos A\cos B} + \dfrac{\cos A\sin B}{\cos A\cos B}}{\dfrac{\cos A\cos B}{\cos A\cos B} - \dfrac{\sin A\sin B}{\cos A\cos B}}$

$$= \frac{\dfrac{\sin A}{\cos A} + \dfrac{\sin B}{\cos B}}{1 - \dfrac{\sin A}{\cos A}\cdot\dfrac{\sin B}{\cos B}}$$

> Cancelling $\cos A$s and $\cos B$s.

i.e. $\tan(A + B) = \dfrac{\tan A + \tan B}{1 - \tan A\tan B}.$

Exactly the same technique works to give

$$\tan(A - B) = \frac{\tan A - \tan B}{1 + \tan A\tan B}.$$

Again, these compound angle formulae may be combined as

i.e. $\tan(A \pm B) = \dfrac{\tan A \pm \tan B}{1 \mp \tan A\tan B}$

because of their similarities. Be careful with the pattern of the $+/-$ signs.

Always take care when applying any of the six compound angle formulae. It is unlikely that you will remember them all. If you are unsure always check back.

Example 4.5

Simplify $\tan(90° - \theta)$.

This expression needs care because $\tan 90° \to \infty$ (infinity). Also $\dfrac{1}{\infty} \to 0$.

This reciprocal is more useful because we have control over a tendency to 0 rather than to ∞.

Comparing $\tan(90° - \theta)$ with $\tan(A - B)$ we substitute for $A = 90°$ and $B = \theta$ to get

$$\tan(90° - \theta) = \frac{\tan 90° - \tan \theta}{1 + \tan 90° \tan \theta}.$$

Because of the problem with $\tan 90°$ tending to infinity we divide each term by $\tan 90°$. This means that ∞ is placed in various denominators and we can apply $\dfrac{1}{\infty} \to 0$.

Then $\tan(90° - \theta) = \dfrac{\dfrac{\tan 90°}{\tan 90°} - \dfrac{\tan \theta}{\tan 90°}}{\dfrac{1}{\tan 90°} + \dfrac{\tan 90° \tan \theta}{\tan 90°}}$

$$= \frac{1 - \left(\dfrac{1}{\infty}\right) \tan \theta}{\dfrac{1}{\infty} + \tan \theta}$$

> Dividing by $\tan 90°$.

> $\dfrac{1}{\infty} \to 0$ and so $\left(\dfrac{1}{\infty}\right) \tan \theta \to 0$.

i.e. $\tan(90° - \theta) = \dfrac{1}{\tan \theta}$.

Again, we remind you that these formulae work for any angle.

EXERCISE 4.2

Rewrite and simplify the following expressions using the compound angle formulae.

1 $\cos(90° + \theta)$
2 $\sin(180° - \theta)$
3 $\cos(180° - \theta)$
4 $\cos(270° + \theta)$
5 $\sin(360° - \theta)$

6 $\cos(90° - \theta)$
7 $\sin(\theta + 360°)$
8 $\tan(90° + \theta)$
9 $\tan(180° - \theta)$
10 $\cos(\theta + 360°)$

There are two interesting triangles that we should look at. The first one comes from bisecting an equilateral triangle of side 2 units. This creates the 30°, 60°, 90° triangle in Fig. 4.1.

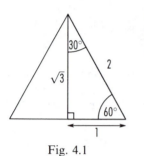

We find the length of the bisector using Pythagoras' theorem. Notice we write it as $\sqrt{3}$ rather than as a decimal. The $\sqrt{}$ of $\sqrt{3}$ means this length is in surd form.

Fig. 4.1

Then

$$\sin 30° = \frac{1}{2} = \cos 60°,$$

$$\sin 60° = \frac{\sqrt{3}}{2} = \cos 30°,$$

$$\tan 30° = \frac{1}{\sqrt{3}},$$

$$\tan 60° = \sqrt{3}.$$

The second one comes from bisecting a square of side 1 unit along a diagonal. This creates the 45°, 45°, 90° (isosceles) triangle in Fig. 4.2. We find the length of the diagonal using Pythagoras' theorem. Again we write it in surd form as $\sqrt{2}$.

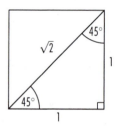

Then

$$\sin 45° = \frac{1}{\sqrt{2}} = \cos 45°,$$

$$\tan 45° = 1.$$

Fig. 4.2

In Examples 4.6 we use these results together with the compound angle formulae.

Examples 4.6

Without a calculator find, in surd form, the values of

i) $\sin 75°$,   ii) $\cos 15°$,   iii) $\sin 105°$.

i) We may split $\sin 75°$ into $\sin(45° + 30°)$ and use both our triangle results. Applying the compound angle formula for $\sin(A + B)$ with $A = 45°$ and $B = 30°$ we get

$$\sin 75° = \sin(45° + 30°)$$

$$= \sin 45° \cos 30° + \cos 45° \sin 30°$$

$$= \frac{1}{\sqrt{2}} \times \frac{\sqrt{3}}{2} + \frac{1}{\sqrt{2}} \times \frac{1}{\sqrt{2}}$$

$$= \frac{\sqrt{3} + 1}{2\sqrt{2}}.$$

ii) We may split $\cos 15°$ into either $\cos(45° - 30°)$ or $\cos(60° - 45°)$. Choosing the latter option and the compound angle formula for $\cos(A - B)$ we substitute for $A = 60°$ and $B = 45°$ to get

$$\cos 15° = \cos(60° - 45°)$$
$$= \cos 60° \cos 45° + \sin 60° \sin 45°$$
$$= \frac{1}{2} \times \frac{1}{\sqrt{2}} + \frac{\sqrt{3}}{2} \times \frac{1}{\sqrt{2}}$$
$$= \frac{1 + \sqrt{3}}{2\sqrt{2}}.$$

This demonstrates our earlier result with $\theta = 15°$ that

$$\cos \theta = \sin(90° - \theta)$$

i.e. $\cos 15° = \sin(90° - 15°) = \sin 75°$.

iii) We may split $\sin 105°$ into $\sin(60° + 45°)$. Applying the compound angle formula for $\sin(A + B)$ we substitute for $A = 60°$ and $B = 45°$ to get

$$\sin 105° = \sin(60° + 45°)$$
$$= \sin 60° \cos 45° + \cos 60° \sin 45°$$
$$= \frac{\sqrt{3}}{2} \times \frac{1}{\sqrt{2}} + \frac{1}{2} \times \frac{1}{\sqrt{2}}$$
$$= \frac{\sqrt{3} + 1}{2\sqrt{2}}.$$

Alternatively we may use our first result for $\sin 75°$,

$$\sin 105° = \sin(180° - 75°)$$
$$= \sin 180° \cos 75° - \cos 180° \sin 75°$$
$$= (0) \cos 75° - (-1) \times \left(\frac{\sqrt{3} + 1}{2\sqrt{2}} \right)$$
$$= \frac{\sqrt{3} + 1}{2\sqrt{2}} \qquad \text{as before.}$$

We can extend these ideas to angles based on general right-angled triangles.

Example 4.7

Simplify $\sin(60° - x) + \cos(x + 30°)$.

Separately for the sine and cosine we apply the appropriate compound angle formula to give

$$\sin(60° - x) + \cos(x + 30°)$$
$$= \sin 60° \cos x - \cos 60° \sin x + \cos x \cos 30° - \sin x \sin 30°$$

$$= \frac{\sqrt{3}}{2}\cos x - \frac{1}{2}\sin x + \cos x \left(\frac{\sqrt{3}}{2}\right) - \sin x \left(\frac{1}{2}\right)$$

$$= 2 \times \frac{\sqrt{3}}{2}\cos x - 2 \times \frac{1}{2}\sin x$$

$$= \sqrt{3}\cos x - \sin x.$$

We could reduce this still further by using phasor addition from the previous chapter. Then the expression would reduce to a single sine wave with a phase angle.

Example 4.8

Given $\sin A = \frac{4}{5}$ and $\cos B = \frac{5}{13}$ write $\sin(A + B)$ in fractional form.

In Figs. 4.3 we have drawn right-angled triangles for angles A and B.

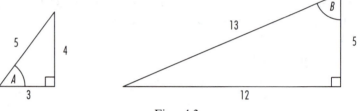

Figs. 4.3

We find the third side in each case by Pythagoras' theorem.

Then using the compound angle formula we write

$$\sin(A + B) = \sin A \cos B + \cos A \sin B$$

and substitute to get

$$\sin(A + B) = \frac{4}{5} \times \frac{5}{13} + \frac{3}{5} \times \frac{12}{13}$$

$$= \frac{20 + 36}{65}$$

$$= \frac{56}{65}.$$

Example 4.9

We are given $\sin A = -\frac{4}{5}$ and $\cos B = \frac{5}{13}$, where A lies between $180°$ and $270°$. Write $\sin(A - B)$ in fractional form.

This is similar to Example 4.8. This time A is in the 3rd quadrant where $\cos A = -\frac{3}{5}$. As before $\sin B = \frac{12}{13}$.

Using the compound angle formula we write

$$\sin(A - B) = \sin A \cos B - \cos A \sin B$$

and substitute to get

$$\sin(A - B) = -\frac{4}{5} \times \frac{5}{13} - -\frac{3}{5} \times \frac{12}{13}$$

$$= \frac{-20 + 36}{65}$$

$$= \frac{16}{65}.$$

We practise all these ideas and applications in Exercise 4.3.

EXERCISE 4.3

Use the compound angle formulae and our two special triangles to find values, in surd form, of the trigonometrical functions in Questions 1 to 10.

1   $\sin 15°$

2   $\sin 165°$

3   $\cos 75°$

4   $\tan 15°$

5   $\cos 345°$

6   $\cos(x - 45°)$

7   $\sin(30° - x)$

8   $\cos(x - 45°) + \cos(45° - x)$

9   $\sin(x + 60°) + \cos(x - 30°)$

10   $\cos(x + 45°) - \sin(x - 45°)$

For Questions 11 to 15 you are given $\sin A = \frac{4}{5}$, $\cos B = \frac{5}{13}$, $\cos C = -\frac{8}{17}$ and $\sin D = \frac{24}{25}$. Draw triangles to represent angles A, B, C and D, where C is obtuse. In each case use Pythagoras' theorem to find the third side of the triangle. Using these triangles and the compound angle formulae find, in fractional form, the following trigonometrical functions.

11   $\sin(D - A)$

12   $\cos(B + C)$

13   $\sin(A - D)$

14   $\tan(A + B)$

15   $\tan(A - D)$

Double angle formulae

We apply some of our compound angle formulae with $B = A$.

Then $\quad \sin(A + B) = \sin A \cos B + \cos A \sin B$

becomes $\quad \sin(A + A) = \sin A \cos A + \cos A \sin A$

i.e. $\quad\quad \mathbf{\sin 2A = 2 \sin A \cos A.}$

Also $\quad \cos(A + B) = \cos A \cos B - \sin A \sin B$

becomes $\quad \cos(A + A) = \cos A \cos A - \sin A \sin A$

i.e. $\quad\quad \mathbf{\cos 2A = \cos^2 A - \sin^2 A.}$

This formula has two other variations. We recall that $\sin^2 A + \cos^2 A = 1$ and substitute to get

$$\cos 2A = \cos^2 A - (1 - \cos^2 A)$$

$$\boxed{\sin^2 A = 1 - \cos^2 A.}$$

$$= \cos^2 A - 1 + \cos^2 A$$

i.e. $\mathbf{\cos 2A = 2\cos^2 A - 1.}$

Also $\cos 2A = (1 - \sin^2 A) - \sin^2 A$

$$\boxed{\cos^2 A = 1 - \sin^2 A.}$$

i.e. $\mathbf{\cos 2A = 1 - 2\sin^2 A.}$

Finally $\tan(A + B) = \dfrac{\tan A + \tan B}{1 - \tan A \tan B}$

becomes $\tan(A + A) = \dfrac{\tan A + \tan A}{1 - \tan A \tan A}$

i.e. $\mathbf{\tan 2A = \dfrac{2\tan A}{1 - \tan^2 A}.}$

Again, all these double angle formulae work for all angles. We simply need the same pattern of angles, e.g. $2A$ and $4A$, $1.5A$ and $3A$, $0.5A$ and A, and so on.

In the next set of examples we use specimen angles to show that these formulae are true.

▰▰▰ Examples 4.10 ▰▰▰

i)   Letting $A = 60°$ and using the formula for $\sin 2A$ we have

$$\sin 2A = \sin(2 \times 60°)$$
$$= \sin 120°$$
$$= 0.866.$$

Also $2\sin A \cos A = 2\sin 60° \cos 60°$

$$= 2 \times 0.866 \times 0.5$$
$$= 0.866 \qquad \text{as expected.}$$

ii)   Letting $A = 145°$ and using the formula for $\cos 2A$ we have

$$\cos 2A = \cos(2 \times 145°)$$
$$= \cos 290°$$
$$= 0.342.$$

Also $\cos^2 A - \sin^2 A = \cos^2 145° - \sin^2 145°$

$$= (-0.819\ldots)^2 - (0.573\ldots)^2$$
$$= 0.671 - 0.329$$
$$= 0.342 \qquad \text{as expected.}$$

iii) Letting $A = 217°$ and using the formula for $\tan 2A$ we have

$$\tan 2A = \tan(2 \times 217°)$$
$$= \tan 434°$$
$$= 3.487.$$

Also
$$\frac{2 \tan A}{1 - \tan^2 A} = \frac{2 \tan 217°}{1 - \tan^2 217°}$$
$$= \frac{2 \times 0.75355\ldots}{1 - (0.75355\ldots)^2}$$
$$= 3.487 \qquad \text{as expected.}$$

Examples 4.11

Find the value of $\sin(2A + B)$ in fractional form where $\sin A = \dfrac{3}{5}$ and $\cos B = \dfrac{12}{13}$.

We draw right-angled triangles for A and B in Figs. 4.4. Using Pythagoras' theorem, we find each third side.

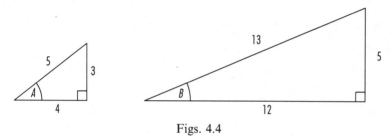

Figs. 4.4

i) First we look at $\sin(2A + B)$ with A and B as acute angles. Using the compound angle formula we write

$$\sin(2A + B) = \sin 2A \cos B + \cos 2A \sin B$$

and substitute for $\sin 2A$ and $\cos 2A$ to get

$$\sin(2A + B) = (2 \sin A \cos A) \cos B + (\cos^2 A - \sin^2 A) \sin B.$$

Rather than expand the brackets we substitute from our triangles to get

$$\sin(2A + B) = \left[2 \times \frac{3}{5} \times \frac{4}{5} \right] \frac{12}{13} + \left[\left(\frac{4}{5} \right)^2 - \left(\frac{3}{5} \right)^2 \right] \frac{5}{13}$$
$$= \frac{288}{325} + \frac{35}{325}$$
$$= \frac{323}{325}.$$

This time we look at $\sin(2A + B)$ with A as an obtuse angle ($90° < A < 180°$) and B as an acute angle. This means that $\cos A$ is negative, i.e. $\cos A = -\dfrac{4}{5}$, using the $\dfrac{S \mid A}{T \mid C}$ diagram from Volume 1, Chapter 5. We substitute into

$$\sin(2A + B) = (2 \sin A \cos A) \cos B + (\cos^2 A - \sin^2 A) \sin B$$

to get

$$\sin(2A + B) = \left[2 \times \frac{3}{5} \times \left(\frac{-4}{5}\right)\right]\frac{12}{13} + \left[\left(\frac{-4}{5}\right)^2 - \left(\frac{3}{5}\right)^2\right]\frac{5}{13}$$

$$= -\frac{288}{325} + \frac{35}{325}$$

$$= -\frac{253}{325}.$$

■■■■■ EXERCISE 4.4 ■■■■■

For Questions 1 to 10 you are given $\sin A = \dfrac{4}{5}$, $\cos B = \dfrac{5}{13}$, $\sin C = \dfrac{8}{17}$ and $\cos D = \dfrac{24}{25}$. Draw triangles to represent angles A, B, C and D. In each case use Pythagoras' theorem to find the third side of the triangle. Using these triangles and the compound angle formulae find, in fractional form, the following trigonometrical functions.

1	$\sin 2C$	**7**	$\tan(2A + D)$
2	$\tan 2A$	**8**	$\sin 4A$
3	$\cos 2C$	**9**	$\sin 3A$ written as
4	$\tan 2B$		$\sin(2A + A)$
5	$\sin(2C - A)$	**10**	$\cos 3A$ written as
6	$\tan(2A - C)$		$\cos(2A + A)$

11 For $A = 140°$ check that $\sin 2A = 2 \sin A \cos A$.

12 Test that $\cos 2x = 2\cos^2 x - 1$ where $x = 50°$.

13 Substitute for $y = 310°$ to check that $\tan 2y = \dfrac{2 \tan y}{1 - \tan^2 y}$

14 Substitute for $\sin 2A$ and $\cos 2A$ in $\dfrac{1 - \cos 2A}{\sin 2A}$. Simplify your expression to get $\tan A$.

15 Show that $\sin(x + 120°) + \sin(x + 60°) = q \cos x$ where q is a value for you to find.

Half angle formulae

It is possible to re-write our double angle formulae and keep our patterns, i.e.

$$\sin A = 2\sin\frac{A}{2}\cos\frac{A}{2},$$

$$\cos A = \cos^2\frac{A}{2} - \sin^2\frac{A}{2} \text{ and others,}$$

> Replacing A with $\frac{A}{2}$ and $2A$ with A.

$$\tan A = \frac{2\tan\frac{A}{2}}{1 - \tan^2\frac{A}{2}}.$$

However, there are alternative forms using $t = \tan\frac{A}{2}$. We can think of $t = \frac{t}{1}$ and draw the right-angled triangle in Fig. 4.5. We find the length of the hypotenuse using Pythagoras' theorem.

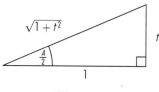

Fig. 4.5

Then $\sin\frac{A}{2} = \frac{t}{\sqrt{1 + t^2}}$, and $\cos\frac{A}{2} = \frac{1}{\sqrt{1 + t^2}}$. We substitute these into our above identities to get

$$\sin A = 2 \times \frac{t}{\sqrt{1 + t^2}} \times \frac{1}{\sqrt{1 + t^2}}$$

i.e. $$\sin A = \frac{2t}{1 + t^2}.$$

Also, $$\cos A = \left(\frac{1}{(\sqrt{1 + t^2})}\right)^2 - \left(\frac{t}{(\sqrt{1 + t^2})}\right)^2$$

i.e. $$\cos A = \frac{1 - t^2}{1 + t^2}.$$

Finally, $$\tan A = \frac{2t}{1 - t^2}.$$

> Simple substitution in the formula for $\tan\frac{A}{2}$.

These formulae work for all angles. Again we could test them for specimen values as we have done for other formulae.

▰▰▰▰ **Example 4.12** ▰▰▰▰▰▰▰▰▰▰▰▰▰▰▰▰▰▰

The angle at the circumference of a circle is half that at the centre, as shown in Fig. 4.6. If A is an obtuse angle and $\tan\dfrac{A}{2}=\dfrac{12}{5}$ find $\cos A$.

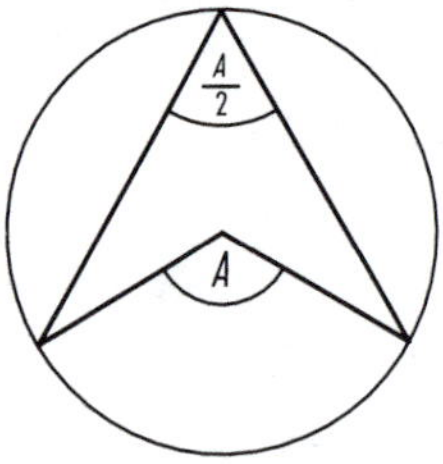

Using $t=\tan\dfrac{A}{2}$ we write $t=\dfrac{12}{5}$ so that

$$\cos A = \frac{1-t^2}{1+t^2}$$

becomes

$$\cos A = \frac{1-\left(\dfrac{12}{5}\right)^2}{1+\left(\dfrac{12}{5}\right)^2}$$

$$= \frac{\dfrac{25-144}{25}}{\dfrac{25+144}{25}}$$

$$= -\frac{119}{169}.$$

Fig. 4.6

> Cancelling common denominators of 25.

Notice the negative value for $\cos A$ where A is an obtuse angle.

▰▰▰▰ **EXERCISE 4.5** ▰▰▰▰▰▰▰▰▰▰▰▰▰▰▰▰▰▰▰

1   If $A=60°$ show that $\cos A=\cos^2\dfrac{A}{2}-\sin^2\dfrac{A}{2}$.

2   Check that $\tan A=\dfrac{2\tan\dfrac{A}{2}}{1-\tan^2\dfrac{A}{2}}$, where $A=150°$.

3   For $t=\tan A$ show that $\dfrac{1-\cos A}{\sin A}=t$.

4   If $A=120°$ and $t=\tan\dfrac{A}{2}$ what is the value of t? Check that this agrees with $\sin A=2\sin\dfrac{A}{2}\cos\dfrac{A}{2}$.

5   For $t=\tan\dfrac{A}{2}$ and $\sec^2\dfrac{A}{2}=1+\tan^2\dfrac{A}{2}$ find an expression for $\sec\dfrac{A}{2}$.

6   Write down $\tan x-\sin x$ in terms of t, where $t=\tan\dfrac{x}{2}$.

7   $y=\dfrac{2}{1+\cos x}$ and $\tan\dfrac{x}{2}=t$. Express y in terms of t as simply as possible.

8 $t = \tan\dfrac{A}{2}$ and $y = \dfrac{1 + \cos A}{1 - \cos A}$. Find y in terms of t as simply as possible.

9 If $t = \tan\dfrac{A}{2}$ write down $\sin 2A$ in terms of $\sin A$ and $\cos A$. Hence write it down as simply as possible in terms of t.

10 Write down $\tan 2A$ in terms of $\tan A$ and then in terms of t where $t = \tan\dfrac{A}{2}$.

Sums and differences

We deduce these four formulae using the same technique. Using our formulae for $\sin(A + B)$ and $\sin(A - B)$, let $x = A + B$
and $y = A - B$.

We can re-arrange these expressions to get $A = \dfrac{x + y}{2}$ and $B = \dfrac{x - y}{2}$.

If we add our formulae for $\sin(A + B)$ and $\sin(A - B)$, i.e.

$$\sin(A + B) = \sin A \cos B + \cos A \sin B$$

and $$\sin(A - B) = \sin A \cos B - \cos A \sin B$$

we get $\quad \sin(A + B) + \sin(A - B) = 2\sin A \cos B$

i.e. $$\sin x + \sin y = 2\sin\left(\frac{x + y}{2}\right)\cos\left(\frac{x - y}{2}\right).$$

> Substituting for A and B.

Similarly, if we subtract our formulae for $\sin(A + B)$ and $\sin(A - B)$ we get

$$\sin(A + B) - \sin(A - B) = 2\cos A \sin B$$

i.e. $$\sin x - \sin y = 2\cos\left(\frac{x + y}{2}\right)\sin\left(\frac{x - y}{2}\right).$$

> Substituting for A and B.

If we add our formulae for $\cos(A + B)$ and $\cos(A - B)$, i.e.

$$\cos(A + B) = \cos A \cos B - \sin A \sin B$$

and $$\cos(A - B) = \cos A \cos B + \sin A \sin B$$

we get $\quad \cos(A + B) + \cos(A - B) = 2\cos A \cos B$

i.e. $$\cos x + \cos y = 2\cos\left(\frac{x + y}{2}\right)\cos\left(\frac{x - y}{2}\right).$$

> Substituting for A and B.

Similarly if we subtract our formulae for $\cos(A + B)$ and $\cos(A - B)$ we get

$$\cos(A + B) - \cos(A - B) = -2\sin A \sin B$$

i.e.
$$\cos x - \cos y = -2\sin\left(\frac{x + y}{2}\right)\sin\left(\frac{x - y}{2}\right).$$

> Substituting for A and B.

▰▰▰▰ Example 4.13 ▰▰▰▰

Simplify $\dfrac{\sin 8\alpha + \sin 4\alpha}{\cos 8\alpha + \cos 4\alpha}$.

We have the addition of two sines in the numerator and of two cosines in the denominator. In both cases the patterns of angles (8α and 4α) are the same. We use our sums formulae, substituting for $x = 8\alpha$ and $y = 4\alpha$.

Then
$$\frac{x + y}{2} = \frac{8\alpha + 4\alpha}{2} = \frac{12\alpha}{2} = 6\alpha$$

and
$$\frac{x - y}{2} = \frac{8\alpha - 4\alpha}{2} = \frac{4\alpha}{2} = 2\alpha.$$

Now
$$\frac{\sin 8\alpha + \sin 4\alpha}{\cos 8\alpha + \cos 4\alpha} = \frac{2\sin 6\alpha \cos 2\alpha}{2\cos 6\alpha \cos 2\alpha}$$

$$= \frac{\sin 6\alpha}{\cos 6\alpha}$$

> Cancelling 2 and $\cos 2\alpha$.

$$= \tan 6\alpha.$$

▰▰▰ EXERCISE 4.6 ▰▰▰

1   For $x = 225°$ and $y = 45°$ check that
$$\cos x + \cos y = 2\cos\left(\frac{x + y}{2}\right)\cos\left(\frac{x - y}{2}\right).$$

2   Check that $\sin x - \sin y = 2\cos\left(\dfrac{x + y}{2}\right)\sin\left(\dfrac{x - y}{2}\right)$ when $x = 30°$ and $y = 120°$.

3   Simplify   i) $\dfrac{\sin 4A + \sin 2A}{\cos 4A - \cos 2A}$,   ii) $\dfrac{\cos 4A + \cos 2A}{\sin 4A - \sin 2A}$.

4   i)   Use the sum formula for $\cos x + \cos y$ to show that
$$\cos A + \cos(-A) = 2\cos A.$$
   ii)   Use the difference formula for $\sin x - \sin y$ to simplify $\sin A - \sin(-A)$.

5   Split the middle terms in the numerator and denominator and then simplify $\dfrac{\sin \alpha + 2\sin 3\alpha + \sin 5\alpha}{\cos 3\alpha + 2\cos 5\alpha + \cos 7\alpha}$.

$R\sin(\omega t \pm \alpha)$

From our work with compound angle formulae we know about $\sin(A + B)$ and $\sin(A - B)$. Now let us look at $R\sin(\omega t + \alpha)$ where R is the amplitude and α is the phase angle. In this simplification technique we aim to find values for R and α. We aim to express $a\sin\omega t + b\cos\omega t$ as a single waveform $R\sin(\omega t + \alpha)$. Let us compare $\sin(\omega t + \alpha)$ with $\sin(A + B)$ so that $A = \omega t$ and $B = \alpha$. Then

$$R\sin(\omega t + \alpha) \equiv R\{\sin\omega t\cos\alpha + \cos\omega t\sin\alpha\}$$
$$\equiv (R\cos\alpha)\sin\omega t + (R\sin\alpha)\cos\omega t.$$

We are going to compare $a\sin\omega t + b\cos\omega t$ with $R\sin(\omega t + \alpha)$ and hence with $(R\cos\alpha)\sin\omega t + (R\sin\alpha)\cos\omega t$. When we compare trigonometric terms in ωt we need $a = R\cos\alpha$ and $b = R\sin\alpha$. Dividing we get

$$\frac{R\sin\alpha}{R\cos\alpha} = \frac{b}{a}$$

i.e. $\qquad\qquad \tan\alpha = \dfrac{b}{a}.$ | Cancelling the Rs. |

Our next step is to square our equations for a and b so that

$$R\cos\alpha = a \qquad \text{and} \qquad R\sin\alpha = b$$

become $\quad R^2\cos^2\alpha = a^2 \qquad$ and $\qquad R^2\sin^2\alpha = b^2.$

Adding these squared equations we get

$$R^2\cos^2\alpha + R^2\sin^2\alpha = a^2 + b^2$$

i.e. $\qquad R^2(\cos^2\alpha + \sin^2\alpha) = a^2 + b^2$ | $\cos^2\alpha + \sin^2\alpha = 1.$ |

$$R^2 = a^2 + b^2$$
$$\therefore \qquad\qquad R = \sqrt{a^2 + b^2}.$$

Now we can write that $a\sin\omega t + b\cos\omega t$ is identical to $R\sin(\omega t + \alpha)$ with these values of R and α.

Exactly the same idea follows for $R\sin(\omega t - \alpha)$ as we see in Example 4.14.

Example 4.14

Express $2\sin t - 4\cos t$ in the form $R\sin(\omega t - \alpha)$. Find values for R and α.

Immediately we see that $\omega = 1$ and so write

$$R\sin(t - \alpha) \equiv R\{\sin t\cos\alpha - \cos t\sin\alpha\}$$
$$\equiv (R\cos\alpha)\sin t - (R\sin\alpha)\cos t.$$

Comparing like terms for $\sin t$ and $\cos t$ between this expression and our original question we have $2 = R\cos\alpha$ and $4 = R\sin\alpha$. Then we divide to get

$$\frac{R\sin\alpha}{R\cos\alpha} = \frac{4}{2}$$

i.e. $\tan\alpha = 2$ | Cancelling the Rs. |

∴ $\alpha = 1.107$ rad or $63.43°$.

Squaring and adding our equations

$$R\cos\alpha = 2 \qquad \text{and} \qquad R\sin\alpha = 4$$

we get $R^2\cos^2\alpha = 4$ and $R^2\sin^2\alpha = 16$

so that $R = \sqrt{4 + 16} = \sqrt{20}$

i.e. $R = 4.47$.

∴ $2\sin t - 4\cos t \equiv 4.47\sin(t - 1.107)$

or in degrees, $2\sin t - 4\cos t \equiv 4.47\sin(t - 63.43°)$.

ASSIGNMENT

Let us look at our parallel a.c. voltages $v_1 = 15\sin 5t$ and $v_2 = 8\cos 5t$. The sum of these voltages is $15\sin 5t + 8\cos 5t$. We are going to express this sum as $R\sin(\omega t + \alpha)$ where $\omega = 5$. Then

$$R\sin(5t + \alpha) \equiv R\{\sin 5t\cos\alpha + \cos 5t\sin\alpha\}$$
$$= (R\cos\alpha)\sin 5t + (R\sin\alpha)\cos 5t.$$

Comparing like terms for $\sin 5t$ and $\cos 5t$ between this expression and our voltage sum we have $15 = R\cos\alpha$ and $8 = R\sin\alpha$. Then we divide to get

$$\frac{R\sin\alpha}{R\cos\alpha} = \frac{8}{15}$$

i.e. $\tan\alpha = 0.5\overline{3}$ | Cancelling the Rs. |

∴ $\alpha = 0.49$ rad.

We use radians because degrees have no real meaning for voltages.

Squaring and adding our equations

$$R\cos\alpha = 15 \qquad \text{and} \qquad R\sin\alpha = 8$$

we get $R^2\cos^2\alpha = 225$ and $R^2\sin^2\alpha = 64$

so that $R = \sqrt{225 + 64} = \sqrt{289}$

i.e. $R = 17$.

This means we can write our voltage sum as $17\sin(5t + 0.49)$ to agree with our previous methods.

Also we can find the maximum voltage and when it first occurs. The maximum value of any sine is 1. This allows us to write $v_{max} = 17 \times 1 = 17$ V. The maximum sine value occurs at $\frac{\pi}{2}$, i.e. $\sin(5t + 0.49) = 1$

gives $5t + 0.49 = \dfrac{\pi}{2} = 1.57$

$5t = 1.57 - 0.49 = 1.08$

$\therefore$ $t = \dfrac{1.08}{5} = 0.22\,\text{s}.$

Gathering our answers, the maximum voltage of 17 V occurs first after 0.22 s.

EXERCISE 4.7

1   Express $6\sin 2x - 4\cos 2x$ in the form $R\sin(2x - \alpha)$. Find values for R and α.

2   What are the values of R and α that allow us to write $12\cos x + 12\sin x$ in the form $R\sin(x + \alpha)$?

3   A unit mass is subjected to forces $7\sin 3t$ and $-3.5\cos 3t$. Write down the sum of these forces and express it as $R\sin(3t - \alpha)$. Calculate values for R and α.

4   Find the total current, $i_1 + i_2$, where $i_1 = 4\sin 2t$ and $i_2 = 3.5\cos 2t$. Simplify this in the form $I\sin(2t + \alpha)$, finding values for I and α. Write down the maximum total current.

5   The sum of two voltages is $v = v_1 + v_2$. Write this as $V\sin(10\pi t + \alpha)$ where $v_1 = 5\sin 10\pi t$ and $v_2 = 15\cos 10\pi t$. Find values for V and α. What is the maximum total voltage?

Trigonometrical equations

Our first type of equation is $a\sin x + b = 0$ where a and b are numbers. We can replace $\sin x$ with any simple trigonometrical function, as you will see in the next few examples. Generally we solve in the range $0°$ to $360°$ or $-180°$ to $180°$ or over 1 cycle. Also we can use radians just as easily as degrees. You need to revise the $\dfrac{S\ |\ A}{T\ |\ C}$ diagram from Volume 1, Chapter 5, and our work on the inverses of sine, cosine and tangent.

Example 4.15

In the range $-180°$ to $180°$ solve the equation $13\cos x - 5 = 0$ for x.

We need to make $\cos x$, and then x, the subject of this equation. -5 is less closely attached to x than anything else. We move it to the right hand side, i.e. $13\cos x = 5$.

From here we solve this equation in two different ways.

i) We plot $y = 13\cos x$ and $y = 5$ in Fig. 4.7. Where the graphs cross are our solutions, i.e. at $x = 67°$ and $293°$. Notice the symmetry in our diagram. Look at the distance of the first solution after the first peak value. It is the same as that of the second solution before the second peak value.

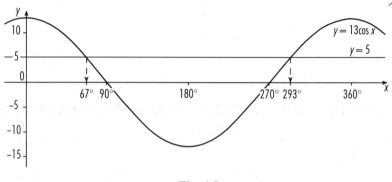

Fig. 4.7

ii) Alternatively we make $\cos x$ the subject of our equation,

$$\cos x = \frac{5}{13} = 0.3846\ldots$$

> Dividing by 13.

Using the calculator we find that $x = 67.38°$. Also from the $\dfrac{S \mid A}{T \mid C}$ diagram we have $x = 360° - 67.38° = 292.62°$.

Example 4.16

Over 1 cycle from $x = 0°$ solve the equation $4\sin x + 2.5 = 0$.

We need to make $\sin x$, and then x, the subject of this equation. 2.5 is less closely attached to x than anything else. We move it to the right hand side, i.e. $4\sin x = -2.5$.

From here we solve this equation in two different ways.

i) We plot $y = 4\sin x$ and $y = -2.5$ in Fig. 4.8. Where the graphs cross are our solutions, i.e. at $x = 219°$ and $321°$.

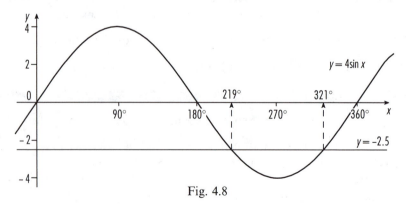

Fig. 4.8

ii) Alternatively we make sin x the subject of our equation,

$$\sin x = -\frac{2.5}{4} = -0.625$$

> Dividing by 4.

Using the calculator we get $x = -38.68°$. From the $\begin{array}{c|c} S & A \\ \hline T & C \end{array}$ diagram we have

$$x = 180° + 38.68° \text{ and } 360° - 38.68°$$

i.e. $x = 218.68°$ and $321.32°$.

For both sine and cosine we know that their maximum value is 1 and their minimum value is -1. All trigonometric equations involving them only have a solution where this is true.

■ EXERCISE 4.8 ■

1 In the range $0° < x < 360°$ solve the equation $3 \sin x = 2.76$ for x.

2 What values of x satisfy the equation $7 \cos x + 12 = 0$ in the range $-180° < x < 180°$?

3 Where $\tan x - 3 = 1.65$ solve for x in the range $0° < x < 360°$.

4 In the range $0 < y < 2\pi$ radians solve $4 \tan y + 6 = 2.5$.

5 Given that $0.1 \sin t + 0.02 = 0.046$ solve for t in radians over 1 cycle from $t = 0$.

▒ Example 4.17 ▒

In the range $0°$ to $360°$ solve $3 \tan 2x - 5.40 = 0$ for x.

We need to make $\tan 2x$, then $2x$ and x, the subject of this equation. -5.40 is less closely attached to x than anything else. We move it to the right hand side, i.e. $3 \tan 2x = 5.40$.

From here we solve this equation in two different ways.

i) We plot $y = 3 \tan 2x$ and $y = 5.40$ in Fig. 4.9. Where the graphs cross are our solutions, i.e. at $x = 30°$, $120°$, $210°$ and $300°$.

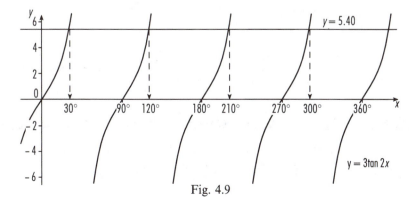

Fig. 4.9

ii) Alternatively, we make $\tan 2x$ the subject of our equation,

$$\tan 2x = \frac{5.40}{3} = 1.80.$$

> Dividing by 3.

Because of 2 in $\tan 2x$ we know this graph has twice as many cycles as the simpler $\tan x$. Our method follows the usual pattern except that we go around the $\dfrac{S \mid A}{T \mid C}$ diagram twice to 720°. Then the same factor of 2 reduces all our answers into our range of 0° to 360°. Using the calculator we find that $2x = 60.945\ldots°$. Using the $\dfrac{S \mid A}{T \mid C}$ diagram we have

$$2x = 60.945\ldots°, \ 180° + 60.945\ldots°, \ 360° + 60.945\ldots°,$$
$$540° + 60.945\ldots°$$

> $540° = 360° + 180°.$

i.e. $2x = 60.945\ldots°, \ 240.945\ldots°, \ 420.945\ldots°, \ 600.945\ldots°$

∴ $x = 30.47°, \ 120.47°, \ 210.47°, \ 300.47°.$

ASSIGNMENT

This time we look separately at our two voltages. In each case we set the pattern of answers rather than complete five cycles.

First we see when the second voltage is a maximum, i.e. when

$$8\cos 5t = 8$$
$$\cos 5t = 1$$

> Dividing by 8.

∴ $5t = 0, \ 2\pi, \ 4\pi, \ \ldots$

$$t = 0, \ \frac{2\pi}{5}, \ \frac{4\pi}{5}, \ \ldots$$

> Times in seconds.

By way of example we look at when the first voltage is halved,

i.e. $15\sin 5t = 7.5$

$$\sin 5t = \frac{7.5}{15} = 0.5$$

> Dividing by 15.

∴ $5t = 0.523\ldots, \ \pi - 0.523\ldots, \ 2\pi + 0.523\ldots,$
$$3\pi - 0.523\ldots$$

> $\dfrac{S \mid A}{T \mid C}$

$$t = 0.10, \ 0.52, \ 1.36, \ 1.78, \ \ldots$$

> Times in seconds.

Again we could obtain these solutions graphically.

EXERCISE 4.9

1   In the range $0° < x < 360°$ solve the equation $6\sin 2x = 4.5$ for x.

2   What values of x satisfy the equation $4\cos 3x - 1.2 = 0$ in the range $-180° < x < 180°$?

3 Where $7\cos 2x - 2.50 = 1.65$ solve for x in the range $0° < x < 360°$.

4 In the range $0 < t < 2\pi$ radians where $i = 4\sin 2t - 3.6$ find when the current is $-2\,\text{A}$.

5 Given that $0.5\sin 2.5t + 0.2 = 0.46$ solve for t in radians over 1 cycle from $t = 0$.

Example 4.18

Solve the equation $2\sin t - 4\cos t = 3.5$ in the range $0°$ to $360°$.

From Example 4.14 we have

$$2\sin t - 4\cos t = \sqrt{20}\sin(t - 63.43°).$$

Then $\qquad 2\sin t - 4\cos t = 3.5$

becomes $\quad \sqrt{20}\sin(t - 63.43°) = 3.5$

i.e. $\qquad\qquad \sin(t - 63.43°) = \dfrac{3.5}{\sqrt{20}} = 0.7826\ldots$

$\therefore \qquad\qquad t - 63.43° = 51.50°, 128.50°$

$$\begin{array}{c|c} {}^{*}\text{S} & \text{A}^{*} \\ \hline \text{T} & \text{C} \end{array}$$

i.e. $\qquad\qquad t = 51.50° + 63.43°, 128.50° + 63.43°$

$\qquad\qquad\qquad t = 114.93°, 191.93°.$

Example 4.19

Solve the equation $2\sin t + 4\cos t = 3.5$ in the range $0°$ to $360°$.

This example is similar to Example 4.18, with a change of $-/+$ sign so that

$$2\sin t + 4\cos t = \sqrt{20}\sin(t + 63.43°).$$

Then $\qquad 2\sin t + 4\cos t = 3.5$

becomes $\quad \sqrt{20}\sin(t + 63.43°) = 3.5$

i.e. $\qquad\qquad \sin(t + 63.43°) = \dfrac{3.5}{\sqrt{20}} = 0.7826\ldots$

$\therefore \qquad\qquad t + 63.43° = 51.50°, 128.50°.$

$$\begin{array}{c|c} {}^{*}\text{S} & \text{A}^{*} \\ \hline \text{T} & \text{C} \end{array}$$

Now when we move $63.43°$ our first value is out of range. This means we need to use $360° + 51.50°$ from the next cycle

i.e. $\qquad\qquad t + 63.43° = 51.50°, 128.50°, 411.50°$

so that $\qquad\qquad t = 51.50° - 63.43°, 128.50° - 63.43°,$

$\qquad\qquad\qquad\qquad 411.50° - 63.43°$

i.e. $\qquad\qquad t = 65.07°, 348.07°.$ $\quad\boxed{-11.93° \text{ is out of range.}}$

▰▰ ASSIGNMENT ▰▰

For the final look at our Assignment we find when the total voltage is three-quarters of its maximum, i.e.

$$15 \sin 5t + 8 \cos 5t = \frac{3}{4} \times 17$$

i.e. $\qquad \sin(5t + 0.49) = 0.75$

$\therefore \qquad\qquad 5t + 0.49 = 0.85, \ldots$

$\qquad\qquad\qquad\quad 5t = 0.36$

$\qquad\qquad\qquad\quad\, t = 0.07,$

> Dividing by 17.
> Using radians.
> Subtracting 0.49.

i.e. after 0.07 seconds the total voltage is three-quarters its maximum for the first time.

As an alternative we may attempt a graphical solution. In Fig. 4.10 we have plotted the first cycle and shown our solution of $t = 0.07\,\text{s}$.

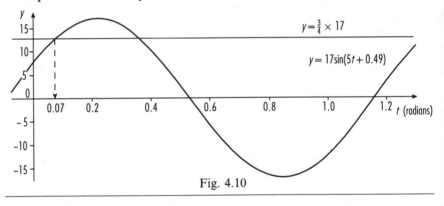

Fig. 4.10

▰▰ EXERCISE 4.10 ▰▰

1   Express $6 \sin 2x + 4 \cos 2x$ in the form $R \sin(2x + \alpha)$. Find values for R and α. Now solve the equation $6 \sin 2x + 4 \cos 2x = 6.5$ in the range $0° \leqslant x \leqslant 360°$.

2   What are the values of R and α that allow us to write $5 \sin 2x - 5 \cos 2x$ in the form $R \sin(2x - \alpha)$? Hence solve the equation $5(\sin 2x - \cos 2x) = 4.5$ in the range $-180° \leqslant x \leqslant 180°$.

3   A unit mass is subjected to forces $3.5 \sin 3t$ and $5 \cos 3t$. Write down the sum of these forces and express it as $R \sin(3t + \alpha)$. Calculate values for R and α. Find the first and second times when the total force is half of its maximum.

4   Find when the total current, $i_1 + i_2$, is first $2.5\,\text{A}$ where $i_1 = 3.5 \sin 2t$ and $i_2 = 3 \cos 2t$.

5   The sum of two voltages is $v = v_1 + v_2$ where $v_1 = 5 \sin 10\pi t$ and $v_2 = 5 \cos 10\pi t$. For the first occasion during what time is the total voltage over 80% of its maximum?

Our final selection of examples and exercise uses some of our solution ideas and formulae. We need to substitute using the necessary formulae before attempting a solution. Alternatively you can still get a graphical solution as we have often done in this chapter.

████████ **Example 4.20** ████████████████████████████████████

For $0° \leqslant x \leqslant 360°$ solve the trigonometric equation $5 \sin x = 2 + \cos 2x$.

Our equation has a mixture of angles (x and $2x$) and trigonometric functions (sine and cosine). We reduce the angles to x only by substituting for $\cos 2x$. Because of $\sin x$ in our original equation we choose the substitution $\cos 2x = 1 - 2 \sin^2 x$. Now we have only angles of x and sines,

i.e. $\qquad\qquad 5 \sin x = 2 + (1 - 2 \sin^2 x)$

i.e. $\quad 2 \sin^2 x + 5 \sin x - 3 = 0$

$\qquad (2 \sin x - 1)(\sin x + 3) = 0$ | This quadratic equation factorises.

$\therefore \qquad 2 \sin x - 1 = 0 \qquad$ or $\qquad \sin x + 3 = 0$

i.e. $\qquad \sin x = 0.5 \qquad$ or $\qquad \sin x = -3.$

$\therefore \qquad\qquad x = 30°, 150°. \qquad$ Impossible because we need $-1 \leqslant \sin x \leqslant 1$.

We could have solved this equation graphically by plotting $y = 5 \sin x$ and $y = 2 + \cos 2x$. As usual the solutions occur where the graphs intersect.

Not all our quadratic equations need to factorise. We know how to solve quadratic equations using the usual formula.

████████ **Example 4.21** ████████████████████████████████████

Solve the equation $1 - \cos 2x - \cos 2x \tan^2 x = 0.5$ for x in the range $0°$ to $360°$.

Again we have a mixture of angles (x and $2x$) and trigonometric functions (cosine and tangent). We substitute $\tan^2 x = \dfrac{\sin^2 x}{\cos^2 x}$.

We have three choices for $\cos 2x$, our choices using a mixture to reach a quick solution. Other choices will achieve exactly the same result by a slightly different route.

Then $\qquad 1 - (1 - 2 \sin^2 x) - (2 \cos^2 x - 1)\dfrac{\sin^2 x}{\cos^2 x} = 0.5$

i.e. $\quad 1 - 1 + 2 \sin^2 x - 2 \cos^2 x \times \dfrac{\sin^2 x}{\cos^2 x} + 1 \times \dfrac{\sin^2 x}{\cos^2 x} = 0.5$

$\qquad\qquad 2 \sin^2 x - 2 \sin^2 x + \dfrac{\sin^2 x}{\cos^2 x} = 0.5$

i.e. $\qquad\qquad\qquad\qquad\qquad \dfrac{\sin^2 x}{\cos^2 x} = 0.5.$

In a last substitution we return to $\tan^2 x$ with

$$\tan^2 x = 0.5$$

i.e. $\tan x = \pm\sqrt{0.5} = \pm 0.707$

Because we have positive and negative values for the tangent we get solutions in all the quadrants.

$\therefore$ $x = 35.26°,\ 180° - 35.26°,\ 180° + 35.26°,\ 360° - 35.26°$

$x = 35.26°,\ 144.74°,\ 215.26°,\ 324.74°.$

It is important to notice how we simplify a trigonometric equation. We do not cancel through by division where a trigonometric function may be zero. Remember we *cannot* divide by zero in Mathematics.

▆▆▆ EXERCISE 4.11 ▆▆▆

1   Solve the equation $\sin 2x = \sin x$ for x in the range $0° \leqslant x \leqslant 360°$.

2   Between $0°$ and $360°$ find the values of x that are solutions of $\sin(x + 45°) + \sin(135° + x) = 0.45$.

3   In the range $-180° \leqslant x \leqslant 180°$ solve the equation $1 - 2\sin^2 x + \cos^2 x = 0$.

4   Solve $3\tan 2\theta = 2\tan \theta$ for θ between $0°$ and $360°$.

5   Over the first cycle from $x = 0°$ solve the equation $\sec^2 x - \tan x - 7 = 0$.

6   For $\tan x + \sin x = 0$ solve for x over 1 cycle from $x = 0$ radians.

7   In the range $0°$ to $360°$ solve $\cos y + \cos(-y) = 0.75$ for y.

8   What values of x satisfy the equation $4\sin^2 x + \cos x = 1$ in the range $0° \leqslant x \leqslant 360°$?

9   Solve the equation $\dfrac{1 - \cos 2x}{\sin 2x} = 2.85$ for x between $0°$ and $360°$.

10   Factorise $\cos^4 t - \sin^4 t$ and hence solve the equation $\cos^4 t - \sin^4 t = 0.64$ for t in the range $0° < t < 360°$.

5 The Binomial Theorem

The objectives of this chapter are to:

1 Expand expressions of the form $(a + x)^n$ for a small positive integer n using Pascal's triangle.
2 Expand expressions of the form $(a + x)^n$ for a positive integer n using the binomial theorem.
3 State the general form for the binomial coefficients for a positive integer.
4 Identify appropriate patterns relating Pascal's triangle to the binomial theorem.
5 Expand expressions of the form $(1 + x)^n$ where n takes positive, negative or fractional (or decimal) powers.
6 State the range of values of x for which the series is convergent.
7 State the binomial approximation.
8 Calculate the effect on the subject of a formula when one or more independent variables is subject to a small change or error.

Introduction

Let us look at the word **binomial** of **binomial theorem**. A binomial expression has **two (bi)** terms added or subtracted. The binomial theorem allows us to expand the brackets around the expression without tedious multiplication. We introduce the idea using Pascal's triangle. This sets up patterns for us to generalise later.

Examples 5.1

The following are some binomial expressions

i) $1 + x$, ii) $(2 - x)^4$,

iii) $(3x + 0.5)^{-2}$, iv) $(x^2 - 0.1)^n$.

> i.e. $(1 + x) = (1 + x)^1$.
> n is any number.

ASSIGNMENT

In this Assignment we will see the binomial theorem applied twice to an engineering company producing cylindrical rods. We give no precise dimensions for their standard rod. They are unnecessary for our calculations. Our first

123

investigation looks at possible faulty rods in a sample batch. Our second investigation looks at how dimensional errors affect the volume.

Pascal's triangle

In our build-up to the binomial theorem we will look for patterns.

We know $(a + x)^0 = 1$ because raising an expression to the power of zero is always 1.

Also $(a + x)^1 = a + x$ as we generally omit writing the power of one.

Now $(a + x)^2 = (a + x)(a + x)$

$$= a^2 + x^2 + ax + xa$$
$$= a^2 + 2ax + x^2.$$

Then $(a + x)^3 = (a + x)(a + x)^2$

$$= (a + x)(a^2 + 2ax + x^2).$$

a from the first bracket multipies each term in the second bracket. Similarly, x from that bracket multiplies each term in the second bracket. Hence we get

$$= a^3 + 2a^2x + ax^2 + xa^2 + 2ax^2 + x^3$$
$$= a^3 + 3a^2x + 3ax^2 + x^3.$$

We follow on with

$$(a + x)^4 = (a + x)(a + x)^3$$
$$= (a + x)(a^3 + 3a^2x + 3ax^2 + x^3).$$

We shall look at this result shortly. Following this method we can multiply out any number of brackets. As an exercise for yourself, you should multiply out these brackets. We group together our results to look for some patterns. You can check your own answers against them.

$$(a + x)^0 = \qquad\qquad\qquad 1$$
$$(a + x)^1 = \qquad\qquad\quad a \quad + \quad x$$
$$(a + x)^2 = \qquad\qquad a^2 \;+\; 2ax \;+\; x^2$$
$$(a + x)^3 = \qquad\quad a^3 \;+\; 3a^2x \;+\; 3ax^2 \;+\; x^3$$
$$(a + x)^4 = \quad\; a^4 \;+\; 4a^3x \;+\; 6a^2x^2 \;+\; 4ax^3 \;+\; x^4$$
$$(a + x)^5 = a^5 \;+\; 5a^4x \;+\; 10a^3x^2 \;+\; 10a^2x^3 \;+\; 5ax^4 \;+\; x^5$$

and so the pattern continues.

You can think of the first value of 1 at the top of this triangle as either a^0 or x^0.

We concentrate on the last line, the expansion of $(a + x)^5$, to highlight some patterns that apply generally. They are

1. The powers of 5 of the first term (a^5) and the last term (x^5) agree with the original expression $(a + x)^5$.
2. From left to right the powers of a decrease by 1 each term.
3. From left to right the powers of x increase by 1 each term.
4. The coefficients are symmetrical about the mid-point.
5. The coefficients may be deduced from the previous line. We show you how this is done below.
6. The coefficients of the first and last terms are both 1.
7. The number of terms is always one more than the original power.

From the previous triangle of terms we separate out all the coefficients and display them as **Pascal's triangle**.

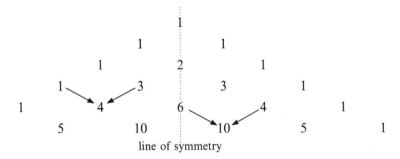

line of symmetry

The first and last coefficients are 1.

Add the two neighbouring (**adjacent**) values (**coefficients**) to give the value (coefficient) directly below

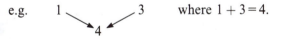

e.g.   1   3   where $1 + 3 = 4$.
       4

e.g.   6   4   where $6 + 4 = 10$.
      10

The coefficients are symmetrical about a vertical line through the original 1.

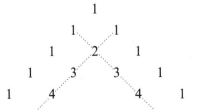

These entries increase by 1. In each case they have the same value as the original power.

━━━━ **Examples 5.2** ━━━━

Using Pascal's triangle write down the coefficients of

i) $(a + x)^6$,   ii) $(a + x)^7$.

i) We use the coefficients of $(a + x)^5$,

   1   5   10   10   5   1.

For our new line of coefficients the first and last values are 1. The other values are found by addition.

1 ↘ 5,   5 ↘↙ 10,   10 ↘↙ 10,   10 ↘↙ 5,   5 ↘↙ 1
↘ 6   15   20   15   6 ↙

Hence the complete row of coefficients is

   1   6   15   20   15   6   1.

ii) Similarly, we use this result to get

       1   7   21   35   35   21   7   1,

where 1 ↘ 6,   6 ↘ 15   and so on.
        ↘ 7 ↙   21 ↙

━━━━ **Example 5.3** ━━━━

Using the established patterns write down the expansion of $(a + x)^6$.

From the previous example we know about the coefficients. Also the bracket is raised to the power 6 so the first term is a^6 and the last term is x^6. In between these terms the powers of a decrease by 1 and the powers of x increase by 1. Hence

$$(a + x)^6 = a^6 + 6a^5x + 15a^4x^2 + 20a^3x^3 + 15a^2x^4 + 6ax^5 + x^6.$$

powers of a decreasing ⟶

powers of x increasing ⟶

━━━━ **Example 5.4** ━━━━

Find the expansion of $(2b - 3c)^6$.

We have done most of the preparatory work in the previous examples. By now we know about the coefficients and what happens to the powers. When we compare $(2b - 3c)^6$ with our previous example of $(a + x)^6$ we use $2b$ in place of a and $-3c$ in place of x.

Remember $2b - 3c = 2b + (-3c)$. This means we substitute directly into the expansion of $(a + x)^6$ to get

$$(2b - 3c)^6 = (2b)^6 + 6(2b)^5(-3c) + 15(2b)^4(-3c)^2 + 20(2b)^3(-3c)^3 +$$
$$15(2b)^2(-3c)^4 + 6(2b)(-3c)^5 + (-3c)^6.$$

We need to tidy up the signs and numbers. When we do this most of our patterns will be lost. The alternate $+/-$ signs do remain because

$(-)^{\text{even power}} = +$ and $(-)^{\text{odd power}} = -$. Remember, when we raise a bracket to a power everything in that bracket needs some attention. We look at these terms in turn.

$$(2b)^6 \qquad = 2^6 b^6 \qquad\qquad = 64b^6.$$

$$6(2b)^5(-3c) \quad = 6 \times 2^5 b^5 \times -3c$$
$$= 6 \times 32b^5 \times -3c \quad = -576b^5 c.$$

$$15(2b)^4(-3c)^2 = 15 \times 2^4 b^4 \times (-)^2 3^2 c^2$$
$$= 15 \times 16b^4 \times 9c^2 \quad = 2160b^4 c^2.$$

$$20(2b)^3(-3c)^3 = 20 \times 2^3 b^3 \times (-)^3 3^3 c^3$$
$$= 20 \times 8b^3 \times -27c^3 \quad = -4320b^3 c^3.$$

$$15(2b)^2(-3c)^4 = 15 \times 2^2 b^2 \times (-)^4 3^4 c^4$$
$$= 15 \times 4b^2 \times 81c^4 \quad = 4860b^2 c^4.$$

$$6(2b)(-3c)^5 \quad = 6 \times 2b \times (-)^5 3^5 c^5$$
$$= 6 \times 2b \times -243c^4 \quad = -2916bc^4.$$

$$(-3c)^6 \qquad = (-)^6 3^6 c^6 \qquad = 729c^6.$$

Gathering together all our terms we get

$$(2b - 3c)^6 \quad = 64b^6 - 576b^5 c + 2160b^4 c^2 - 4320b^3 c^3 + 4860b^2 c^4$$
$$- 2916bc^5 + 729c^6.$$

Notice how tedious this example has been with only minor numerical amendments. The next short exercise makes only the slightest alterations to the general $(a + x)^n$. Afterwards we look to a method that is less complicated yet still based on a pattern.

▰▰▰ ASSIGNMENT ▰▰▰

The company decides to test samples of five rods at a time from the production process. It has a range of tests the sample must pass before the entire batch is despatched. There is a further range of follow-up tests should one of the batch fail. We need not worry ourselves with these.

Now each of the five rods may be perfect (p) or faulty (q). We list below all the possible combinations of p and q.

5ps	4ps, 1q	3ps, 2qs	2ps, 3qs	1p, 4qs	5qs
ppppp	ppppq	pppqq	ppqqq	pqqqq	qqqqq
	pppqp	ppqpq	pqpqq	qpqqq	
	ppqpp	pqppq	pqqpq	qqpqq	
	pqppp	qpppq	pqqqp	qqqpq	
	qpppp	ppqqp	qppqq	qqqqp	
		pqpqp	qpqpq		
		qppqp	qpqqp		
		pqqpp	qqppq		
		qpqpp	qqpqp		
		qqppp	qqqpp		
1	5	10	10	5	1

Notice how the total numbers for each combination tally with a line from Pascal's triangle. They are associated with $(a + x)^5$. In our case, with p and q, they must be associated with $(p + q)^5$, i.e.

$$(p + q)^5 = p^5 + 5p^4q + 10p^3q^2 + 10p^2q^3 + 5pq^4 + q^5.$$

■ EXERCISE 5.1 ■

Use Pascal's triangle and the patterns we have already worked out to expand the following.

1 $(a + x)^8$

2 $(a - x)^5$

3 $(a - x)^6$

4 $(2r + s)^4$

5 $(x - 3y)^5$

Factorial numbers

We use the symbol **!** in mathematics to mean **factorial**. You may have a calculator key with this symbol. If not, do not worry because the calculation is very easy.

Usually we can find only the factorial of a positive integer (positive whole number). For example 4! (said '4 factorial') is

$$4! = 4 \times 3 \times 2 \times 1 \quad = 24,$$
$$5! = 5 \times 4 \times 3 \times 2 \times 1 = 120.$$

| 5 factorial. |

Notice that $5! = 5 \times (4!)$.

We can keep working out values like 6!, 7! and so on.

For the smaller numbers we have

$$2! = 2 \times 1 = 2,$$
$$1! = 1$$

and strangely $0! = 1$. This is the only exception to the original number being positive.

Let us look again at $5! = 5 \times 4 \times 3 \times 2 \times 1$. Notice how we start the multiplication with the original 5. Then we reduce the multiplying numbers in succession until the final number is 1. More generally

$$n! = n \times (n - 1) \times (n - 2) \times \ldots \times 3 \times 2 \times 1.$$

You can try working out some factorial numbers with your calculator. The factorial key is $\underline{x!}$. Examples 5.5 show you the simple order of key operations.

Use your calculator to find the values of   i) 5!,   ii) 9!.

i) For 5! the order of key operations is $\underline{5}|\ \underline{x!}|$ to display 120 as expected.

ii) For 9! the order of key operations is $\underline{9}|\ \underline{x!}|$ to display 362880.

Your calculator cannot cope with large values that exceed its capacity. For example it may evaluate 69! as $1.7112\ldots \times 10^{98}$, but display an error message, -E-, for 70! due to its excessive size.

The binomial theorem

The general expression for the **binomial theorem** is

$$(a+x)^n = a^n + \frac{na^{n-1}x}{1!} + \frac{n(n-1)a^{n-2}x^2}{2!} + \frac{n(n-1)(n-2)a^{n-3}x^3}{3!} \ldots + x^n.$$

This form of the theorem is **valid for all positive integers (positive whole numbers) of** n.

For large values of n this is easier to use than Pascal's triangle. The patterns we discussed for Pascal's triangle still apply here. Sometimes they may not be as obvious as they were previously. More importantly, let us look at the powers in a particular term. For example $\dfrac{n(n-1)a^{n-2}x^2}{2!}$ is the 3rd term. Notice the 2s in the term and that they are 1 less than 3. This original pattern is true for any term so, for example, the 6th term is $\dfrac{n(n-1)(n-2)(n-3)(n-4)a^{n-5}x^5}{5!}$. Again notice the 5s, $(6-1=5)$. We will look at this in detail in Example 5.8.

Use the binomial theorem to find the expansions of

i) $(a+x)^6$,   ii) $(a-x)^5$,   iii) $(2b-3c)^4$.

i) Comparing $(a+x)^6$ with the general $(a+x)^n$ we have $n=6$ to give

$$(a+x)^6 = a^6 + \frac{6a^{6-1}x}{1!} + \frac{6 \times 5a^{6-2}x^2}{2!} + \frac{6 \times 5 \times 4a^{6-3}x^3}{3!}$$

$$+ \frac{6 \times 5 \times 4 \times 3a^{6-4}x^4}{4!} + \frac{6 \times 5 \times 4 \times 3 \times 2a^{6-5}x^5}{5!}$$

$$+ \frac{6 \times 5 \times 4 \times 3 \times 2a^{6-6}x^6}{6!}.$$

By following the patterns we know to stop here.
$6 \times 5 \times 4 \times 3 \times 2 \times 1$ can go no further, as the next figure would need to be 0. Also $a^{6-6} = a^0 = 1$ and we have reached x^6. Our expansion simplifies to

$$(a + x)^6 = a^6 + \frac{6a^5x}{1} + \frac{6 \times 5a^4x^2}{2 \times 1} + \frac{6 \times 5 \times 4a^3x^3}{3 \times 2 \times 1}$$

$$+ \frac{6 \times 5 \times 4 \times 3a^2x^4}{4 \times 3 \times 2 \times 1} + \frac{6 \times 5 \times 4 \times 3 \times 2a^1x^6}{5 \times 4 \times 3 \times 2 \times 1}$$

$$+ \frac{6 \times 5 \times 4 \times 3 \times 2 \times 1a^0x^6}{6 \times 5 \times 4 \times 3 \times 2 \times 1}$$

$$= a^6 + 6a^5x + 15a^4x^2 + 20a^3x^3 + 15a^2x^4 + 6ax^5 + a^6$$

as before.

ii) Comparing $(a - x)^5$ with the original $(a + x)^n$ we have $n = 5$ and replace x with $-x$ to give

$$(a - x)^5 = \frac{a^5 + 5a^{5-1}(-x)}{1!} + \frac{5 \times 4a^{5-2}(-x)^2}{2!}$$

$$+ \frac{5 \times 4 \times 3a^{5-3}(-x)^3}{3!} + \frac{5 \times 4 \times 3 \times 2a^{5-4}(-x)^4}{4!}$$

$$+ \frac{5 \times 4 \times 3 \times 2 \times 1a^{5-5}(-x)^5}{5!}$$

$$= a^5 - \frac{5a^4x}{1} + \frac{5 \times 4a^3x^2}{2 \times 1} - \frac{5 \times 4 \times 3a^2x^3}{3 \times 2 \times 1}$$

$$+ \frac{5 \times 4 \times 3 \times 2a^1x^4}{4 \times 3 \times 2 \times 1} - \frac{5 \times 4 \times 3 \times 2 \times 1a^0x^5}{5 \times 4 \times 3 \times 2 \times 1}$$

$$= a^5 - 5a^4x + 10a^3x^2 - 10a^2x^3 + 5ax^4 - x^5.$$

Notice how we replace x with $-x$ and the signs now alternate $+/-$.

iii) Comparing $(2b - 3c)^4$ with the general $(a + x)^n$ we have $n = 4$. We replace a with $2b$ and x with $-3c$ to give

$$(2b - 3c)^4 = (2b)^4 + \frac{4(2b)^{4-1}(-3c)}{1!} + \frac{4 \times 3(2b)^{4-2}(-3c)^2}{2!}$$

$$+ \frac{4 \times 3 \times 2(2b)^{4-3}(-3c)^3}{3!}$$

$$+ \frac{4 \times 3 \times 2 \times 1(2b)^{4-4}(-3c)^4}{4!}$$

$$= (2b)^4 + \frac{4(2b)^3(-3c)}{1} + \frac{4 \times 3(2b)^2(-3c)^2}{2 \times 1}$$

$$+ \frac{4 \times 3 \times 2(2b)^1(-3c)^3}{3 \times 2 \times 1} + \frac{4 \times 3 \times 2 \times 1(2b)^0(-3c)^4}{4 \times 3 \times 2 \times 1}$$

$$= 16b^4 - 4 \times 8b^3 \times 3c + 6 \times 4b^2 \times 9c^2 - 4 \times 2b \times 27c^3$$

$$+ 81c^4$$

$$= 16b^4 - 96b^3c + 216b^2c^2 - 216bc^3 + 81c^4.$$

Some of the patterns do remain in this third example. However, most of them are lost because the original expression is complicated slightly.

As the power of n increases the expansion becomes too long to write out in full. Often we give only the early terms.

▓▓▓▓▓ Examples 5.7 ▓▓▓▓▓▓▓▓▓▓▓▓▓▓▓▓▓▓▓▓▓▓▓▓▓▓▓▓▓▓

Use the binomial theorem to find the expansions of

i) $(a + 2b)^{12}$ for the first 5 terms,

ii) $\left(z - \dfrac{1}{z}\right)^9$ for the first 4 terms.

i) Comparing $(a + 2b)^{12}$ with the general $(a + x)^n$ we have $n = 12$ and replace x with $2b$.

$$(a + 2b)^{12} = a^{12} + \frac{12a^{11}(2b)}{1!} + \frac{12 \times 11a^{10}(2b)^2}{2!}$$
$$+ \frac{12 \times 11 \times 10a^9(2b)^3}{3!} + \frac{12 \times 11 \times 10 \times 9a^8(2b)^4}{4!} \cdots$$
$$= a^{12} + 12a^{11} \times 2b + 66a^{10} \times 4b^2 + 220a^9 \times 8b^3$$
$$+ 495a^8 \times 16b^4 \ldots$$
$$= a^{12} + 24a^{11}b + 264a^{10}b^2 + 1760a^9b^3 + 7920a^8b^4 \ldots$$

ii) Comparing $\left(z - \dfrac{1}{z}\right)^9$ with the general $(a + x)^n$ we have $n = 9$. We replace a with z and x with $-\dfrac{1}{z}$ to give

$$\left(z - \frac{1}{z}\right)^9 = z^9 + \frac{9z^8}{1!}\left(\frac{-1}{z}\right) + \frac{9 \times 8z^7}{2!}\left(\frac{-1}{z}\right)^2$$
$$+ \frac{9 \times 8 \times 7z^6}{3!}\left(\frac{-1}{z}\right)^3 \cdots$$
$$= z^9 - \frac{9z^8}{z} + \frac{36z^7}{z^2} - \frac{84z^6}{z^3} \cdots$$
$$= z^9 - 9z^7 + 36z^5 - 84z^3 \ldots$$

Suppose we need to find a particular term in an expansion. We could work out the terms in order from the first term. Alternatively, and more quickly, we could follow the patterns. The next example uses those patterns.

████ **Example 5.8** ████

What is the 5th term in the expansion of $(a - 3x)^{11}$?

Earlier, for the general $(a + x)^n$ we noted the 3rd term to be $\dfrac{n(n - 1)a^{n-2}x^2}{2!}$. Values are 1 less than the 3 in that '3rd term'.

   n and $(n - 1)$ appear as the product of 2 terms,

   2 appears somewhere in all the powers,

   2 appears in the denominator as 2!

For $(a - 3x)^{11}$ we have $n = 11$ and replace x with $-3x$. We are interested in the 5th term. This means that 4 (as $5 - 1 = 4$) is important to us.

   n, $(n - 1)$, $(n - 2)$ and $(n - 3)$ appear as the product of 4 terms,

   4 appears somewhere in all the powers,

   4 appears in the denominator as 4!

$$\text{The 5th term is} \quad \frac{11 \times 10 \times 9 \times 8a^{11-4}(-3x)^4}{4!}$$

$$= \frac{11 \times 10 \times 9 \times 8a^7 \times 81x^4}{4 \times 3 \times 2 \times 1}$$

$$= 26730a^7x^4.$$

████ **ASSIGNMENT** ████

Previously we looked at combinations of p and q. We related them to $(p + q)^5$ and Pascal's triangle. In this section we look at expanding $(p + q)^5$ using the binomial theorem, i.e.

$$(p + q)^5 = p^5 + \frac{5p^{5-1}q}{1!} + \frac{5 \times 4p^{5-2}q^2}{2!} + \frac{5 \times 4 \times 3p^{5-3}q^3}{3!}$$

$$+ \frac{5 \times 4 \times 3 \times 2p^{5-4}q^4}{4!} + \frac{5 \times 4 \times 3 \times 2 \times 1p^{5-5}q^5}{5!}$$

$$= p^5 + 5p^4q + \frac{5 \times 4p^3q^2}{2 \times 1} + \frac{5 \times 4 \times 3p^2q^3}{3 \times 2 \times 1}$$

$$+ \frac{5 \times 4 \times 3 \times 2p^1q^4}{4 \times 3 \times 2 \times 1} + \frac{5 \times 4 \times 3 \times 2 \times 1p^0q^5}{5 \times 4 \times 3 \times 2 \times 1}$$

$$= p^5 + 5p^4q + 10p^3q^2 + 10p^2q^3 + 5pq^4 + q^5 \text{ as before.}$$

■■■■ EXERCISE 5.2 ■■■■■■■■■

Use the binomial theorem in the following questions.

Fully expand the expressions

1 $(a + b)^5$

2 $(1 + b)^5$

3 $(1 - b)^5$

4 $(x + 4y)^4$

5 $(2x - y)^6$

6 $(3x - 2y)^5$

7 $(\frac{1}{2}a + 2b)^4$

8 $(2 - x)^7$

9 $\left(\dfrac{1}{a} + a\right)^8$

10 $\left(a^2 - \dfrac{1}{a^2}\right)^5$

Find the first 4 terms of

11 $(2x + y)^8$

12 $(x - 3y)^{15}$

13 $(3x + 2y)^6$

Find the first 5 terms of

14 $(a + b)^{12}$

15 $(a - b)^{11}$

16 $(a - 2b)^{14}$

Find the first 6 terms of

17 $(1 + b)^{16}$

18 $\left(x - \dfrac{1}{x}\right)^{10}$

19 Find the 5th term of $(b - a)^{10}$

20 Find the 7th term of $(4a + b)^9$

The binomial series

We introduced the binomial theorem for $(a + x)^n$ where n had to be a positive whole number (positive integer).

There is an alternative form where n is negative or fractional (including decimal). It also applies where n is both negative and fractional. The form is

$$(1 + x)^n = 1 + \frac{nx}{1!} + \frac{n(n-1)x^2}{2!} + \frac{n(n-1)(n-2)x^3}{3!} \cdots$$

and can be called the **binomial series**. It has an infinite number of terms. We may use as many as we need to show a pattern or to meet a level of accuracy. Notice we have had to replace a with 1, otherwise the patterns are similar to those from before.

This expansion is only valid for $-1 < x < 1$,
i.e. x is greater than -1 and less than 1,
i.e. x lies between -1 and 1.

▓▓▓▓▓ **Examples 5.9** ▓▓▓▓▓▓▓▓▓▓▓▓▓▓▓▓▓▓▓▓▓▓▓▓▓▓▓▓▓▓▓▓▓▓▓▓

We can use the binomial series to expand

i) $\sqrt{1+x}$, ii) $\dfrac{1}{1+2x}$, iii) $\dfrac{1}{(1+4x)^3}$.

In each case, for what range of values of x is the series valid?

We return to our general $(1+x)^n$ and look at $1+x$. In each case we compare against this section.

i) $\sqrt{1+x}$ includes $1+x$ which compares exactly with our general $1+x$. Hence this series is also valid for $-1 < x < 1$.

ii) $\dfrac{1}{1+2x}$ includes $1+2x$ to compare against our general $1+x$. We have replaced x with $2x$ so our series is valid for $-1 < 2x < 1$, i.e. $-\frac{1}{2} < x < \frac{1}{2}$.

> Dividing by 2.

iii) $\dfrac{1}{(1+4x)^3}$ includes $1+4x$ to compare against our general $1+x$. We have replaced x with $4x$ so our series is valid for $-1 < 4x < 1$, i.e. $-\frac{1}{4} < x < \frac{1}{4}$.

> Dividing by 4.

▓▓▓▓▓ **Examples 5.10** ▓▓▓▓▓▓▓▓▓▓▓▓▓▓▓▓▓▓▓▓▓▓▓▓▓▓▓▓▓▓▓▓▓▓▓

Use the binomial series to expand

i) $\sqrt{1+x}$ up to and including x^4,

ii) $\dfrac{1}{1+2x}$ for the first 4 terms,

iii) $\dfrac{5}{1+2x}$ for the first 4 terms,

iv) $\dfrac{1}{(1-4x)^3}$ for the first 4 terms,

v) $\dfrac{1}{\sqrt[4]{1+2x}}$ up to and including x^3.

i) We re-write $\sqrt{1+x}$ as $(1+x)^{\frac{1}{2}}$. Comparing it with the general $(1+x)^n$ we have $n = \dfrac{1}{2}$. This means that

$$(1+x)^{\frac{1}{2}} = 1 + \frac{\frac{1}{2}x}{1!} + \frac{\frac{1}{2}(\frac{1}{2}-1)x^2}{2!} + \frac{\frac{1}{2}(\frac{1}{2}-1)(\frac{1}{2}-2)x^3}{3!}$$
$$+ \frac{\frac{1}{2}(\frac{1}{2}-1)(\frac{1}{2}-2)(\frac{1}{2}-3)x^4}{4!} \cdots$$

$$= 1 + 0.5x + \frac{0.5(-0.5)x^2}{2} + \frac{0.5(-0.5)(-1.5)x^3}{6}$$
$$+ \frac{0.5(-0.5)(-1.5)(-2.5)x^4}{24} \cdots$$

$$= 1 + 0.5x - 0.125x^2 + 0.0625x^3 - 0.0391x^4 \ldots$$

ii) We rewrite $\dfrac{1}{1+2x}$ as $1(1+2x)^{-1}$ or simply $(1+2x)^{-1}$. Comparing it with the general $(1+x)^n$ we have $n = -1$ and replace x with $2x$. This means that

$$(1+2x)^{-1} = 1 + \frac{(-1)2x}{1!} + \frac{(-1)(-1-1)(2x)^2}{2!}$$
$$+ \frac{(-1)(-1-1)(-1-2)(2x)^3}{3!} \cdots$$

$$= 1 - 2x + \frac{(-1)(-2)4x^2}{2} + \frac{(-1)(-2)(-3)8x^3}{6} \cdots$$

$$= 1 - 2x + 4x^2 - 8x^3 \ldots$$

iii) We rewrite $\dfrac{5}{(1+2x)}$ as $5(1+2x)^{-1}$. This means we can apply our previous result, multiplying by 5 to get

$$5(1+2x)^{-1} = 5(1 - 2x + 4x^2 - 8x^3 \ldots)$$
$$\text{or } 5 - 10x + 20x^2 - 40x^3 \ldots$$

iv) We rewrite $\dfrac{1}{(1-4x)^3}$ as $(1-4x)^{-3}$. Comparing it with the general $(1+x)^n$ we have $n = -3$ and replace x with $-4x$. This means that

$$(1-4x)^{-3} = 1 + \frac{(-3)(-4x)}{1!} + \frac{(-3)(-3-1)(-4x)^2}{2!}$$
$$+ \frac{(-3)(-3-1)(-3-2)(-4x)^3}{3!} \cdots$$

$$= 1 + 12x + \frac{(-3)(-4)16x^2}{2} + \frac{(-3)(-4)(-5)(-64x^3)}{6} \cdots$$

$$= 1 + 12x + 96x^2 + 640x^3 \ldots$$

v) We rewrite $\dfrac{1}{\sqrt[4]{1+2x}}$ in stages as $\dfrac{1}{(1+2x)^{\frac{1}{4}}}$ and then $(1+2x)^{-\frac{1}{4}}$ or $(1+2x)^{-0.25}$. Comparing it with our general $(1+x)^n$ we have $n = -0.25$ and replace x with $2x$. This means that

$$(1 + 2x)^{-0.25} = 1 + \frac{(-0.25)2x}{1!} + \frac{(-0.25)(-0.25 - 1)(2x)^2}{2!}$$

$$+ \frac{(-0.25)(-0.25 - 1)(-0.25 - 2)(2x)^3}{3!} \cdots$$

$$= 1 - 0.5x + \frac{(-0.25)(-1.25)4x^2}{2}$$

$$+ \frac{(-0.25)(-1.25)(-2.25)8x^3}{6} \cdots$$

$$= 1 - 0.5x + 0.625x^2 - 0.9375x^3 \ldots$$

Example 5.11

Using the binomial series find the first 4 terms of $\dfrac{1}{(4 - x)^3}$.

Immediately $(4 - x)$ does *not* compare favourably with $(1 + x)$. The problem is caused by the 4. We must start with 1. To achieve this we falsely remove a common factor of 4 so that $4 - x = 4\left(1 - \dfrac{x}{4}\right)$.

Then $\dfrac{1}{(4 - x)^3} = \dfrac{1}{\left(4\left(1 - \dfrac{x}{4}\right)\right)^3} = \dfrac{1}{4^3\left(1 - \dfrac{x}{4}\right)^3}$ or $\dfrac{1}{64\left(1 - \dfrac{x}{4}\right)^3}$.

We apply the binomial series to the bracketed section, i.e.

$$\frac{1}{(4 - x)^3} = \frac{1}{64}\left(1 - \frac{x}{4}\right)^{-3}$$

$$= \frac{1}{64}\left[1 + \frac{(-3)}{1!}\left(\frac{-x}{4}\right) + \frac{(-3)(-3 - 1)}{2!}\left(-\frac{x}{4}\right)^2\right.$$

$$\left. + \frac{(-3)(-3 - 1)(-3 - 2)}{3!}\left(-\frac{x}{4}\right)^3 \cdots\right]$$

$$= \frac{1}{64}\left[1 + \frac{3x}{4} + \frac{(-3)(-4)}{2}\left(\frac{x^2}{16}\right)\right.$$

$$\left. + \frac{(-3)(-4)(-5)}{6}\left(-\frac{x^3}{64}\right) \cdots\right]$$

$$= \frac{1}{64}\left(1 + 0.75x + 0.375x^2 + 0.15625x^3 \ldots\right)$$

We could multiply through by $\dfrac{1}{64}$ but choose not to do so.

EXERCISE 5.3

Use the binomial series to write down series in the following questions.

1 $\dfrac{1}{1+x}$ for the first 5 terms.

2 $\dfrac{1}{(1+b)^5}$ up to and including b^3.

3 $\dfrac{3}{(1-b)^4}$ for the first 4 terms.

4 $\dfrac{2}{(1+3x)^3}$ up to and including x^4.

5 $\sqrt{1+2x}$ up to and including x^3.

6 $\sqrt{1-2x}$ for the first 4 terms.

7 $(1-x)^{\frac{1}{3}}$ up to and including x^2.

8 $(1+3x)^{\frac{1}{4}}$ for the first 3 terms.

9 $\dfrac{1}{\sqrt{1+2x}}$ for the first 4 terms.

10 $\dfrac{1}{\sqrt[4]{1-2x}}$ up to and including x^3.

11 $(1+x)^{-\frac{1}{3}}$ up to and including x^2.

12 $\dfrac{1}{\sqrt{1+4x}}$ for the first 5 terms.

13 $2(1-6x)^{-\frac{1}{2}}$ up to and including x^3.

14 $\dfrac{1}{\sqrt{4+x}}$ for the first 4 terms.

15 $\dfrac{3}{(2+x)^4}$ up to and including x^3.

Example 5.12

Using the binomial series expand $\dfrac{2+x}{1+2x}$ up to and including the x^3 term.

We may rewrite $\dfrac{2+x}{1+2x}$ as $(2+x)(1+2x)^{-1}$ and apply the binomial series to $(1+2x)^{-1}$. From Examples 5.10(ii) we have

$$(1+2x)^{-1} = 1 - 2x + 4x^2 - 8x^3 \ldots$$

Then $\dfrac{2+x}{1+2x} = (2+x)(1+2x)^{-1}$

$$= (2+x)(1 - 2x + 4x^2 - 8x^3 \ldots).$$

We multiply out these brackets. Each term in the second bracket is multiplied by 2 and then by x. We add together these results, requiring only the terms up to and including x^3. This is the accuracy of our earlier binomial series application. We get

$$2 - 4x + 8x^2 - 16x^3 \ldots$$
$$+x - 2x^2 + 4x^3 \ldots$$
$$= 2 - 3x + 6x^2 - 12x^3 \ldots$$

Example 5.13

Using the binomial series find the first 3 terms of $\dfrac{5\sqrt{1+x}}{1+2x}$.

We use earlier results from Examples 5.10, for $\dfrac{5}{1+2x}$ and for $\sqrt{1+x}$,

i.e. $\dfrac{5\sqrt{1+x}}{1+2x} = (5 - 10x + 20x^2 \ldots)(1 + 0.5x - 0.125x^2 \ldots)$

We quote our results to include the x^2 terms. This is also the limit of our accuracy when we multiply each term in the second bracket by each term in the first bracket. We get

$$5 + 2.5x - 0.625x^2 \ldots$$
$$- 10x - 5x^2 \ldots$$
$$+ 20x^2 \ldots$$
$$= 5 - 7.5x + 14.375x^2 \ldots$$

EXERCISE 5.4

Apply the binomial series to write down series in each question. (Hint: Use your results from previous exercises.) Complete each question by multiplying together the brackets and gathering together like terms.

1 $\dfrac{2x}{1+x}$ for the first 5 terms.

2 $\dfrac{\sqrt{1+2x}}{1+x}$ up to and including x^3.

3 $\dfrac{1-x}{\sqrt{1+2x}}$ for the first 4 terms.

4 $\dfrac{3(x-1)}{(2+x)^4}$ up to and including x^3.

5 $\sqrt{\dfrac{1+2x}{1+4x}}$ up to and including x^2.

The binomial series – small changes

We know that we can improve our accuracy with the binomial series by using more terms. In contrast, suppose we want a quick approximation. We do this by using fewer terms, usually just the first and second ones,

i.e. $(1+x)^n \approx 1 + nx$.

$\approx$ means 'approximately equals'.

This is the **binomial approximation**. For this approximation we assume all the other terms are small enough to be neglected. This assumes that x and n are small. As x and n increase so the approximation worsens. In the next examples we look at the approximation worsening as x increases.

Examples 5.14

We investigate the general relationship $(1+x)^n \approx 1 + nx$ for $n=4$ and different values of x. Each attempt is checked against the calculator value.

For $x = 0.025$, $(1+x)^4$ is $(1+0.025)^4 \approx 1 + 4 \times 0.025$

$$\approx 1.100.$$

The calculator value of $(1+0.025)^4$ is 1.104.
The difference between these values is 0.004.

We repeat this calculation for different values of x and draw a table.

x	Binomial approx.	Calculator value	Difference
0.025	1.100	1.104	0.004
0.050	1.200	1.216	0.016
0.075	1.300	1.335	0.035
0.100	1.400	1.464	0.064
0.150	1.600	1.749	0.149

As x increases the differences increase, but at a greater rate. In our table, x increases by 6 times from its first to last value. The increase in the differences is over 37 times.

As an exercise you can choose a value of x (e.g. $x = 0.075$) and test the binomial approximation against the calculator for different values of n. Again, as n increases so will the differences, at a greater rate.

The remaining examples in this section look at how this approximation affects formulae.

Example 5.15

The radius, r, of a sphere is overestimated by 3.5%. What effect does this have on the   i) surface area and   ii) volume?

3.5% is $\dfrac{3.5}{100}$, i.e. 0.035, and is taken as positive because it is an over-estimate. This change alters the radius, r, to $(1+0.035)r$.

i) The formula for the surface area, A, of a sphere is $A = 4\pi r^2$. Our new area is $A + \delta A$ where δA is the change from the original area, A. We apply these changes to our formula to get

$$A + \delta A = 4\pi[(1 + 0.035)r]^2$$
$$= 4\pi r^2(1 + 0.035)^2$$
$$= A(1 + 0.035)^2$$
$$\approx A(1 + (2 \times 0.035))$$

> $A = 4\pi r^2$.
> Binomial approx.

i.e. $A + \delta A \approx A(1 + 0.070) = A + 0.07A$

i.e. $\delta A \approx 0.07A$

i.e.   the surface area is 7% $\left(0.07 = \dfrac{7}{100}\right)$ too large.

ii) The formula for the volume, V, of a sphere is $V = \dfrac{4}{3}\pi r^3$.

Our new volume is $V + \delta V$ where δV is the change from the original volume, V. We apply these changes to get

$$V + \delta V = \frac{4}{3}\pi[(1 + 0.035)r]^3$$
$$= \frac{4}{3}\pi r^3(1 + 0.035)^3$$
$$= V(1 + 0.035)^3$$
$$\approx V(1 + (3 \times 0.035))$$

> $V = \dfrac{4}{3}\pi r^3$.
> Binomial approx.

i.e. $V + \delta V \approx V(1 + 0.105) = V + 0.105V$

i.e. $\delta V \approx 0.105V$

i.e.   the volume is 10.5% $\left(0.105 = \dfrac{10.5}{100}\right)$ too large.

Example 5.16

In an electrical circuit the formula for voltage magnification, Q, is given by $Q = \dfrac{1}{R}\sqrt{\dfrac{L}{C}}$. R is the resistance, L is the inductance and C is the capacitance, as usual. Find the change in Q (δQ) when

i) R is overestimated by 4%   (i.e. 0.04),

ii) L is overestimated by 3%   (i.e. 0.03),

iii) C is underestimated by 6%   (i.e. −0.06),

iv) all the misestimates occur together.

We interpret an overestimate as positive and an underestimate as negative.

i) The overestimate of R is written as $(1 + 0.04)R$. Our new voltage magnification is $Q + \delta Q$ where δQ is the change from the original Q. We apply these alterations to our formula to get

$$Q + \delta Q = \frac{1}{(1 + 0.04)R} \sqrt{\frac{L}{C}}$$

$$= \frac{1}{R} \sqrt{\frac{L}{C}} (1 + 0.04)^{-1}$$

$$= Q(1 + 0.04)^{-1}$$ — Substituting for Q.

$$\approx Q(1 + (-1 \times 0.04))$$ — Binomial approx.

i.e. $Q + \delta Q \approx Q(1 - 0.04) = Q - 0.04Q$

i.e. $\delta Q \approx -0.04Q$ — Negative is an underestimate.

i.e.   the voltage magnification is 4% too small.

ii) The overestimate of L is written as $(1 + 0.03)L$. Our new voltage magnification is $Q + \delta Q$ where δQ is the change from the original Q. We apply these alterations to our formula to get

$$Q + \delta Q = \frac{1}{R} \sqrt{\frac{(1 + 0.03)L}{C}}$$

$$= \frac{1}{R} \sqrt{\frac{L}{C}} (1 + 0.03)^{0.5}$$

$$\approx Q(1 + (0.5 \times 0.03))$$ — Substituting for Q. Binomial approx.

i.e. $Q + \delta Q \approx Q(1 + 0.015) = Q + 0.015Q$

i.e. $\delta \approx 0.015Q$

i.e.   the voltage magnification is 1.5% too large.

iii) The underestimate of C is written as $(1 - 0.06)C$. Our new voltage magnification is $Q + \delta Q$ where δQ is the change from the original Q. We apply these alterations to our formula to get

$$Q + \delta Q = \frac{1}{R} \sqrt{\frac{L}{(1 - 0.06)C}}$$

$$= \frac{1}{R} \sqrt{\frac{L}{C}} \sqrt{\frac{1}{1 - 0.06}}$$ — Substituting for Q. Binomial approx.

$$= Q(1 - 0.06)^{-0.5}$$

$$\approx Q(1 - (-0.5 \times 0.06))$$

i.e. $Q + \delta Q \approx Q(1 + 0.03) = Q + 0.03Q$

i.e. $\delta Q \approx 0.03Q$

i.e.   the voltage magnification is 3% too large.

iv) Now we bring together all the misestimates as

$$Q + \delta Q = \frac{1}{(1 + 0.04)R} \sqrt{\frac{(1 + 0.03)L}{(1 - 0.06)C}}$$

$$= \frac{1}{R}\sqrt{\frac{L}{C}}(1 + 0.04)^{-1}(1 + 0.03)^{0.5}(1 - 0.06)^{-0.5}$$

$$\approx Q(1 - 0.04)(1 + 0.015)(1 + 0.03)$$

> Substituting for Q.
> Binomial approx.

We expand these brackets in stages, neglecting terms that are too small. Because we have used the binomial approximation we *cannot* achieve greater accuracy now.

$$\therefore \quad Q + \delta Q \approx Q(\overbrace{1 - 0.04)(1 + 0.01}5)(1 + 0.03)$$

$$\approx Q(1 + 0.015 - 0.04\ldots)(1 + 0.03)$$

> Neglecting -0.04×0.015.

$$\approx (\overbrace{1 - 0.025)(1 + 0.03})$$

$$\approx Q(1 + 0.03 - 0.025\ldots)$$

> Neglecting -0.025×0.03.

i.e. $Q + \delta Q \approx Q(1 + 0.005) = Q + 0.005Q$

i.e. $\delta Q \approx 0.005Q$

i.e. the voltage magnification is 0.5% too large.

Notice how the positions of R, L and C in the numerator and denominator have affected the overall misestimation. Some alterations have compensated for others to reduce the overall error. Sometimes, in other cases, they can amplify the errors rather than reduce them.

■■■ ASSIGNMENT ■■■

We take a final look at our cylindrical rods, this time concentrating on the volume, V, where $V = \pi r^2 h$. r is the radius and h is the height (or length). There can be many measurement errors so we will use some specimen ones.

i) Suppose the radius is overestimated by 3% and the height underestimated by 2%. What effect does this have on the volume?

The overestimate of r is written as $(1 + 0.03)r$ and the underestimate of h as $(1 - 0.02)h$. Our new volume is $V + \delta V$ where δV is the change from the original V. We apply these alterations to our formula to get

$$V + \delta V = \pi\{(1 + 0.03)r\}^2(1 - 0.02)h$$
$$= \pi r^2 h(1 + 0.03)^2(1 - 0.02)$$
$$= V(1 + 0.03)^2(1 - 0.02) \qquad \text{Substituting for } V.$$
$$\approx V(1 + (2 \times 0.03))(1 - 0.02) \qquad \text{Binomial approx.}$$
$$\approx V\overbrace{(1 + 0.06)(1 - 0.02)}$$
$$\approx V(1 - 0.02 + 0.06\ldots)$$

i.e. $\quad V + \delta V \approx V(1 + 0.04) = V + 0.04V$

i.e. $\qquad \delta V \approx 0.04V$

i.e. the volume is 4% too large.

ii) Suppose the radius is underestimated by 3% and the height overestimated by 6%. What effect does this have on the volume?

This time we have the underestimate of r written as $(1 - 0.03)r$ and the overestimate of h as $(1 + 0.06)h$. Our new volume is $V + \delta V$ where δV is the change from the original V. We apply these alterations to our formula to get

$$V + \delta V = \pi\{(1 - 0.03)r\}^2(1 + 0.06)h$$
$$= \pi r^2 h(1 - 0.03)^2(1 + 0.06)$$
$$= V(1 - 0.03)^2(1 + 0.06) \qquad \text{Substituting for } V.$$
$$\approx V(1 - (2 \times 0.03))(1 + 0.06) \qquad \text{Binomial approx.}$$
$$\approx V\overbrace{(1 - 0.06)(1 + 0.06)}$$
$$\approx V(1 + 0.06 - 0.06\ldots)$$

i.e. $\quad V + \delta V \approx V$

i.e. $\qquad \delta V \approx 0$

i.e. there is no approximate change in the volume.

■ EXERCISE 5.5 ■

For all these questions apply the binomial approximation as appropriate.

1 What is the effect on the area, A, of a circle if the radius, r, is overestimated by 2.5%? $\qquad [A = \pi r^2]$

2 The radius, r, of a sphere is underestimated by 3%. What is the change in its volume, V? $\qquad \left[V = \dfrac{4}{3}\pi r^3\right]$

3 For a cone the radius, r, is measured 1.5% too large and the height, h, 2.5% too small. What is the percentage change in the volume, V? $$\left[V = \frac{1}{3}\pi r^2 h\right]$$

4

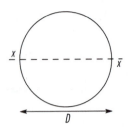

The second moment of area about the line XX is given by $I_{xx} = \dfrac{\pi D^4}{64}$. When D is measured 1.5% too small what is the effect on I_{xx}?

5 In an electrical circuit the voltage, V, and resistance, R, are related to power, P, by $P = \dfrac{V^2}{R}$. V is underestimated by 3.75% and R is overestimated by 2.50%. Calculate the percentage change in P.

6

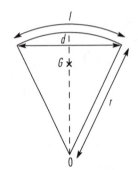

G is the centre of gravity of the sector where OG is $\bar{y} = \dfrac{2dr}{3l}$.

d is measured 4% too large and l 2% too small. Calculate the percentage change in $\bar{y}$.

7 A standard beam of length l carries a uniformly distributed load, W. Its maximum deflection, y, is given by $y = \dfrac{5Wl^3}{384EI}$. If the beam is 1.5% too long what is the increase to the maximum deflection?

8 The capacitive reactance, X_c, in an a.c. circuit is given by $X_c = \dfrac{1}{2\pi f C}$. f is the frequency and C is the capacitance. If f is 48 Hz instead of 50 Hz, what is the percentage error of f? Along with this error C is underestimated by 1.5%. Calculate the overall percentage change in X_c.

9 The periodic time, T, of a simple pendulum is related to its length, l, by $T = 2\pi\sqrt{\dfrac{l}{g}}$. g is the acceleration due to gravity. If l is measured 3.5% too large while g remains constant, what happens to T?

10 The resonant frequency, f_r, in an electrical circuit is given by $f_r = \dfrac{1}{2\pi\sqrt{LC}}$. L is the inductance and C is the capacitance. L is measured 5% too large and C 3% too large. Calculate the effect on f_r.

11 For $\bar{x} = np$ and $\sigma = \sqrt{npq}$, p and q are inaccurately measured. p is overestimated by 2% and q is underestimated by 2%. What are the effects on $\bar{x}$ and σ?

12 A vehicle's speed, v, on a curved horizontal road is related to the radius of the curve, r, by $v = \sqrt{g\mu r}$. g is the acceleration due to gravity and μ is the coefficient of friction. Calculate the change in v due to a 7.5% change in μ.

13 A body falling from rest has a velocity, v, given by $v = \sqrt{2gh}$. g is the acceleration due to gravity and h is the height fallen. If h is measured 5% too large what effect does this have on v? By moving to another position on the earth's surface there is a 1% change in g. For both an increase and a decrease in g together with the change in h calculate the effects on v.

14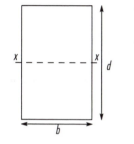

The second moment of area about the line XX is given by $I_{xx} = \dfrac{bd^3}{12}$.

i) What is the effect on I_{xx} when d is 2.5% too large and b is 3.5% too small?

ii) d is 1.5% too large. What change is needed in b so that there is no overall change in I_{xx}?

15 In a drilling machine the drill diameter, D, is connected to the volume of metal removed, V, by $V = \dfrac{\pi D^2 f N}{4}$. f is the feed and N is the rotational speed. D is increased by 2.5% and f by 1.5%. N is decreased by 2%. What is the overall effect on V?

6 The Exponential Function

The objectives of this chapter are to:

1 Understand the origin of e.

2 Evaluate e.

3 State the expansion of e^x in a power series.

4 Deduce the expansion of e^{-x} in a power series.

5 Understand the expansions are convergent for all x.

6 Generalise the expansions in the form Ae^{ax} where A and a are constant multipliers.

7 Consider the graphical similarities of exponential functions and their gradients.

Introduction

We introduced the exponential function, e^x, in Volume 1, Chapters 10 and 11. Now we look back further to the origin of this function. Also we write it in terms of a power series. We complete the chapter by looking at the gradient of the exponential function. We consider this graphically, leaving more details to Chapter 7, on differentiation. There we will extend our introductory chapter on differentiation in Volume 1.

■■■■■ ASSIGNMENT ■■■■■

The Assignment for this chapter looks closely at the exponential function and its gradient. We are going to do this in two ways. First, we look at them graphically. Then we look at them term by term in the power series. In Chapter 7, Differentiation, we include the exponential function in detail. There we continue our discussion of gradient but without needing to resort to graphical sketches each time.

The origin of e

In Chapter 5 we looked at the binomial theorem and developed the binomial series. Of the various forms let us concentrate on the series $(1 + x)^n$. Remember, the series is valid only for x lying between -1 and 1.

We replace n with $1/x$. With the aid of a calculator we are going to test $(1 + x)^{1/x}$ for various values of x. In the following table we start with $x = 0.5$ and reduce this progressively. In each row we give our final value correct to 4 decimal places.

x	$1 + x$	$\dfrac{1}{x}$	$(1 + x)^{1/x}$
0.5	1.5	2	$1.5^2 = 2.2500$
0.25	1.25	4	$1.25^4 = 2.4414$
0.10	1.10	10	$1.10^{10} = 2.5937$
0.05	1.05	20	$1.05^{20} = 2.6533$
0.01	1.01	100	$1.01^{100} = 2.7048$
0.005	1.005	200	$1.005^{200} = 2.7115$
0.001	1.001	1 000	$1.001^{1000} = 2.7169$
0.0005	1.0005	2 000	$1.0005^{2000} = 2.7176$
0.0001	1.0001	10 000	$1.0001^{10000} = 2.7181$
0.00005	1.00005	20 000	$1.00005^{20000} = 2.7182$

We could continue with this table. We could also use more decimal places to refine our final values on each row.

You can see that we appear to be approaching a value of approximately 2.718... We may write that we are **converging towards 2.718**... You can check this on your calculator using the keys $\boxed{1}\ \boxed{e^x}$ to display 2.7182818. The table leads us to a definition of e. We define e to be the limit of $(1 + x)^{1/x}$ as $x \to 0$,

i.e. $e = \underset{x \to 0}{\text{Lim}}\, (1 + x)^{1/x}.$

Obviously $x \neq 0$ because $\dfrac{1}{x}$ would *not* be defined. Remember, division by zero is not allowed in Mathematics.

The exponential series

It is possible to write the exponential function, e^x, as a power series, i.e.

$$e^x = 1 + x + \frac{x^2}{2!} + \frac{x^3}{3!} + \frac{x^4}{4!} + \cdots$$

Remember factorials from Chapter 5.

This is a simple power series. We will generalise it later. First, let us use it to find the value of e, confirming our earlier value of 2.718... Comparing e^x with e (or e^1) we put $x = 1$ in the power series. This gives us

$$e = e^1 = 1 + 1 + \frac{1^2}{2!} + \frac{1^3}{3!} + \frac{1^4}{4!} + \frac{1^5}{5!} + \frac{1^6}{6!} + \cdots$$

$$= 1 + 1 + 0.5 + 0.1\bar{6} + 0.041\bar{6} + 0.008\bar{3} + 0.00138\ldots$$

$$= 2.718\ldots$$

When you try this calculation for yourself use the calculator memory to avoid early approximation. Failure to do this will lead to a less accurate value for *e*.

Notice how after the first few terms the values get smaller. This will always happen. The more terms we use the more accurate will be our answer. The exponential series is convergent, i.e. it approaches a particular value.

Example 6.1

Write out e^2 as a power series. Hence, using the first 8 terms find the value of $3e^2$.

Using our power series for e^x we substitute for $x = 2$ to get

$$e^2 = 1 + 2 + \frac{2^2}{2!} + \frac{2^3}{3!} + \frac{2^4}{4!} + \frac{2^5}{5!} + \frac{2^6}{6!} + \frac{2^7}{7!} + \cdots$$

Then
$$3e^2 = 3\left(1 + 2 + \frac{2^2}{2!} + \frac{2^3}{3!} + \frac{2^4}{4!} + \frac{2^5}{5!} + \frac{2^6}{6!} + \frac{2^7}{7!} + \cdots\right)$$

$$= 3(1 + 2 + 2 + 1.\bar{3} + 0.\bar{6} + 0.2\bar{6} + 0.0\bar{8} + 0.025\ldots)$$

$$= 3 \times 7.38095\ldots$$

$$= 22.14.$$

These values compare favourably with the calculators ones; $7.389056\ldots$ to our $7.38095\ldots$ and $22.167\ldots$ to our 22.14. There is a pattern in working out this power series. We can spot this by looking at consecutive terms, e.g. $\frac{2^3}{3!}$ and $\frac{2^4}{4!}$.

We can write them out as $\frac{2 \times 2 \times 2}{3 \times 2 \times 1}$ and $\frac{2 \times 2 \times 2 \times 2}{4 \times 3 \times 2 \times 1}$. You can see the second of these fractions contains the first fraction,

i.e. $\frac{2^4}{4!} = \frac{2 \times (2 \times 2 \times 2)}{4 \times (3 \times 2 \times 1)} = \frac{2 \times 2^3}{4 \times 3!}$. We can use our value of $\frac{2^3}{3!}$ and multiply by $\frac{2}{4}$ to find the value of $\frac{2^4}{4!}$.

ASSIGNMENT

From Volume 1, Chapter 11, we know about the shape of the exponential function, $y = e^x$. Rather than just quoting the curve we can build up its shape. We do this using terms from the power series. The more terms we use, the closer the graphs. Our series of sketches in Figs. 6.1 shows the build up. We start with the horizontal straight line, $y = 1$. It is common to $y = e^x$ only at one point, $(0, 1)$. As we include more terms from the power series we increase the common central portion. This shows how we improve the graphical accuracy by using more terms from the power series.

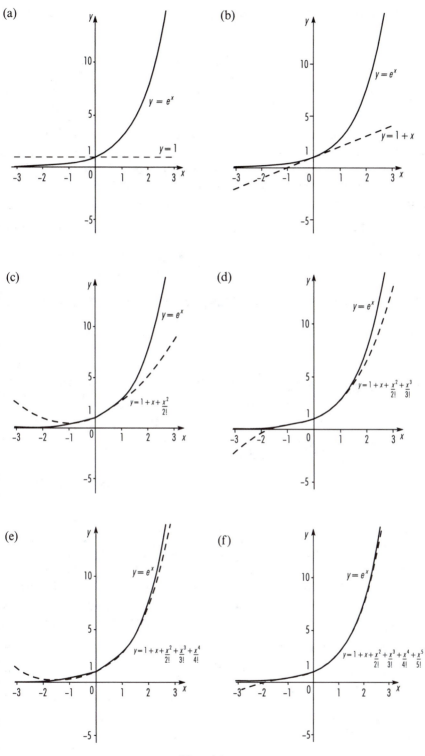

Figs. 6.1

Let us return to our power series for e^x. Suppose we use $-x$ in place of x so that

$$e^{-x} = 1 + (-x) + \frac{(-x)^2}{2!} + \frac{(-x)^3}{3!} + \frac{(-x)^4}{4!} + \cdots$$

i.e. $\qquad e^{-x} = 1 - x + \frac{x^2}{2!} - \frac{x^3}{3!} + \frac{x^4}{4!} \cdots$

Remember, when we raise a negative value to an even power it becomes positive. When we raise a negative value to an odd power it remains negative. Notice the differences and similarities between the power series for e^x and e^{-x}. The terms are so similar. The series only differ with alternate $+/-$ signs. Having discussed the series for e^{-x} we need a word of warning. Recall that $\frac{1}{e^x} = e^{-x}$. We *cannot* usefully expand e^x as a power series in the denominator. We always do so from the numerator position. This creates *no* difficulty. We have seen it work well for e^{-x}.

▨▨▨ Example 6.2 ▨▨▨

Write out $\frac{1}{e^{1.5}}$ as a power series. Hence, using the first 6 terms find the value of $2e^{-1.5}$.

Using our power series for e^x we substitute for $x = -1.5$ to get

$$\frac{1}{e^{1.5}} = e^{-1.5} = 1 - 1.5 + \frac{1.5^2}{2!} - \frac{1.5^3}{3!} + \frac{1.5^4}{4!} - \frac{1.5^5}{5!} + \cdots$$

Then $\qquad 2e^{-1.5} = 2(1 - 1.5 + \frac{1.5^2}{2!} - \frac{1.5^3}{3!} + \frac{1.5^4}{4!} - \frac{1.5^5}{5!} + \cdots)$

$$= 2(1 - 1.5 + 1.125 - 0.5625 + 0.2109\ldots - 0.0632\ldots)$$

$$= 2 \times 0.210\ldots$$

$$= 0.42.$$

We contrast the quality of our approximation with the one we achieved in Example 6.1. These values compare *un*favourably with the calculator ones; $0.223130\ldots$ to our $0.210\ldots$ This means our answer of 0.42 is rather *in*accurate. It is due to the reduced number of terms we used from the power series. Again, there is a pattern in working out this power series. We can identify this by looking at consecutive terms, e.g. $\frac{1.5^4}{4!}$ and $\frac{-1.5^5}{5!}$.

You can see the second fraction contains the first fraction, i.e. $\frac{-1.5 \times (1.5 \times 1.5 \times 1.5 \times 1.5)}{5 \times (4 \times 3 \times 2 \times 1)}$. We can use our value of $\frac{1.5^4}{4!}$ and multiply by $\frac{-1.5}{5}$ to find the value of $\frac{-1.5^5}{5!}$.

ASSIGNMENT

From Volume 1, Chapter 11, we know about the shape of the exponential function, $y = e^{-x}$. Again, rather than just quoting the curve we can build up its shape. We do this by using terms from the power series. Our series of sketches in Figs. 6.2 shows the build up. Again we start with the horizontal straight line, $y = 1$. It is common to $y = e^{-x}$ only at one point, $(0,1)$. As we include more terms from the power series we increase the common central portion. You have seen this done for $y = e^x$ and so should attempt it for yourself for $y = e^{-x}$. Use the series of sketches in Figs. 6.2 as a check.

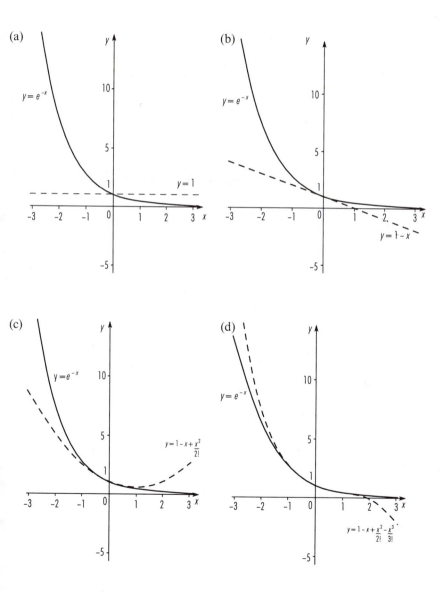

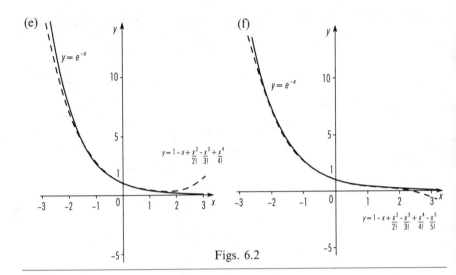

Figs. 6.2

In the following exercise look at the convergence of the exponentials. You should find those with small-sized powers converge more quickly than those with larger-sized powers. You can check this with your calculator.

▇▇▇▇ EXERCISE 6.1 ▇▇▇▇▇▇▇▇▇▇▇▇▇

Find the values of the following exponentials. Use the first 6 terms. Check each answer for accuracy using your calculator.

1   $e^{0.5}$

2   $e^{1.5}$

3   e^{-1}

4   $\dfrac{1}{e^2}$

5   $e^{-0.5}$

6   $2e^{0.5}$

7   $0.5e^{1.5}$

8   $\dfrac{e^{-1}}{2}$

9   $\dfrac{3}{e^2}$

10 $\dfrac{3}{4e^{0.5}}$

We move on further with our generalisation. This time we replace x with ax in the power series so that

$$e^{ax} = 1 + ax + \frac{(ax)^2}{2!} + \frac{(ax)^3}{3!} + \frac{(ax)^4}{4!} + \ldots$$

or $$e^{ax} = 1 + ax + \frac{a^2x^2}{2!} + \frac{a^3x^3}{3!} + \frac{a^4x^4}{4!} + \ldots$$

a is a constant. It may be positive or negative. It may be a whole number or a decimal or a fraction. The series we quote contains various patterns. However, once we start fiddling with values for a we do tend to lose them.

Examples 6.3

Write down the power series for i) e^{3x}, ii) e^{-2x}, iii) $e^{-x/2}$.

i) For e^{3x} we let $a = 3$ in our general power series so that

$$e^{3x} = 1 + 3x + \frac{(3x)^2}{2!} + \frac{(3x)^3}{3!} + \frac{(3x)^4}{4!} + \ldots$$

$$= 1 + 3x + \frac{9x^2}{2} + \frac{27x^3}{6} + \frac{81x^4}{24} + \ldots$$

$$= 1 + 3x + 4.5x^2 + 4.5x^3 + 3.375x^4 + \ldots$$

ii) For e^{-2x} we let $a = -2$ in our general power series. Remember that because the power is negative the signs of the terms will eventually alternate $+/-$. Now

$$e^{-2x} = 1 + (-2x) + \frac{(-2x)^2}{2!} + \frac{(-2x)^3}{3!} + \frac{(-2x)^4}{4!} + \ldots$$

$$= 1 + (-2x) + \frac{4x^2}{2} + \frac{(-8x^3)}{6} + \frac{16x^4}{24} \ldots$$

$$= 1 - 2x + 2x^2 - \frac{4x^3}{3} + \frac{2x^4}{3} \ldots$$

ii) For $e^{-x/2}$ we let $a = -\frac{1}{2}$ or -0.5 so that

$$e^{-x/2} = 1 + (-0.5x) + \frac{(-0.5x)^2}{2!} + \frac{(-0.5x)^3}{3!} \ldots$$

$$= 1 - 0.5x + 0.125x^2 - 0.0208\bar{3}x^3 \ldots$$

We choose to use decimals rather than fractions for simplicity.

Examples 6.4

Write down the power series for i) e^{2x}, ii) xe^{2x}, iii) $(3 + x)e^{2x}$.

i) For e^{2x} we let $a = 2$ in our general power series so that

$$e^{2x} = 1 + 2x + \frac{(2x)^2}{2!} + \frac{(2x)^3}{3!} + \frac{(2x)^4}{4!} + \ldots$$

$$= 1 + 2x + \frac{4x^2}{2} + \frac{8x^3}{6} + \frac{16x^4}{24} + \ldots$$

$$= 1 + 2x + 2x^2 + \frac{4x^3}{3} + \frac{2x^4}{3} + \ldots$$

ii) For xe^{2x} we use our previous result and multiply each term by x, i.e.

$$xe^{2x} = x\left(1 + 2x + 2x^2 + \frac{4x^3}{3} + \frac{2x^4}{3} + \ldots\right)$$

$$= x + 2x^2 + 2x^3 + \frac{4x^4}{3} + \frac{2x^5}{3} + \ldots$$

Notice the power of x in each term is increased by 1.

iii)   For $(3 + x)e^{2x}$ we multiply all the terms in our first power series by both 3 and x. Then we add the results, i.e.

$$(3 + x)e^{2x} = 3\left(1 + 2x + 2x^2 + \frac{4x^3}{3} + \frac{2x^4}{3} + \ldots\right)$$

$$+x\left(1 + 2x + 2x^2 + \frac{4x^3}{3} + \frac{2x^4}{3} + \ldots\right)$$

$$= 3 + 6x + 6x^2 + 4x^3 + 2x^4 + \ldots$$

$$+ \ x + 2x^2 + 2x^3 + \frac{4x^4}{3} + \ldots$$

Notice how we stop both multiplications at the x^4 term. This is because we worked out our power series to this term only. Hence we are limited to this accuracy when we combine the two sections of the calculation. Adding together like terms we finally get

$$(3 + x)e^{2x} = 3 + 7x + 8x^2 + 6x^3 + \frac{10x^4}{3} + \ldots$$

▬▬▬ EXERCISE 6.2 ▬▬▬

In Questions 1–8 write down the exponential function as a power series. The number of terms requested is given in each question. Fully simplify each term in your series.

1	e^{4x}	(4 terms)		**5**	$4xe^{x/2}$	(4 terms)
2	$e^{x/2}$	(4 terms)		**6**	$(1 + 4x)e^{x/2}$	(4 terms)
3	e^{-3x}	(5 terms)		**7**	$(2 - 3x)e^{1.5x}$	(to x^3)
4	$2e^{-4x}$	(6 terms)		**8**	$\dfrac{1 + e^{6x}}{e^{3x}}$	(to x^4)

9   Using the results from the chapter find power series for the following exponential functions. In each case use terms up to and including the term in x^6. Fully simplify your answers.

i) $\frac{1}{2}(e^{2x} + e^{-2x})$,   ii) $\frac{1}{2}(e^{2x} - e^{-2x})$.

10   Only attempt this question if you have studied Chapter 2 (Complex Numbers). As usual, j indicates an imaginary value. Find power series for the following exponential functions up to and including the term in x^6. Fully simplify your answers.

i) e^{jx},   ii) e^{-jx},   iii) $\frac{1}{2}(e^{jx} + e^{-jx})$,   iv) $\frac{1}{2}(e^{jx} - e^{-jx})$.

Gradients of exponential functions

In Volume 1, Chapter 18, we quoted and applied the result that for

$$y = e^x,$$

then $\dfrac{dy}{dx} = e^x.$

Now we are going to show that this is true. We write e^x as a power series and differentiate term by term, i.e.

$$y = e^x = 1 + x + \frac{x^2}{2!} + \frac{x^3}{3!} + \frac{x^4}{4!} + \frac{x^5}{5!} + \cdots$$

then $\dfrac{dy}{dx} = \dfrac{d}{dx}(e^x) = \quad 1 + \dfrac{2x}{2!} + \dfrac{3x^2}{3!} + \dfrac{4x^3}{4!} + \dfrac{5x^4}{5!} + \cdots$

$$= \quad 1 + \frac{2x}{2 \times 1} + \frac{3x^2}{3 \times 2 \times 1} + \frac{4x^3}{4 \times 3 \times 2 \times 1}$$

$$+ \frac{5x^4}{5 \times 4 \times 3 \times 2 \times 1} \cdots$$

$$= \quad 1 + \frac{x}{1} + \frac{x^2}{2 \times 1} + \frac{x^3}{3 \times 2 \times 1}$$

> Notice how each term cancels.

$$+ \frac{x^4}{4 \times 3 \times 2 \times 1} \cdots$$

$$= \quad 1 + x + \frac{x^2}{2!} + \frac{x^3}{3!} + \frac{x^4}{4!} + \cdots$$

i.e. $\dfrac{d}{dx}(e^x) = e^x.$

We can generalise this term-by-term differentiation using the power series for e^{ax}. Using

$$e^{ax} = 1 + ax + \frac{a^2x^2}{2!} + \frac{a^3x^3}{3!} + \frac{a^4x^4}{4!} + \frac{a^5x^5}{5!} + \cdots$$

then $\dfrac{d}{dx}(e^{ax}) = \quad a + \dfrac{2a^2x}{2!} + \dfrac{3a^3x^2}{3!} + \dfrac{4a^4x^3}{4!} + \dfrac{5a^5x^4}{5!} + \cdots$

$$= \quad a + a^2x + \frac{a^3x^2}{2!} + \frac{a^4x^3}{3!} + \frac{a^5x^4}{4!} + \cdots$$

> Cancelling as before.

Notice how the power of a is always greater than the power of x by 1. We remove a from each term as a common factor to get

$$\frac{d}{dx}(e^{ax}) = a(1 + ax + \frac{a^2x^2}{2!} + \frac{a^3x^3}{3!} + \frac{a^4x^4}{4!} + \ldots)$$

i.e. $$\frac{d}{dx}(e^{ax}) = ae^{ax}.$$

This is the general rule we use in Chapter 7.

◼◼◼◼ ASSIGNMENT ◼◼◼◼

Now let us look at the gradient of a specific exponential function, $y = e^{2x}$.
We know now that $\frac{dy}{dx} = 2e^{2x}$ where $a = 2$.

In Fig. 6.3 we show the original function, $y = e^{2x}$, and the graph of the gradient, $2e^{2x}$. As predicted, you can see the gradient is greater than its original function. However, by just looking you would not expect always to see it as twice as large. You should use Fig. 6.3 as a check for yourself. Accurately plot the graph of $y = e^{2x}$ from $x = -1.5$ to $x = 1.5$ at intervals of 0.25. Then draw tangents at intervals of x and measure each gradient. Now plot the values of these gradients against the corresponding x values. If you do this carefully your attempt ought to closely resemble our gradient graph. Obviously you can label your original graph $y = e^{2x}$. Also, you can label the gradient graph as a graph in its own right, i.e. $y = 2e^{2x}$.

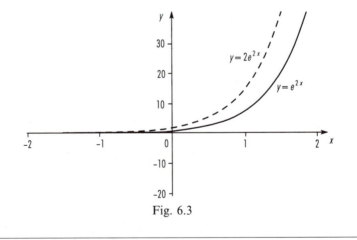

Fig. 6.3

EXERCISE 6.3

In each question write the exponential function as a power series. Fully simplify your series up to and including the term in x^5. Now differentiate it term by term. Check that your expression for $\dfrac{dy}{dx}$ agrees with the general rule $\dfrac{d}{dx}(e^{ax}) = ae^{ax}$.

1   e^{4x}

2   e^{-x}

3   $2e^{-x}$

4   e^{-2x}

5   e^{3x}

6   $e^{x/2}$

7   $e^{-x/2}$

8   e^{-4x}

9   $2e^x$

10   $3e^{2x}$

7 | Differentiation

The objectives of this chapter are to:

1 Use the derivatives of the functions ax^n, $\sin ax$, $\cos ax$, $\tan ax$, $\ln ax$ and e^{ax}.

2 State the basic rules for the derivatives of sum, function of a function, product and quotient.

3 Determine the derivatives of combinations of any 2 functions in **1**.

4 Evaluate the derivatives at given points.

Introduction

In Volume 1, Chapter 18, we introduced the idea of differentiation by looking at graphical change. Before going on with this chapter you should reread part of that chapter. Start with 'Differentiation of Algebraic Functions' and continue through to the end. Generally, but *not* always, in that chapter we looked at y as a function of x and then $\dfrac{dy}{dx}$.

Mathematically we use $y=f(x)$ to mean 'y **is a function of x**'. Notice the positions of y and x in the Mathematics and in the English. f **stands for function**. That function may involve x to some power, addition/subtraction, multiplication/division, trigonometry, logarithms or exponentials. The exact function will be given in each example.

Remember that $\dfrac{dy}{dx}$ has various interpretations:

i) the change in y due to the change in x,

ii) the differentiation of y with respect to x,

iii) the first differential,

iv) the first derivative,

v) the gradient of a tangent to a curve at any point and

vi) $\tan \alpha$ where α is the angle of inclination of the tangent to the horizontal.

We can use $f'(x)$ (said 'f **dashed**' or 'f **dashed of x**') in place of $\dfrac{dy}{dx}$. The $'$ means we have differentiated $f(x)$ once.

ASSIGNMENT

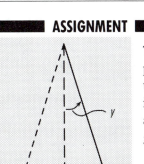

The Assignment for this chapter is a simple pendulum performing small oscillations about the vertical. This takes place in a fluid, so involving a resistive force to the motion. The angular displacement, y, from the vertical is given in terms of time, t, by the formula
$$y = 0.05e^{-0.5t}\sin 3t.$$

Later in this chapter we will look at how y varies depending upon time.

Fig. 7.1

Differentiation by rule

Let us revise some of the features of our previous work on differentiation. We have a general result.

For $y = ax^n$ we have $\dfrac{dy}{dx} = anx^{n-1}$ where a and n are constants.

There is the alternative notation of
$$\frac{d}{dx}(ax^n) = anx^{n-1}.$$

Remember that it is important to follow the pattern of differentiation. The old power n comes forward to multiply and the power is reduced by 1. Also remember that *not* all things change, i.e. some things are constant.

This leads to the general result that if $y = $ constant then $\dfrac{dy}{dx} = 0$.

Examples 7.1

In this set of examples we refresh your memory with some basic differentiation. We start with y in terms of x and find a general expression for $\dfrac{dy}{dx}$.

i) $y = 2.7$, a constant, $\dfrac{dy}{dx} = 0$.

ii) $y = 3.5x$, $\dfrac{dy}{dx} = 3.5 \times 1x^{1-1} = 3.5.$

$$\boxed{\begin{array}{l} x^1 = x. \\ x^0 = 1. \end{array}}$$

iii) $y = \dfrac{6}{5x^3} = \dfrac{6}{5}x^{-3}$, $\dfrac{dy}{dx} = \dfrac{6}{5} \times (-3)x^{-3-1} = -\dfrac{18}{5}x^{-4}.$

iv) $y = 7x^{\frac{3}{2}}$, $\dfrac{dy}{dx} = 7 \times \dfrac{3}{2}x^{\frac{3}{2}-1} = \dfrac{21}{2}x^{\frac{1}{2}}.$

v) $y = \dfrac{2}{x^{\frac{1}{2}}} = 2x^{-\frac{1}{2}}$, $\dfrac{dy}{dx} = 2 \times (-\dfrac{1}{2})x^{-\frac{1}{2}-1} = -x^{-\frac{3}{2}}.$

From Volume 1 we recall that we can differentiate an expression involving addition and/or subtraction of terms. We do this for each term in turn.

███████ **Example 7.2** ████████████████████████████

Before we can differentiate $s = \dfrac{3t - t^3}{t^{\frac{1}{2}}} + \dfrac{1}{t}$ we simplify the algebra.

This gives
$$s = \frac{3t}{t^{\frac{1}{2}}} - \frac{t^3}{t^{\frac{1}{2}}} + \frac{1}{t}$$

i.e.
$$s = 3t^{0.5} - t^{2.5} + t^{-1}$$

> Accurate decimals are as good as fractions.

Differentiating term by term we get

$$\frac{ds}{dt} = 1.5t^{-0.5} - 2.5t^{1.5} - t^{-2}.$$

The first exercise in this chapter is for revision. It allows you to practise again skills you first learned in Volume 1.

███████ **EXERCISE 7.1** ████████████████████████████

Differentiate the following expressions by rule.

1 $y = 3 - 3x + 4x^2$

2 $s = 7t^4 - 8t + 4$

3 $y = 9.5 - 3x^{3.5} - 4x^{-2}$

4 $y = 3x^2 + x^{-3.5} - x^{-4.2}$

5 $s = 6t^{-1.6} + 2t^{-2.4}$

6 $s = \dfrac{14}{t^{2.4}}$

7 $y = \dfrac{16}{x^{2.5}} + \dfrac{20}{x^{1.5}}$

8 $y = 3x - \dfrac{2}{x}$

9 $y = \dfrac{2}{x^{3.5}} - 3x^3$

10 $y = \dfrac{3x^{\frac{1}{2}} + 2x^2}{x}$

11 $y = \dfrac{x - x^3}{x^4} + 3.5x^{0.5}$

12 $y = 8x^{\frac{1}{3}} + \dfrac{1}{3}x^{\frac{1}{8}}$

13 $y = \sqrt{x^3} - \sqrt[3]{x^2} + 3$

14 $s = 14t^{0.2} - 12t^{0.4}$

15 $s = 4\sqrt{t} - \dfrac{1}{4\sqrt{t}}$

Differentiation using function of a function or by substitution

This is one of the methods we use when the functions become more complicated. We know from Volume 1 that the first differential of $\sin\theta$ is cos θ, provided θ is in radians. Also the differential of cos θ is − sin θ. Suppose we now have a more complicated angle in place of the simple θ.

Let us do this by looking at the form of $y = \sin(3\theta + 7)$ before we think about any differentiation. Suppose we need to know the calculator value of y for a given value of θ. Firstly we multiply θ by 3 and then add 7. We need to do all this before thinking about the <u>sin</u> button on our calculator. In fact $(3\theta + 7)$ is a function. Then $\sin(3\theta + 7)$ is a sine function of $(3\theta + 7)$, i.e. we have a [sine] function of a function. In our method we make a substitution, say u, for $3\theta + 7$. This means we split $y = \sin(3\theta + 7)$ into

$$y = \sin u \qquad \text{where } u = 3\theta + 7.$$

Because $y = \sin u$ gives y as a function of u we find $\dfrac{dy}{du}$.

Also $u = 3\theta + 7$ gives u as a function of θ so we find $\dfrac{du}{d\theta}$.

When we multiply these results together as $\dfrac{dy}{du} \times \dfrac{du}{d\theta}$ it looks like the du terms cancel, leaving $\dfrac{dy}{d\theta}$.

Hence for y as a function of x we can write down the general result

$$\frac{dy}{dx} = \frac{dy}{du} \times \frac{du}{dx}.$$

Example 7.3

Differentiate $y = \sin(3\theta + 7)$ with respect to θ.

Making a substitution we have

$$y = \sin u \qquad \text{where } u = 3\theta + 7.$$

Differentiating in both cases we get

$$\frac{dy}{du} = \cos u \qquad \text{and } \frac{du}{d\theta} = 3.$$

In our general result we use θ in place of x so that

$$\frac{dy}{d\theta} = \frac{dy}{du} \times \frac{du}{d\theta}$$

gives $\quad \dfrac{dy}{d\theta} = \cos u \times 3$

$$= 3\cos u = 3\cos(3\theta + 7). \qquad \boxed{\text{Substituting for } u = 3\theta + 7.}$$

Our original question is in terms of θ and we have found $\dfrac{dy}{d\theta}$.

For consistency we quote our final result in terms of θ.

<hr>

███████ **Example 7.4** ███████████████████████████████████████

Find the first derivative of $y = 4\cos(2 + 5\theta)$.

This example follows a similar pattern to Example 7.3. Notice the order within the brackets is changed. This will *not* upset our method.

Making a substitution we have

$$y = 4\cos u \qquad \text{where} \quad u = 2 + 5\theta.$$

Differentiating in both cases we get

$$\frac{dy}{du} = -4\sin u \qquad \text{and} \quad \frac{du}{d\theta} = 5.$$

We substitute into

$$\frac{dy}{d\theta} = \frac{dy}{du} \times \frac{du}{d\theta}$$

to get $\quad \dfrac{dy}{d\theta} = -4\sin u \times 5$

$$= -20\sin u = -20\sin(2 + 5\theta). \qquad \boxed{\text{Substituting for } u = 2 + 5\theta.}$$

More generally we have the differentiation rules for

$$y = \sin(a\theta + b) \qquad \text{and} \qquad y = \cos(a\theta + b)$$

to give $\quad \dfrac{dy}{d\theta} = a\cos(a\theta + b) \quad$ **and** $\quad \dfrac{dy}{d\theta} = -a\sin(a\theta + b)$

where a and b are constants.

As usual, when differentiating trigonometric functions θ (or some other letter) is in radians.

<hr>

███████ **Example 7.5** ███████████████████████████████████████

Find the first derivative of $y = \sin^3\theta$.

We have already discussed the meaning of expressions like $\sin^3\theta$. Remember $\sin^3\theta$ is the way we write $(\sin\theta)^3$.

Thinking of $y = \sin^3\theta$ as $y = (\sin\theta)^3$ our substitution gives

$$y = u^3 \qquad \text{where } u = \sin\theta.$$

Differentiating in both cases we get

$$\frac{dy}{d\theta} = 3u^2 \qquad \text{and } \frac{du}{d\theta} = \cos\theta.$$

As usual we substitute into

$$\frac{dy}{d\theta} = \frac{dy}{du} \times \frac{du}{d\theta}$$

to get $\quad \dfrac{dy}{d\theta} = 3u^2 \times \cos\theta$

$$= 3\sin^2\theta\cos\theta \quad \text{or} \quad 3\cos\theta\sin^2\theta.$$

Because the order of trigonometric multiplication does *not* affect the result you may choose either version.

Example 7.6

Differentiate $y = (\cos 5\theta)^2$ with respect to θ.

This example brings together the two previous example types.

We make our substitution to give

$$y = u^2 \qquad \text{where} \quad u = \cos 5\theta.$$

Differentiating in both cases we get

$$\frac{dy}{du} = 2u \qquad \text{and} \quad \frac{du}{d\theta} = -5\sin 5\theta.$$

As usual we substitute into

$$\frac{dy}{d\theta} = \frac{dy}{du} \times \frac{du}{d\theta}$$

to get $\dfrac{dy}{d\theta} = 2u \times (-5\sin 5\theta)$

$$= -10u\sin 5\theta = -10\cos 5\theta \sin 5\theta \quad \text{or} \quad -10\sin 5\theta \cos 5\theta.$$

The following exercise allows you to practise further this trigonometric differentiation.

EXERCISE 7.2

Find the first differential in each question.

1 $y = \sin 2\theta$

2 $y = \cos(2x + 3)$

3 $y = 4\sin(3 - 2\theta)$

4 $y = \sin^4 \theta$

5 $y = (\cos x)^{\frac{1}{2}}$

6 $y = \dfrac{1}{\sqrt{\sin x}}$

7 $y = \sin(4\theta - 7) + \cos(7 - 4\theta)$

8 $y = \cos^2 x - \sin^2 x$

9 $y = 4\sqrt{\sin \theta} - 2\sin \theta$

10 $y = \sin^3(6 - 3x)$

We can apply the function of a function rule to natural logarithms in a similar way to the trigonometric functions.

Examples 7.7

Differentiate the following logarithmic functions with respect to x

i) $y = \ln(2x^3 - 5)$, ii) $y = \ln(2x + 9)^3$, iii) $y = \ln\left(\dfrac{2x^3 - 5}{7x + 6}\right)$.

i) Making a substitution we have

$$y = \ln u \qquad \text{where} \quad u = 2x^3 - 5.$$

Differentiating in both cases we get

$$\frac{dy}{du} = \frac{1}{u} \qquad \text{and} \quad \frac{du}{dx} = 6x^2.$$

We substitute into

$$\frac{dy}{dx} = \frac{dy}{du} \times \frac{du}{dx}.$$

to get $\dfrac{dy}{dx} = \dfrac{1}{u} \times 6x^2$

$$= \frac{6x^2}{u} = \frac{6x^2}{2x^3 - 5}.$$

> Substituting for $u = 2x^3 - 5$.

ii) In this example notice the different position for the power 3. We use one of our laws of logarithms to write $y = 3\ln(2x + 9)$ and then make a substitution so that

$$y = 3\ln u \qquad \text{where} \quad u = 2x + 9.$$

Differentiating in both cases we get

$$\frac{dy}{du} = \frac{3}{u} \qquad \text{and} \qquad \frac{du}{dx} = 2.$$

We substitute into

$$\frac{dy}{dx} = \frac{dy}{du} \times \frac{du}{dx}.$$

to get $\dfrac{dy}{dx} = \dfrac{3}{u} \times 2$

$$= \frac{6}{u} = \frac{6}{2x + 9}.$$

> Substituting for $u = 2x + 9$.

iii) This example uses another of the laws of logarithms to give $y = \ln(2x^3 - 5) - \ln(7x + 6)$. We differentiate both logarithmic functions. The first one we have just attempted and the second one is left as an exercise for you. The expression for y in terms of x involved two functions separated by a minus sign. This means our two differentials are also separated by a minus sign,

i.e. $\dfrac{dy}{dx} = \dfrac{6x^2}{2x^3 - 5} - \dfrac{7}{7x + 6}.$

Now look back at each logarithmic function, paying particular attention to the function in brackets. Our answer for $\dfrac{dy}{dx}$ has that function (generally $f(x)$) in the denominator and its differential (generally $f'(x)$) in the numerator. We may write this as a general rule.

For $y = \ln f(x),$ $\dfrac{dy}{dx} = \dfrac{f'(x)}{f(x)}.$

Many cases are simple like $y = \ln ax$ to give $\dfrac{dy}{dx} = \dfrac{1}{x}$ where a is a constant.

EXERCISE 7.3

For each logarithmic function find $\dfrac{dy}{dx}$.

1 $y = \ln(9x - 11)$

2 $y = 2\ln(9x - 11)$

3 $y = \ln(9x + 4)^2$

4 $y = \ln(4x^2 - 3x + 6)$

5 $y = \ln(2x + 3)(4 - x)$

6 $y = \ln\left(\dfrac{3x - 7}{x^2 + 1}\right)$

7 $y = \ln\sqrt{x}$

8 $y = \ln(\cos x)$

9 $y = \frac{1}{2}\ln x$

10 $y = \ln\left(\dfrac{1}{\sqrt{x}}\right)$

We can apply our function of a function rule again, this time to exponential functions.

Examples 7.8

Find the first differentials of the exponential functions

i) $y = 4e^{3x}$, ii) $y = 4e^{3x-7}$, iii) $y = \dfrac{1}{e^{\sin\theta}}$.

i) Making a substitution we have

$$y = 4e^u \quad \text{where} \quad u = 3x.$$

Differentiating in both cases we get

$$\frac{dy}{du} = 4e^u \qquad \text{and} \qquad \frac{du}{dx} = 3.$$

We substitute into

$$\frac{dy}{dx} = \frac{dy}{du} \times \frac{du}{dx}$$

to get $\dfrac{dy}{dx} = 4e^u \times 3$

$$= 12e^u = 12e^{3x}. \qquad \boxed{\text{Substituting for } u = 3x.}$$

ii) This example follows a pattern similar to the previous example with

$$y = 4e^u \qquad \text{where} \quad u = 3x - 7.$$

The working is left as an exercise for you to get

$$\frac{dy}{dx} = 12e^u = 12e^{3x-7}.$$

iii) We may rewrite $y = \dfrac{1}{e^{\sin\theta}}$ as $y = e^{-\sin\theta}$.

Making a substitution we have

$$y = e^u \qquad \text{where} \quad u = -\sin\theta.$$

Differentiating in both cases we get

$$\frac{dy}{du} = e^u \qquad \text{and} \quad \frac{du}{d\theta} = -\cos\theta.$$

We substitute into

$$\frac{dy}{d\theta} = \frac{dy}{du} \times \frac{du}{d\theta}$$

to get $\quad \dfrac{dy}{d\theta} = e^u \times (-\cos\theta)$

$$= -\cos\theta \times e^u$$

$$= -\cos\theta . e^{-\sin\theta}.$$

We have used . for multiplication between the cos and exponential.

If we attempt to substitute some values we need to be careful. Separately we need to work out $\cos\theta$ and $e^{-\sin\theta}$, before multiplying those answers together to get a result.

For the first two of this set of three examples we have a general rule.

For $\qquad y = e^{ax+b}, \qquad\qquad \dfrac{dy}{dx} = ae^{ax+b}.$

Notice how the power of the exponential remains the same. We give this rule in the summary at the end of the chapter.

■■■■ EXERCISE 7.4 ■■■■■■

For each exponential function find the first differential.

1 $\quad y = 3e^{4x}$

2 $\quad y = 3e^{4x+1}$

3 $\quad y = 2.5e^{-2x}$

4 $\quad y = \dfrac{4}{e^{2x}}$

5 $\quad y = \dfrac{4}{e^{2x+5}}$

6 $\quad y = 7.5e^{1-3x}$

7 $\quad y = 7.5x + 2e^{4+x}$

8 $\quad y = 5e^{2x^3}$

9 $\quad y = 3e^{4\sin\theta}$

10 $\quad y = \sin\theta + 2e^{\cos\theta}$

We come to our final application of the function of a function rule. It involves more complicated expressions for x within a bracket raised to a power. Again, we use the rule in the same way as before.

▓▓▓▓▓ **Examples 7.9** ▓▓

Find the first differential of

i) $y = (5 + 2x)^4$, ii) $y = 7(5 + 2x)^4$, iii) $y = \dfrac{1}{(x^3 + 1)^5}$, iv) $y = \sqrt{6 - x^2}$.

i) There is an alternative to raising this bracket to the power 4. We might think of multiplying,

   i.e. $(5 + 2x)(5 + 2x)(5 + 2x)(5 + 2x)$.

   However we cannot recommend this method because of the chances of algebraic errors before we attempt the differentiation. We will use the function of a function rule from the beginning.

   Making a substitution we have

   $$y = u^4 \qquad \text{where} \quad u = 5 + 2x.$$

   Differentiating in both cases we get

   $$\frac{dy}{du} = 4u^3 \qquad \text{and} \frac{du}{dx} = 2.$$

   We substitute into

   $$\frac{dy}{dx} = \frac{dy}{du} \times \frac{du}{dx}$$

   to get $\dfrac{dy}{dx} = 4u^3 \times 2$

   $$= 8u^3 = 8(5 + 2x)^3.$$

ii) For $y = 7(5 + 2x)^4$ we have the previous example multiplied by 7.

   Making a substitution we have

   $$y = 7u^4 \qquad \text{where} \quad u = 5 + 2x.$$

   The rest of this example is left as an exercise for you to get

   $$\frac{dy}{dx} = 7 \times 8(5 + 2x)^3 = 56(5 + 2x)^3.$$

iii) $y = \dfrac{1}{(x^3 + 1)^5}$ needs to be rewritten so that all the x terms are in the numerator, i.e. $y = (x^3 + 1)^{-5}$. Notice that only the power of 5 changes its sign to -5. Now we can use our method, making a substitution so that

   $$y = u^{-5} \qquad \text{where} \quad u = x^3 + 1.$$

   Differentiating in both cases we get

   $$\frac{dy}{du} = -5u^{-5-1} \qquad \text{and} \quad \frac{du}{dx} = 3x^2.$$

   $$= -5u^{-6}$$

We substitute into

$$\frac{dy}{dx} = \frac{dy}{du} \times \frac{du}{dx}$$

to get $\dfrac{dy}{dx} = -5u^{-6} \times 3x^2$

$$= -15x^2u^{-6} = -15x^2(x^3 + 1)^{-6}.$$

iv)   We rewrite the $\sqrt{}$ as the power $\frac{1}{2}$ to give $y = (6 - x^2)^{\frac{1}{2}}$. Making a substitution we have

$$y = u^{\frac{1}{2}} \qquad \text{where} \quad u = 6 - x^2.$$

Differentiating in both cases we get

$$\frac{dy}{du} = \frac{1}{2}u^{\frac{1}{2}-1} \qquad \text{and} \quad \frac{du}{dx} = -2x.$$

$$= \frac{1}{2}u^{-\frac{1}{2}}.$$

We substitute into

$$\frac{dy}{dx} = \frac{dy}{du} \times \frac{du}{dx}$$

to get $\dfrac{dy}{dx} = \dfrac{1}{2}u^{-\frac{1}{2}} \times (-2x)$ $\boxed{\dfrac{1}{2} \times (-2) = -1.}$

$$= -xu^{-\frac{1}{2}}$$

$$= -x(6 - x^2)^{-\frac{1}{2}} \quad \text{or} \quad \frac{-x}{(6 - x^2)^{\frac{1}{2}}} \quad \text{or} \quad \frac{-x}{\sqrt{6 - x^2}}.$$

There is a pattern in these examples. Notice how the original power comes forward to multiply and that power is reduced by 1. Also notice the differential of the bracket is involved in the multiplication. For reference, the pattern is written mathematically in the summary at the end of this chapter.

▬▬ EXERCISE 7.5 ▬▬

Find $\dfrac{dy}{dx}$ in each question.

1   $y = (2x - 5)^4$

2   $y = (2x - 5)^3$

3   $y = 2.5(2x - 5)^4$

4   $y = \dfrac{2.5}{2x - 5}$

5   $y = \dfrac{1}{(2x - 5)^2}$

6   $y = \sqrt{1 + 7x}$

7   $y = \dfrac{6}{\sqrt{1 + 4x^2}}$

8   $y = (4 - x)^{\frac{3}{2}}$

9   $y = 3(1 + x - x^2)^5$

10   $y = \dfrac{3}{(4x + x^3)^2}$

Differentiation of a product

In mathematics, when we write about a product of terms we understand that those terms are multiplied together. A product is a result of multiplication. For example, the product of 2 and 11 is 22 ($=2 \times 11$). We can extend this to more terms. For example the product of 0.5, 4 and 5 is 10 ($=0.5 \times 4 \times 5$). The product of $2x^2$ and $11x$ is $22x^3$. These products are easy to work out. Now suppose we have ($2x^2 + 5x$) and ($11x - 1$). We can multiply the brackets together as ($2x^2 + 5x$)($11x - 1$). If you wish you may complete this for yourself, though we choose not to do so here. At this stage there is no mathematical reason to find the answer.

Suppose we have a product of u and v. If $y = uv$ (or $u \times v$) where u and v are both functions of x, then

$$\frac{dy}{dx} = u\frac{dv}{dx} + v\frac{du}{dx}.$$

This is the **product rule**. In each example we need to identify u and v, differentiate them separately and substitute into the formula.

▓▓▓▓▓▓ Examples 7.10 ▓▓▓▓▓▓▓▓▓▓▓▓▓▓▓▓▓▓▓▓▓▓▓▓▓▓▓▓▓▓▓▓▓▓

Find the first differentials of

i) $y = x^2.e^{3x}$, ii) $y = (1 - x^2)\ln(7x + 6)$, iii) $y = (2x^2 + 5x)(11x - 1)$.

i) In $y = x^2.e^{3x}$ we have the product of x^2 and e^{3x}.

Let $u = x^2$ and $v = e^{3x}$.

Differentiating both parts we get

$$\frac{du}{dx} = 2x \qquad \text{and} \qquad \frac{dv}{dx} = 3e^{3x}.$$

We substitute into

$$\frac{dy}{dx} = u\frac{dv}{dx} + v\frac{du}{dx}$$

to get $\dfrac{dy}{dx} = (x^2)(3e^{3x}) + (e^{3x})(2x)$.

This is an answer, though it can be tidied and simplified algebraically. We look for common factors by comparing the pairs of brackets. From each pair of brackets we may remove x and e^{3x}. This leaves another x and 3 from the first pair. It leaves just 2 from the second pair. Thus we write $\dfrac{dy}{dx} = (xe^{3x})(3x) + (xe^{3x})(2)$.

Because the pairs of brackets are added, then what remains ($3x$ and 2) will also be added,

i.e. $\dfrac{dy}{dx} = xe^{3x}(3x + 2)$ or $x(3x + 2)e^{3x}$.

ii) In $y = (1 - x^2) \ln(7x + 6)$ the $\times$ sign is understood to be between $(1 - x^2)$ and $\ln(7x + 6)$.

Let $u = 1 - x^2$ and $v = \ln(7x + 6)$.

Differentiating both parts we get

$$\frac{du}{dx} = -2x \qquad \text{and} \qquad \frac{dv}{dx} = \frac{7}{7x + 6}. \qquad \boxed{\text{From Examples 7.7(iii).}}$$

We substitute into

$$\frac{dy}{dx} = u\frac{dv}{dx} + v\frac{du}{dx}$$

to get $\dfrac{dy}{dx} = (1 - x^2)\left(\dfrac{7}{7x + 6}\right) + (\ln(7x + 6))(-2x).$

This answer really needs rewriting. Unlike the previous example we *cannot* readily remove common factors. We just write

$$\frac{dy}{dx} = \frac{7(1 - x^2)}{7x + 6} - 2x \ln(7x + 6).$$

In the first pair of brackets we simply gather together the terms. In the second pair of brackets we rearrange the $-2x$ because it is *not* involved with the natural logarithm.

iii) In this example the $\times$ sign is between the brackets as usual so that

$$u = 2x^2 + 5x \qquad \text{and} \quad v = 11x - 1.$$

Differentiating both parts we get

$$\frac{du}{dx} = 4x + 5 \qquad \text{and } \frac{dv}{dx} = 11.$$

We substitute into

$$\frac{dy}{dx} = u\frac{dv}{dx} + v\frac{du}{dx}$$

to get $\dfrac{dy}{dx} = (2x^2 + 5x)(11) + (11x - 1)(4x + 5).$

Again, this answer is correct, but may be better presented with some simplification of the algebra. We multiply out the brackets to get

$$\frac{dy}{dx} = 22x^2 + 55x + 44x^2 + 55x - 4x - 5$$

$$= 66x^2 + 106x - 5.$$

ASSIGNMENT

Our simple pendulum has an angular displacement, y, in terms of time, t, according to $y = 0.05e^{-0.5t} \sin 3t$. Let us look at the graph of y against t in Fig. 7.2.

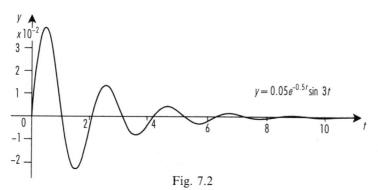

Fig. 7.2

The curve oscillates above and below the horizontal axis. Each successive peak or trough is closer to the axis than the previous one. Physically we would see the pendulum swinging a smaller and smaller distance away from the vertical. These distances would alternate on either side of that vertical. Eventually the pendulum would stop, hanging in line with the central vertical line. Because the graph continually varies in shape, so does its gradient. The most useful way to look at the gradient is through differentiation.

In $y = 0.05e^{-0.5t} \sin 3t$ we have a product of $e^{-0.5t}$ and $\sin 3t$ together with a scaling factor of 0.05.

Let $\qquad u = 0.05e^{-0.5t} \qquad$ and $\quad v = \sin 3t.$

Differentiating both parts we get

$$\frac{du}{dt} = -0.025e^{-0.5t} \quad \text{and} \quad \frac{dv}{dt} = 3\cos 3t.$$

We substitute into

$$\frac{dy}{dt} = u\frac{dv}{dt} + v\frac{du}{dt}$$

to get $\quad \dfrac{dy}{dt} = (0.05e^{-0.5t})(3\cos 3t) + (\sin 3t)(-0.025e^{-0.5t})$

$$= e^{-0.5t}(0.15\cos 3t - 0.025\sin 3t).$$

y is the angular displacement from the vertical measured in radians. t is the time for the motion measured in seconds. This means that $\dfrac{dy}{dt}$ is the change in angular displacement with respect to time or the rate of change of angular displacement, i.e. the angular velocity, measured in rads^{-1}.

◼◼◼◼ EXERCISE 7.6 ◼◼◼◼

Apply the product rule to find the first differential in each question.

1 $y = 3xe^{2x}$

2 $y = (x^3 + 2x)\ln 3x$

3 $y = \theta \sin \theta$

4 $y = (1 - x^2)(4x + 5)$

5 $y = (4x + x^3)e^{3x}$

6 $y = (4 + \theta^2)\cos \theta$

7 $y = \sin \theta \cos \theta$

8 $y = 3\theta e^{4\cos \theta}$

9 $y = \sqrt{x}\ln(3x + 1)$

10 $y = (\sqrt{x} - 7)\ln(3x + 1)$

Differentiation of a quotient

In mathematics, when we write about a quotient we are writing about division. For example $\dfrac{3}{4}$ and $\dfrac{u}{v}$ are quotients. The first example has an obvious numerical value while the second one is a general algebraic expression.

If $y = \dfrac{u}{v}$ where u and v are both functions of x then

$$\frac{dy}{dx} = \frac{v\dfrac{du}{dx} - u\dfrac{dv}{dx}}{v^2}.$$

This is the **quotient rule**. In each example we identify u and v, differentiate them separately and substitute into the formula above.

It is possible to work out the quotient rule. This is done using the product rule applied to the rewritten $y = uv^{-1}$.

▓▓▓▓ Examples 7.11 ▓▓▓▓

Find the first differentials of   i) $y = \dfrac{e^{2x}}{1 + x^2}$,   ii) $y = \dfrac{2\sin \theta}{\theta}$.

i) Comparing our quotient with the general $\dfrac{u}{v}$ we let

$$u = e^{2x} \qquad \text{and} \quad v = 1 + x^2.$$

Differentiating both parts we get

$$\frac{du}{dx} = 2e^{2x} \qquad \text{and} \quad \frac{dv}{dx} = 2x.$$

We substitute into

$$\frac{dy}{dx} = \frac{v\dfrac{du}{dx} - u\dfrac{dv}{dx}}{v^2}$$

to get $\dfrac{dy}{dx} = \dfrac{(1 + x^2)(2e^{2x}) - (e^{2x})(2x)}{(1 + x^2)^2}$

$\qquad\qquad = \dfrac{2(1 - x + x^2)e^{2x}}{(1 + x^2)^2}.$

> $2e^{2x}$ is a common factor.

ii) Comparing our quotient with the general $\dfrac{u}{v}$ we let

$$u = 2\sin\theta \qquad \text{and} \quad v = \theta.$$

Differentiating both parts we get

$$\dfrac{du}{d\theta} = 2\cos\theta \qquad \text{and} \quad \dfrac{dv}{d\theta} = 1.$$

We substitute into

$$\dfrac{dy}{d\theta} = \dfrac{v\dfrac{du}{d\theta} - u\dfrac{dv}{d\theta}}{v^2}$$

to get $\dfrac{dy}{d\theta} = \dfrac{(\theta)(2\cos\theta) - (2\sin\theta)(1)}{\theta^2}$

$\qquad\qquad = \dfrac{2(\theta\cos\theta - \sin\theta)}{\theta^2}$

> 2 is a common factor.

Example 7.12

For $y = \tan\theta$ find $\dfrac{dy}{d\theta}$.

So far we have no rule for differentiating tan. However we do know that tan may be rewritten in terms of sin and cos,

i.e. $\qquad y = \dfrac{\sin\theta}{\cos\theta}.$

Now we have a quotient where we let

$$u = \sin\theta \qquad \text{and} \quad v = \cos\theta.$$

Differentiating both parts we get

$$\dfrac{du}{d\theta} = \cos\theta \qquad \text{and} \quad \dfrac{dv}{d\theta} = -\sin\theta.$$

We substitute into

$$\dfrac{dy}{d\theta} = \dfrac{v\dfrac{du}{d\theta} - u\dfrac{dv}{d\theta}}{v^2}$$

to get $\dfrac{dy}{d\theta} = \dfrac{(\cos\theta)(\cos\theta) - (\sin\theta)(-\sin\theta)}{(\cos\theta)^2}$

$\qquad\qquad = \dfrac{\cos^2\theta + \sin^2\theta}{\cos^2\theta}$

$\qquad\qquad = \dfrac{1}{\cos^2\theta} = \sec^2\theta.$

This example derives the differentiation rule for $\tan\theta$,

i.e. for $\quad y = \tan\theta \qquad$ then $\quad \dfrac{dy}{d\theta} = \sec^2\theta.$

More generally we write that for

$$y = \tan(a\theta + b)$$

then $\quad \dfrac{dy}{d\theta} = a\sec^2(a\theta + b) \qquad$ where a and b are constants.

▬▬▬ ASSIGNMENT ▬▬▬

Let us have another look at our simple pendulum. We can rewrite our equation $y = 0.05e^{-0.5t}\sin 3t$ as $y = \dfrac{0.05\sin 3t}{e^{0.5t}}$

When we re-position the exponential we amend the sign in the power position. Now we have a quotient to differentiate.

Comparing our quotient with the general $\dfrac{u}{v}$ we let

$$u = 0.05\sin 3t \qquad \text{and} \qquad v = e^{0.5t}.$$

Differentiating both parts we get

$$\frac{du}{dt} = 0.15\cos 3t \qquad \text{and} \qquad \frac{dv}{dt} = 0.5e^{0.5t}.$$

We substitute into

$$\frac{dy}{dt} = \frac{v\dfrac{du}{dt} - u\dfrac{dv}{dt}}{v^2}$$

to get $\quad \dfrac{dy}{dt} = \dfrac{(e^{0.5t})(0.15\cos 3t) - (0.05\sin 3t)(0.5e^{0.5t})}{(e^{0.5t})^2}$

$$= \frac{e^{0.5t}(0.15\cos 3t - 0.025\sin 3t)}{(e^{0.5t})^2}$$

$$= \frac{0.15\cos 3t - 0.025\sin 3t}{e^{0.5t}} \qquad \boxed{\begin{array}{l}\text{Cancelling some}\\ \text{exponentials.}\end{array}}$$

$$= e^{-0.5t}(0.15\cos 3t - 0.025\sin 3t), \quad \text{as before.}$$

■ EXERCISE 7.7 ■

Apply the quotient rule to find the first differential in each case.

1 $y = \dfrac{2 + x}{e^x}$

2 $y = \dfrac{\sin x}{x + 1}$

3 $y = \dfrac{\ln x}{x}$

4 $y = \dfrac{\cos t}{e^t}$

5 $y = \dfrac{2x - 1}{1 - x}$

6 $y = \cot \theta$ i.e. $\dfrac{\cos \theta}{\sin \theta}$

7 $y = \dfrac{e^{2x}}{2x - 7}$

8 $y = \dfrac{1 + 3x + x^2}{x - 1}$

9 $y = \dfrac{\ln(2 + x)}{x - 2}$

10 $y = \dfrac{1 + \cos \theta - \sin \theta}{\theta}$

We have looked at differentiation using three rules as necessary, i.e. i) function of a function, ii) product and iii) quotient.

The next exercise is a mixture of questions. You need to decide which of the 3 methods is appropriate to each question.

■ EXERCISE 7.8 ■

Find the first differential in each case.

1 $y = \sin^2 \theta - \cos^2 \theta$

2 $y = 2xe^x$

3 $y = \dfrac{\ln(1 + x)}{1 + x}$

4 $y = \dfrac{3 + 4x}{e^x}$

5 $y = \dfrac{\sin \theta + \cos \theta}{2\theta}$

6 $y = \cos(2\theta - \pi) + \sin(\theta - \pi)$

7 $y = (2 + x) \ln 2x$

8 $y = \sin x . e^{2x}$

9 $y = \ln(2x + 3)^2$

10 $y = \dfrac{\tan t}{e^t}$

11 $y = \theta^2 \cos(\theta + 2\pi)$

12 $y = e^{2\cos \theta}$

13 $y = \dfrac{7}{\sqrt{5 + 2x}}$

14 $y = \dfrac{\ln x}{\sqrt{x}}$

15 $y = \dfrac{\sin \theta}{\tan \theta}$

Gradient at a point

We looked at the gradient at a point in Volume 1, Chapter 18. As before we differentiate the original equation. This gives us a general expression for the first derivative. It is at this stage that we make a numerical substitution from the given values.

Example 7.13

For the curve $y = \frac{1}{2}(e^{2x} - e^{-2x})$ find the value of $\frac{dy}{dx}$ where $x = 0.75$.

We differentiate

$$y = \frac{1}{2}(e^{2x} - e^{-2x})$$

to get

$$\frac{dy}{dx} = \frac{1}{2}(2e^{2x} - (-2)e^{-2x}).$$

This is our general expression for the first derivative.

Now we substitute for $x = 0.75$ so that

$$\frac{dy}{dx} = \frac{1}{2}(2e^{2(0.75)} + 2e^{-2(0.75)})$$

$$\frac{dy}{dx} = \frac{2}{2}(e^{1.5} + e^{-1.5})$$

> We remove a common factor of 2 and tidy up the powers.

$$= 4.4817 + 0.2231$$

$$= 4.70.$$

Example 7.14

Find the gradient of the curve $y = x \sin x$ at the point $\left(\frac{\pi}{4}, \frac{\pi}{4\sqrt{2}}\right)$.

In $y = x \sin x$ we have the product of x and $\sin x$.

Let $u = x$ and $v = \sin x$.

Differentiating both parts we get

$$\frac{du}{dx} = 1 \qquad \text{and} \quad \frac{dv}{dx} = \cos x.$$

We substitute into

$$\frac{dy}{dx} = u\frac{dv}{dx} + v\frac{du}{dx}$$

to get $\dfrac{dy}{dx} = (x)(\cos x) + (\sin x)(1).$

This is our general expression for the first derivative.

From the values at the point $\left(\frac{\pi}{4}, \frac{\pi}{4\sqrt{2}}\right)$ we need $x = \frac{\pi}{4}$ so that

$$\frac{dy}{dx} = \left(\frac{\pi}{4}\right)\left(\cos\frac{\pi}{4}\right) + \left(\sin\frac{\pi}{4}\right)$$

$$= 1.26 \quad \text{is the value of the gradient.}$$

ASSIGNMENT

We take our final look at our simple pendulum.

In the first case we find the angular velocity after 4 seconds. Using our general expression

$$\frac{dy}{dt} = e^{-0.5t}(0.15\cos 3t - 0.025\sin 3t)$$

we substitute for $t = 4$ to get

$$\frac{dy}{dt} = e^{-2}(0.15\cos 12 - 0.025\sin 12)$$

$$= 0.1353(0.1266 + 0.0134)$$

$$= 0.019,$$

i.e. the angular velocity after 4 seconds is $0.019\,\text{rads}^{-1}$.

In the second case we look at when it stops, i.e. when the angular velocity is zero as shown by $\frac{dy}{dt} = 0$. We put our general expression for $\frac{dy}{dt}$ equal to 0,

i.e. $\quad e^{-0.5t}(0.15\cos 3t - 0.025\sin 3t) = 0.$

This exponential decay function can never be 0. We know this from its graph never crossing the horizontal axis.

$$\therefore \qquad 0.15\cos 3t - 0.025\sin 3t = 0.$$

Dividing through by $0.025\cos 3t$ we get

$$\frac{0.15\cos 3t}{0.025\cos 3t} - \frac{0.025\sin 3t}{0.025\cos 3t} = 0$$

i.e. $\qquad\qquad\qquad 6 - \tan 3t = 0$

i.e. $\qquad\qquad\qquad \tan 3t = 6$

so that $\qquad\qquad 3t = \tan^{-1}(6) = 1.4056$

i.e. $\qquad\qquad\qquad t = 0.47.$

This means the pendulum stops after 0.47 seconds.

EXERCISE 7.9

1 Find a general expression for $\frac{dy}{dx}$ given $y = \frac{1}{2}(e^{4x} + e^{-4x})$.

   What is the value of $\frac{dy}{dx}$ where i) $x = -1.5$? ii) $x = 0$? iii) $x = 1.5$?

2 For the curve $y = 1.5xe^{3x}$ find the value of $\frac{dy}{dx}$ where $x = 0.4$.

3 y and θ are connected by the equation $y = \operatorname{cosec}\theta$. We may rewrite this as $y = \frac{1}{\sin\theta}$. Apply the quotient rule to find a general expression for $\frac{dy}{d\theta}$. What is the value of $\frac{dy}{d\theta}$ where $\theta = \frac{3\pi}{4}$ radians?

   Alternatively we may rewrite our original equation as $y = (\sin\theta)^{-1}$.

Apply the function of a function rule to confirm your general expression for $\dfrac{dy}{d\theta}$.

4   If $y = \dfrac{4\cos\theta}{\theta}$ what is the value of $\dfrac{dy}{d\theta}$ at the point where $\theta = \dfrac{\pi}{2}$ radians?

5   Find the gradient to the curve $y = (1 + 2x)\ln x$ at the point where $x = 1.5$.

6   y and θ are connected by the equation $y = \sec\theta$. We may rewrite this as $y = \dfrac{1}{\cos\theta}$. Apply the quotient rule to find a general expression for $\dfrac{dy}{d\theta}$. What is the value of $\dfrac{dy}{d\theta}$ where $\theta = \dfrac{2\pi}{3}$ radians?

Alternatively we may rewrite our original equation as $y = (\cos\theta)^{-1}$. Apply the function of a function rule to confirm your general expression for $\dfrac{dy}{d\theta}$.

7   For a cylinder of height (h) 0.45 m and radius r m the total surface area, A, is given by $A = 2\pi r(r + 0.45)$. What are the units of A?

Apply the product rule to find $\dfrac{dA}{dr}$. Write down in words the meaning of $\dfrac{dA}{dr}$ in this case. Find the value of $\dfrac{dA}{dr}$ where

   i) r and h are equal in length,
   ii) r is half the length of h,
   iii) r is twice the length of h.

8   A charged particle moves in a magnetic field subject to exponential decay. In the y direction on a pair of axes during time, t, the path is given by $y = 10 + 2t + e^{-0.25t}\cos 12t$. Find a general expression for $\dfrac{dy}{dt}$ and its specific value after 1 second.

In the x direction for the particle we have $x = 15 - 2t + e^{-0.25t}\sin 12t$. Find the value of $\dfrac{dx}{dt}$ at that same time.

9   A uniform heavy beam has length $2a$ and weight $2W$. Its bending moment, M, is given by $M = \dfrac{Wx}{2a}(2a - x)$ where x is the distance from one end. Apply the product rule to find $\dfrac{dM}{dx}$. If $W = 50$ N and $a = 4$ m find the value of $\dfrac{dM}{dx}$ where $x = 1.75$ m.

10   A body is subject to damped harmonic oscillations. Its distance, x, from its equilibrium position during time, t, is given by $x = \dfrac{2\cos(3t + \pi/3)}{e^t}$. Find an expression for $\dfrac{dx}{dt}$ and find its value after 2 seconds.

Summary of differentiation rules

y	$\dfrac{dy}{dx}$
ax^n	anx^{n-1}
$\sin(ax+b)$	$a\cos(ax+b)$
$\cos(ax+b)$	$-a\sin(ax+b)$
$\tan(ax+b)$	$a\sec^2(ax+b)$
$\ln ax$	$\dfrac{1}{x}$
$\ln f(x)$	$\dfrac{f'(x)}{f(x)}$
e^{ax+b}	ae^{ax+b}
$[f(x)]^n$	$nf'(x)[f(x)]^{n-1}$

$\left.\begin{array}{l} a\cos(ax+b) \\ -a\sin(ax+b) \\ a\sec^2(ax+b) \end{array}\right\}$ angles in radians

where a, b and n are constants.

Function of a function rule $\qquad \dfrac{dy}{dx} = \dfrac{dy}{du} \times \dfrac{du}{dx}$

Product rule $\qquad \dfrac{dy}{dx} = u\dfrac{dv}{dx} + v\dfrac{du}{dx}$

Quotient rule $\qquad \dfrac{dy}{dx} = \dfrac{v\dfrac{du}{dx} - u\dfrac{dv}{dx}}{v^2}$

8 Further Differentiation

The objectives of this chapter are to:

1 State the notation for second derivatives, e.g. $\dfrac{d^2y}{dx^2}, \dfrac{d^2x}{dt^2}$.

2 Find the second derivative by differentiating a simplified version of the first derivative.

3 Evaluate a second derivative at a given point.

4 For the displacement, s, as a function of time, t, state that $\dfrac{ds}{dt}$ and $\dfrac{d^2s}{dt^2}$ represent velocity and acceleration respectively.

5 Evaluate velocity and acceleration at a given time.

Introduction

In this chapter we are going to apply the techniques we learned in the previous one. Exactly the same differentiation rules continue to apply. We will concentrate on the second derivative, but the ideas can be extended to the third and fourth derivatives and beyond. The mathematics continues to work, only the physical interpretation becomes difficult.

ASSIGNMENT

The Assignment for this chapter looks at the motion of a missile. We are going to work out expressions for the velocity and acceleration. These are based on the displacement, s (metres), in terms of time, t (seconds), given by

$$s = 35t.\ln(1 + t).$$

We have drawn the graph of s against t in Fig. 8.1.

180

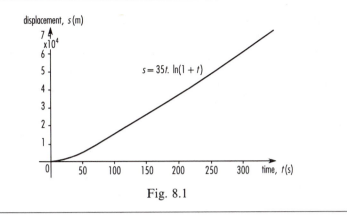

Fig. 8.1

The second derivative

For y as a function of x (i.e. $y = f(x)$) we know that $\dfrac{dy}{dx}$ (or $f'(x)$) is the change in y with respect to x. One interpretation of this is the gradient of the graph. We saw in the previous assignment how this, itself, varied. By differentiating the expression for gradient we can discuss how that gradient in turn changes.

We start with $\qquad\qquad\qquad$ **y** or **f(x)**

and differentiate to get $\qquad\qquad$ $\dfrac{\boldsymbol{dy}}{\boldsymbol{dx}}$ or $\boldsymbol{f'(x)}$.

> First derivative/differential.

We differentiate again and write $\quad \dfrac{\boldsymbol{d}}{\boldsymbol{dx}}\left(\dfrac{\boldsymbol{dy}}{\boldsymbol{dx}}\right)$ or $\boldsymbol{f''(x)}$.

> Second derivative/differential.

$\boldsymbol{f''(x)}$ is said '**f double dashed**' or '**f double dashed of x**'. $\dfrac{\boldsymbol{d}}{\boldsymbol{dx}}\left(\dfrac{\boldsymbol{dy}}{\boldsymbol{dx}}\right)$ is a complete symbol stating that we intend to differentiate with respect to x whatever appears in the bracket. For simplicity we write $\dfrac{d}{dx}\left(\dfrac{dy}{dx}\right)$ as $\dfrac{d^2y}{dx^2}$ (said '**d two y by dx squared**'). Again this is a complete symbol. The positions of the 2's have *no* algebraic significance.

In the following examples we further apply the techniques of the previous chapter.

████████ **Example 8.1** ████████

If $y = 7x^2 + 3x - \dfrac{2}{x}$ find expressions for $\dfrac{dy}{dx}$ and $\dfrac{d^2y}{dx^2}$.

We rewrite the last term in our equation so that

$$y = 7x^2 + 3x - 2x^{-1}.$$

We differentiate once to get

$$\frac{dy}{dx} = 7(2x^{2-1}) + 3(1x^{1-1}) - 2(-1x^{-1-1})$$

$$= 14x + 3 + 2x^{-2}$$

$$\text{or} \quad 14x + 3 + \frac{2}{x^2}$$

Using the expression for $\dfrac{dy}{dx}$ we differentiate again to get

$$\frac{d^2y}{dx^2} = 14(1x^{1-1}) + 0 + 2(-2x^{-2-1})$$

$$= 14 - 4x^{-3}$$

$$\text{or} \quad 14 - \frac{4}{x^3}.$$

> $x^1 = x, \ x^0 = 1.$

████████ **Example 8.2** ████████

Find the first and second derivatives of $y = \sin^3 \theta$.

In Example 7.5 we found the first derivative to be

$$\frac{dy}{d\theta} = 3 \cos \theta \sin^2 \theta.$$

Now we attempt to find the second derivative by differentiating this expression. We have a product of $3 \cos \theta$ and $\sin^2 \theta$ so needing to let

$$u = 3 \cos \theta \qquad \text{and} \qquad v = \sin^2 \theta.$$

Differentiating in both cases we get

$$\frac{du}{d\theta} = -3 \sin \theta \qquad \text{and} \qquad \frac{dv}{d\theta} = 2 \sin \theta \cos \theta.$$

> Use the function of a function rule to check $\dfrac{dv}{d\theta}$.

As usual we substitute into the formula for the product rule. Because we are finding the second derivative the left-hand side is $\dfrac{d^2y}{d\theta^2}$,

i.e. $\dfrac{d^2y}{d\theta^2} = (3 \cos \theta)(2 \sin \theta \cos \theta) + (\sin^2 \theta)(-3 \sin \theta)$

> Using $u\dfrac{dv}{d\theta} + v\dfrac{du}{d\theta}$.

$$= 3 \sin \theta (2 \cos^2 \theta - \sin^2 \theta).$$

> $3 \sin \theta$ is a common factor.

Example 8.3

Find the first and second derivatives of $y = \ln(2x^3 - 5)$.

In Example 7.7 we found the first derivative to be

$$\frac{dy}{dx} = \frac{6x^2}{2x^3 - 5}.$$

To find the second derivative we differentiate this expression using the quotient rule.

Let $\quad u = 6x^2 \quad$ and $\quad v = 2x^3 - 5$.

Differentiating in both cases we get

$$\frac{du}{dx} = 12x \quad \text{and} \quad \frac{dv}{dx} = 6x^2.$$

As usual, we substitute into the formula for the quotient rule. We have $\frac{d^2y}{dx^2}$ on the left-hand side so that

$$\frac{d^2y}{dx^2} = \frac{(2x^3 - 5)(12x) - (6x^2)(6x^2)}{(2x^3 - 5)^2}$$

$$= \frac{24x^4 - 60x - 36x^4}{(2x^3 - 5)^2}$$

$$= \frac{-60x - 12x^4}{(2x^3 - 5)^2}$$

$$= \frac{-12x(5 + x^3)}{(2x^3 - 5)^2}$$

> Using $\dfrac{v\dfrac{du}{dx} - u\dfrac{dv}{dx}}{v^2}$.

> It is usual *not* to expand the denominator.

EXERCISE 8.1

In each case find the first and second derivatives.

1 $\quad y = 3\cos^2 x - 2\sin^2 x$

2 $\quad y = \ln(4x + 5)^2$

3 $\quad y = 4e^{3x-2}$

4 $\quad y = (3 + 2t)^3$

5 $\quad y = \ln(3x^2 + 2)$

6 $\quad y = e^{x^2}$

7 $\quad y = 2\theta \cos \theta$

8 $\quad y = \ln(2\cos \theta)$

9 $\quad y = (x + 5)^{3/2}$

10 $\quad y = \dfrac{\sin \theta}{e^\theta}$

Values of second derivatives

We continue with the differentiation rules and then substitute numerically, as in the next example.

Example 8.4

For $y = 15\ln(3x^2 + 8)$ find the value of $\dfrac{d^2y}{dx^2}$ at the point where $x = 1.5$.

Starting with $y = 15\ln(3x^2 + 8)$ we differentiate once to get

$$\frac{dy}{dx} = 15 \times \frac{6x}{3x^2 + 8}.$$

> By recognition or using the function of a function rule.

Now we have a quotient to differentiate letting

$$u = 6x \qquad \text{and} \qquad v = 3x^2 + 8.$$

Differentiating in both cases we get

$$\frac{du}{dx} = 6 \qquad \text{and} \qquad \frac{dv}{dx} = 6x.$$

As before, we substitute into the formula for the quotient rule to get

$$\frac{d^2y}{dx^2} = 15 \times \frac{(3x^2 + 8)(6) - (6x)(6x)}{(3x^2 + 8)^2}$$

> Using $\dfrac{v\dfrac{du}{dx} - u\dfrac{dv}{dx}}{v^2}$.

$$= 15 \times \frac{(18x^2 + 48 - 36x^2)}{(3x^2 + 8)^2}$$

$$= 15 \times \frac{(48 - 18x^2)}{(3x^2 + 8)^2}$$

This is a general expression for the second derivative. Our next step is to substitute for $x = 1.5$ so that

$$\frac{d^2y}{dx^2} = 15 \times \frac{(48 - 18(1.5)^2)}{(3(1.5)^2 + 8)^2}$$

$$= 15 \times \frac{(48 - 18(2.25))}{(3(2.25) + 8)^2}$$

$$= 15 \times \frac{7.5}{217.56}$$

$$= 0.52.$$

EXERCISE 8.2

1 Find $\dfrac{d^2y}{dx^2}$ for the curve $y = (3 + 7x)^3$ and hence its value where $x = 0.5$.

2 Given $y = 2.5\cos\theta\sin\theta$ find the value of the second derivative at the point $\left(\dfrac{\pi}{4}, \dfrac{5}{4}\right)$.

3 Apply the product rule to find $\dfrac{dy}{dx}$ and $\dfrac{d^2y}{dx^2}$ given that $y = 2xe^{3x}$. Find their values where $x = 0.75$. It is possible for the first and second derivatives to be equal at a particular value of x. Calculate this value.

4 For the curve $y = 2e^{3\sin\theta}$ differentiate to find expressions for $\dfrac{dy}{d\theta}$ and $\dfrac{d^2y}{d\theta^2}$. Hence find the values of these differentials at the point (0, 2).

5 For the curve $y = \dfrac{\ln(x-1)}{x-1}$ find the values of the first and second derivatives where $x = 2.5$.

Velocity and acceleration

We know that $\dfrac{dy}{dx}$ is a change in y due to a change in x. Instead of y being a function of x, suppose that s is a function of t, i.e. $s = f(t)$. For a body moving s metres in a time t seconds we give the displacement as $s = f(t)$.

Now **velocity is defined as the change in displacement, s, due to a change in time, t, i.e. $v = \dfrac{ds}{dt}$ or $f'(t)$.**

For a displacement–time graph (i.e. s against t) v is the gradient at an instant. In Volume 1, Chapter 18, we distinguished between average gradients (chord to the curve) and instantaneous gradients (tangent to the curve).

Acceleration is defined as the change in velocity, v, due to a change in time, t, i.e. $a = \dfrac{dv}{dt}$. Using $v = \dfrac{ds}{dt}$ we may write $a = \dfrac{d}{dt}\left(\dfrac{ds}{dt}\right) = \dfrac{d^2s}{dt^2}$ or $f''(t)$.

Let us look more closely at our definitions. In Volume 1, Chapter 13, we used **displacement** and **velocity** as examples of **vectors**. The corresponding **scalars** are **distance** and **speed**. If we differentiate the displacement to find an expression for velocity, then its size is speed, i.e. **speed** $= \left|\dfrac{ds}{dt}\right|$ **or** $|v|$.

Alternatively **we may define speed as the change in distance due to a change in time.**

Because vectors have both magnitude and direction we need to choose and label a positive direction. A negative displacement is a distance in the opposite direction. A negative velocity is a speed in the opposite direction. Rather differently we interpret a **negative acceleration (deceleration** or **retardation)** to be a slowing down.

Acceleration has no corresponding scalar term. We may treat acceleration as either a vector or a scalar. If an example uses displacement and/or velocity then acceleration is a vector. If it uses distance and/or speed then acceleration is a scalar.

▰▰▰▰ **Example 8.5** ▰▰▰▰▰▰▰▰▰▰▰▰▰▰▰▰▰▰▰▰▰▰▰▰▰▰▰▰▰▰▰▰▰▰▰▰▰

In a particular case the displacement, s, is connected to time, t, by $s = \dfrac{\cos 2t}{2e^t}$. By differentiation find expressions for the velocity and acceleration.

Our equation for s is a quotient in terms of t.

Let $\qquad u = \cos 2t \qquad$ and $\qquad v = 2e^t$.

Differentiating in both cases we get

$$\frac{du}{dt} = -2\sin 2t \quad \text{and} \quad \frac{dv}{dt} = 2e^t.$$

We substitute into

$$\frac{ds}{dt} = \frac{v\dfrac{du}{dt} - u\dfrac{dv}{dt}}{v^2}$$

> Velocity is $\dfrac{ds}{dt}$.

to get $\qquad \dfrac{ds}{dt} = \dfrac{(2e^t)(-2\sin 2t) - (\cos 2t)(2e^t)}{(2e^t)^2}$

$$= \frac{-2e^t(2\sin 2t + \cos 2t)}{(2e^t)^2}$$

i.e. $\qquad \dfrac{ds}{dt} = \dfrac{-(2\sin 2t + \cos 2t)}{2e^t}$

> Cancelling by $2e^t$.

is our expression for velocity.

To find the acceleration we differentiate again using the quotient rule.

Let $\qquad u = -2\sin 2t - \cos 2t \qquad$ and $\qquad v = 2e^t$.

Differentiating in both cases we get

$$\frac{du}{dt} = -4\cos 2t + 2\sin 2t \quad \text{and} \quad \frac{dv}{dt} = 2e^t.$$

Again we substitute into the quotient rule to get

$$\frac{d^2s}{dt^2} = \frac{(2e^t)(-4\cos 2t + 2\sin 2t) - (-2\sin 2t - \cos 2t)(2e^t)}{(2e^t)^2}$$

> Acceleration is a or $\dfrac{d^2s}{dt^2}$.

$$= \frac{(2e^t)(-4\cos 2t + 2\sin 2t + 2\sin 2t + \cos 2t)}{(2e^t)^2}$$

$$= \frac{4\sin 2t - 3\cos 2t}{2e^t}$$

> Cancelling by $2e^t$.

is our general expression for acceleration.

ASSIGNMENT

For our missile we have $s = 35t \ln(1 + t)$. We find an expression for its velocity by differentiation using the product rule.

Let $\qquad u = 35t \qquad$ and $\qquad v = \ln(1 + t)$.

Differentiating in both cases we get

$$\frac{du}{dt} = 35 \qquad \text{and} \qquad \frac{dv}{dt} = \frac{1}{1 + t}.$$

We substitute into

$$\frac{ds}{dt} = u\frac{dv}{dt} + v\frac{du}{dt}$$

to get $\qquad \dfrac{ds}{dt} = (35t)\left(\dfrac{1}{1+t}\right) + (\ln(1 + t))(35)$

$$= 35\left(\frac{t}{1+t} + \ln(1 + t)\right)$$

is our expression for velocity.

Our graph of velocity against time in Fig. 8.2 shows the velocity increasing less rapidly with time.

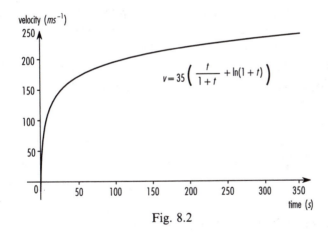

velocity (ms^{-1})

$$v = 35\left(\frac{t}{1+t} + \ln(1 + t)\right)$$

time (s)

Fig. 8.2

We can differentiate again to find an expression for its acceleration. The term $\dfrac{t}{1+t}$ is a quotient and $\ln(1+t)$ is a simple logarithmic function. For the quotient let

$$u = t \qquad \text{and} \qquad v = 1 + t.$$

Differentiating in both cases we get

$$\frac{du}{dt} = 1 \qquad \text{and} \qquad \frac{dv}{dt} = 1.$$

We substitute into the quotient rule and simply differentiate the logarithm to get

$$\frac{d^2s}{dt^2} = 35\left(\frac{(1+t)(1) - (t)(1)}{(1+t)^2} + \frac{1}{1+t}\right)$$

$$= 35\left(\frac{1}{(1+t)^2} + \frac{1}{1+t}\right)$$

$$\text{or } \frac{35(2+t)}{(1+t)^2}$$

is our expression for acceleration.

Again, we could produce a graph of acceleration against time. The gradient of our velocity–time graph is positive, so the acceleration is positive too.

■ EXERCISE 8.3 ■

1 Given that $x = 2 - 5t + 2t^2 - 3t^4$ relates displacement, x, to time, t, find expressions for the i) velocity and ii) acceleration.

2 A body's displacement, x metres, over time, t seconds, is given by $x = 10t\ln(1 + 3t)$. Differentiate, and fully simplify your answers, to get expressions for velocity and acceleration.

3 Find expressions for the velocity and acceleration of a body where the displacement, x, is related to the time of motion, t, according to $x = \dfrac{2 + t^2}{e^{-t}}$.

4 A body moves with simple harmonic motion about an equilibrium point O. Its displacement from O, x metres, is related to time, t seconds, by $x = 1.75\sin\left(\dfrac{\pi}{4}t + 2.75\right)$. By differentiation find expressions for the velocity, $\dfrac{dx}{dt}$, and acceleration, $\dfrac{d^2x}{dt^2}$.

Displacement and acceleration are connected by $\dfrac{d^2x}{dt^2} = -kx$. What is the value of k?

5 The angular displacement of a point on a wheel, θ radians, is given by $\theta = (1 + t^2)(t - 0.5)$ for time, t seconds. Use the product rule to find expressions for the angular velocity, $\dfrac{d\theta}{dt}$, and angular acceleration, $\dfrac{d^2\theta}{dt^2}$.

███████ **Example 8.6** ███████████████████████████████

The displacement of a body, s metres, in time, t seconds, is given by $s = 0.2(4 + 7t - 2t^2 - t^3)$. Find

i)   the velocity after 3 seconds,
ii) when the velocity is zero,
iii) the change in speed during the 4th second,
iv) the acceleration when $t = 1.5$ seconds.

We differentiate our original expression for s to find an expression for velocity $\left(v \text{ or } \dfrac{ds}{dt} \right)$.

$$\therefore \quad \frac{ds}{dt} = 0.2(7 - 4t - 3t^2).$$

i)   When $t = 3$, $v = 0.2(7 - 4(3) - 3(3)^2) = -6.4\,\text{ms}^{-1}$, i.e. a velocity of $6.4\ \text{ms}^{-1}$ in a direction opposite to the original direction.

ii) For $v = 0$, $0 = 7 - 4t - 3t^2$ | Cancelling by 0.2. |

   i.e. $0 = (1 - t)(7 + 3t)$

   i.e. $1 - t = 0, \quad 7 + 3t = 0$

   i.e. $t = 1, \ -\dfrac{7}{3}.$

   Because time is positive (i.e. we *cannot* go back in time) the velocity is zero after 1 second.

iii) The 4th second starts at $t = 3$ and continues until $t = 4$. Already we have substituted for $t = 3$ in our expression for velocity. Also, when $t = 4$, $v = 0.2(7 - 4(4) - 3(4)^2) = -11.4\ \text{ms}^{-1}$. In this time the velocity changes from $-6.4\ \text{ms}^{-1}$ to $-11.4\ \text{ms}^{-1}$, i.e. the size of the change is 5 ms^{-1}, i.e. the change in speed is 5 ms^{-1}.

iv) We differentiate our expression for velocity to find an expression for acceleration $\left(a \text{ or } \dfrac{dv}{dt} \text{ or } \dfrac{d^2s}{dt^2} \right)$,

$$\frac{d^2s}{dt^2} = 0.2(-4 - 6t).$$

When $t = 1.5$, $\dfrac{d^2s}{dt^2} = 0.2(-4 - 6(1.5)) = -2.6\ \text{ms}^{-2}$,

   i.e. the body is slowing down, the retardation (deceleration) indicated by the minus sign.

▆▆▆▆ ASSIGNMENT ▆▆▆▆▆▆▆▆▆▆▆▆▆▆

In this final look at our missile let us calculate some velocity values. We can find its velocity after 1 minute (i.e. 60 seconds) by substituting $t = 60$ into our velocity expression,

i.e. $\dfrac{ds}{dt} = 35\left(\dfrac{t}{1+t} + \ln(1+t)\right)$

becomes $\dfrac{ds}{dt} = 35\left(\dfrac{60}{1+60} + \ln(1+60)\right)$

$\phantom{becomes \dfrac{ds}{dt}} = 35(0.9836 + 4.1109)$

$\phantom{becomes \dfrac{ds}{dt}} = 178.3 \text{ ms}^{-1}.$

We may also wish to know the average velocity during this first minute. This calculation needs to discuss the displacement during the minute.

In $\qquad\qquad\qquad s = 35t\ln(1+t)$

when $t = 0$, $\qquad\quad s_0 = 35(0)\ln(1+0) = 0.$

Also when $t = 60$, $\quad s_{60} = 35(60)\ln(1+60) = 8632.8.$

Then Average velocity $= \dfrac{\text{Displacement}}{\text{Time}}$

$ = \dfrac{8632.8 - 0}{60}$

$ = 143.9 \text{ ms}^{-1}.$

Finally, suppose we wish to know the initial acceleration of the missile, i.e. the acceleration at $t = 0$.

In $\qquad\qquad \dfrac{d^2s}{dt^2} = \dfrac{35(2+t)}{(1+t)^2}$

we substitute for $t = 0$ to get

$\qquad\qquad a_0 = \dfrac{35(2+0)}{(1+0)^2} = 70 \text{ ms}^{-2}$ $\qquad\boxed{a = \dfrac{d^2s}{dt^2}.}$

▆▆▆▆ EXERCISE 8.4 ▆▆▆▆▆▆▆▆▆▆▆▆▆▆

1   A body moves according to the equation $x = 2t^2 + 16t + 4$. x metres is the displacement at time t seconds.

   i)   How far does the body move in the 3rd second?
   ii) Calculate the speed at the beginning and end of that 3rd second. What is the change in speed during this time?

2   A particle oscillates about E, a point of equilibrium. From E in time t seconds its displacement x metres is given by $x = 4\sin\dfrac{\pi t}{6} + 3\cos\dfrac{\pi t}{6}$.

What is the   i)   initial speed?
　　　　　　　 ii)   velocity after 0.5 seconds?
　　　　　　　iii)   acceleration after 0.75 seconds?

3 Cargo is pushed from a military transport aircraft during flight. It falls subject to air resistance and the effects of its parachutes. During its descent the vertical displacement from the aircraft, y metres, is connected to time, t seconds, by $y = 49(t + 0.5e^{-2t})$. Find a general expression for the velocity.

As $t \to \infty$ (i.e. as the time tends to infinity) what happens to the negative exponential? What does this tell you about the velocity?

4 A city-centre tram, operating a shuttle service, travels x kilometres in time t minutes according to $x = 0.025t^2(8 - t)$.

When is its displacement from its initial station zero? What does this tell you about the time for a return trip? What is its velocity after 7.5 minutes? When is its acceleration zero?

5 In a water-treatment installation the height of effluent, y metres, is related to time, t hours. During a period of 24 hours timed from midnight this relationship is

$$y = 0.36\left(2 + \frac{10}{t - 24} + \frac{50}{(t - 24)^2}\right).$$

$\dfrac{dy}{dt}$ represents the rate of change of this height. Find a general expression for $\dfrac{dy}{dt}$ and find its value at 9.30 a.m. and 9.30 p.m.

9 Maximum and Minimum

The objectives of this chapter are to:

1 Define a turning point of a graph.
2 Determine the first derivative of the function of the graph concerned.
3 Find the value of x (independent variable) at a turning point.
4 Determine the nature of a turning point by considering the gradient on either side of that point.
5 Determine and evaluate the second derivative of the function at the turning point.
6 Determine the nature of a turning point by the sign of the second derivative.
7 Find the value of y (dependent variable) at a turning point.
8 Find any points of inflexion on a curve.
9 Solve problems involving maxima and minima relevant to technology.

Introduction

This chapter looks at a mathematical view of **maximum** and **minimum**. These values can occur at particular points on a curve. We are not interested in the absolute biggest and smallest values of the graph. Our maximum and minimum values are associated with the **turning points** of the graph. For $y = f(x)$ the **maximum** and **minimum values** refer to **values of y**, the dependent variable, where x is the independent variable. The plural of **maximum** is **maxima** and of **minimum** is **minima**. Our method always involves looking at the first derivative and generally at the second derivative.

■■■ ASSIGNMENT ■■■

The Assignment for this chapter looks again at the simple pendulum performing small oscillations about the vertical.

Remember this takes place in a fluid so involving a resistive force to the motion. The angular displacement, y, from the vertical is given in terms of time, t, by the formula $y = 0.05e^{-0.5t} \sin 3t$.

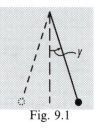

Fig. 9.1

192

Turning points

Fig. 9.2 shows y as some general function of x, i.e. $y = f(x)$. We are interested in points A and B. Here the tangents are parallel to the horizontal axis, i.e. the gradients are zero, i.e. $\dfrac{dy}{dx} = 0$. A and B are called **turning points**. We mentioned turning points in Volume 1, Chapter 18. Now we are interested in finding the value of the independent variable (usually x) at a turning point.

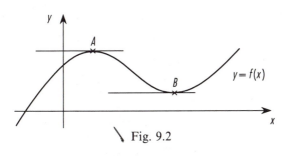

Fig. 9.2

Has the graph of $y = 2x^2 + x + 1$ any turning points? If there are any, at what values of x do they occur?

Fig. 9.3 shows there is a turning point where $x = -0.25$. Indeed the sketch includes a horizontal tangent at this point. We do not always need to draw a graph of a function just to find any turning points. Indeed, we can save ourselves time by not having to draw one. Our method uses differentiation.

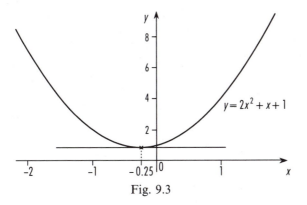

Fig. 9.3

Using $\qquad y = 2x^2 + x + 1$

we differentiate once,

i.e. $\qquad \dfrac{dy}{dx} = 4x + 1.$

We know that $\dfrac{dy}{dx}=0$ at a turning point. This specific condition can be linked to our general expression, $4x + 1$, so that

$$4x + 1 = 0$$

i.e. $x = -0.25.$

This confirms what we discovered in Fig. 9.3, that there is a turning point where $x = -0.25$. Notice that we have only one turning point rather than several. We recall from Volume 1, Chapter 7, that quadratic functions only turn once. Also, when we differentiate a quadratic function (highest powered term being x^2) we get a linear function (highest powered term being x). The ensuing linear equation we create can have only one solution.

Example 9.2

Has the graph of $y = x^3 - 3.5x^2 + 2x - 6$ any turning points? If there are any, at what values of x do they occur?

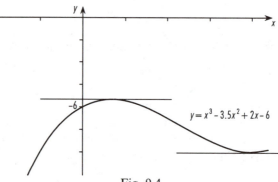

Fig. 9.4

Fig. 9.4 shows there to be two turning points. We have removed the scales from the axes on purpose. This is because the aim of this chapter is to find a turning point using differentiation.

Using $y = x^3 - 3.5x^2 + 2x - 6$

we differentiate once,

i.e. $\dfrac{dy}{dx} = 3x^2 - 7x + 2.$

Using our specific knowledge that $\dfrac{dy}{dx}=0$ at a turning point we have

$$3x^2 - 7x + 2 = 0.$$

This quadratic equation factorises to give

$$(3x - 1)(x - 2) = 0$$

∴ $3x - 1 = 0,\ x - 2 = 0$

i.e. $x = \dfrac{1}{3},\ 2.$

This calculation confirms what we see in our sketch. We have two turning points, one where $x = \frac{1}{3}$ and the other where $x = 2$. Our particular quadratic equation factorised, but if necessary we can use the formula.

> For $ax^2 + bx + c = 0$
>
> $x = \dfrac{-b \pm \sqrt{b^2 - 4ac}}{2a}.$

Notice how we started with a cubic function and differentiated to get a quadratic function. Generally, though *not* always, a quadratic equation has two solutions. These two solutions are consistent with a cubic function that has two turning points. Now this means we need to be able to distinguish between them. We look to do this after the first exercise.

EXERCISE 9.1

In each question find, by differentiation, the value(s) of x at the turning point(s).

1 $y = x^2 - 5x + 3.5$

2 $y = 2.25 + 3x - x^2$

3 $y = 2x^3 - 4.5x^2 + 3x + 8$

4 $y = 4x^3 - 3x^2 - 11x + 2.5$

5 $y = x^3 - 4.4x^2 + 5.6x - 1.6$

Maximum and minimum turning points

Let us redraw our first figure. In Fig. 9.5 we include the sign of the gradient, noticing how it changes.

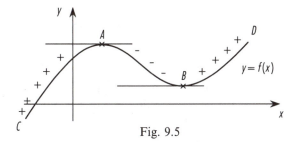

Fig. 9.5

We start from the left, moving to the right as x increases. The gradient changes from positive ($+$) through zero at A to negative ($-$). This type of change shows that A is a **maximum turning point**. It is also called a **local maximum point**. This distinguishes it from points like D which are absolutely bigger. The gradient changes from negative ($-$) through zero at B to positive ($+$). This type of change shows that B is a **minimum turning point**. It is also called a **local minimum point**. This distinguishes it from points like C which are absolutely smaller. These features mean we can test a turning point by examining the gradient closely on either side of it. This is called the **gradient test**.

████ **Example 9.3** ████

Has the graph of $y = 3 + 2x - 5x^2$ a turning point? If there is a turning point decide whether it is a maximum or minimum.

Using
$$y = 3 + 2x - 5x^2$$

we differentiate once,

i.e.
$$\frac{dy}{dx} = 2 - 10x.$$

At a turning point $\frac{dy}{dx} = 0$ so that

$$2 - 10x = 0$$

giving a solution $\qquad x = 0.2.$

So far we know there is a turning point at $x = 0.2$. Our next step is to check the gradients closely on either side of this turning point. We simply choose values of x slightly below and slightly above 0.2.

In $\quad \dfrac{dy}{dx} = 2 - 10x$

we substitute for $x = 0.175$ to get

$$\frac{dy}{dx} = 2 - 10(0.175) = 0.25.$$

Also for $x = 0.225$ we get

$$\frac{dy}{dx} = 2 - 10(0.225) = -0.25.$$

Let us look at our numerical results in order.

x	0.175	0.2	0.225
$\dfrac{dy}{dx}$	0.25 (+)	0	−0.25 (−)

As x increases the gradient changes from positive through zero to negative. This means we have a maximum turning point.

There is an alternative method to the gradient test that uses the second derivative.

We know the gradient changes from positive through zero to negative at a maximum turning point. This means the gradient is decreasing, i.e. the change is negative. Mathematically we write this gradient change as $\dfrac{d}{dx}\left(\dfrac{dy}{dx}\right)$ is negative, i.e. $\dfrac{d^2y}{dx^2} < 0$.

This means our test for a **maximum turning point (local maximum)** is $\dfrac{dy}{dx} = 0$ **and** $\dfrac{d^2y}{dx^2} < 0$.

Also we know the gradient changes from negative through zero to positive at a minimum turning point. This means the gradient is increasing, i.e. the change is positive. Mathematically we write this gradient change as $\frac{d}{dx}\left(\frac{dy}{dx}\right)$ is positive, i.e. $\frac{d^2y}{dx^2} > 0$.

This means our test for a **minimum turning point (local minimum) is** $\frac{dy}{dx} = 0$ and $\frac{d^2y}{dx^2} > 0$.

Example 9.4

Find the turning points of $y = 2x^3 - 9x^2 - 60x + 4$. Using the second derivative discover if these are maximum or minimum turning points.

Using $$y = 2x^3 - 9x^2 - 60x + 4$$
we differentiate once,

i.e. $$\frac{dy}{dx} = 6x^2 - 18x - 60.$$

At a turning point $\frac{dy}{dx} = 0$ so that

$$6x^2 - 18x - 60 = 0$$

i.e. $$6(x^2 - 3x - 10) = 0$$

$$(x + 2)(x - 5) = 0$$

Dividing by 6 and factorising.

$$\therefore \quad x + 2 = 0,\ x - 5 = 0$$

i.e. $$x = -2,\ 5.$$

This shows we have turning points where $x = -2$ and $x = 5$. Our next step is to find the type of each turning point.

Using $$\frac{dy}{dx} = 6x^2 - 18x - 60$$

we differentiate again to get

$$\frac{d^2y}{dx^2} = 12x - 18.$$

This is a general expression for the second derivative. We need to know whether it is positive or negative at our specific values of x. Substituting for $x = -2$,

$$\frac{d^2y}{dx^2} = 12(-2) - 18 = -42, \quad \text{i.e. negative.}$$

This means we have a maximum turning point where $x = -2$.

Also substituting for $x = 5$,

$$\frac{d^2y}{dx^2} = 12(5) - 18 = 42, \quad \text{i.e. positive.}$$

This means we have a minimum turning point where $x = 5$.

For maximum and minimum turning points there are exceptions to the rules for the second derivative. When this fails to be either positive or negative we need to try our first method, the gradient test.

▓▓▓ Example 9.5 ▓▓▓

Find the type (nature) of turning point of the curve $y = x^4 + 2$.

Using $y = x^4 + 2$

we differentiate once,

i.e. $\dfrac{dy}{dx} = 4x^3.$

At a turning point $\dfrac{dy}{dx} = 0$ so that

$$4x^3 = 0$$

i.e. $x^3 = 0$ | Dividing by 4. |

i.e. $x = 0.$

This shows we have a turning point where $x = 0$.

Using $\dfrac{dy}{dx} = 4x^3$

we differentiate again to get

$$\dfrac{d^2y}{dx^2} = 12x^2.$$

Substituting for $x = 0$ into our general expression we get

$$\dfrac{d^2y}{dx^2} = 12(0)^2 = 0.$$

This is neither positive nor negative yet Fig. 9.6 shows we have a minimum turning point.

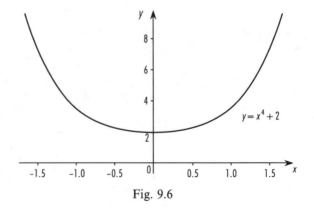

Fig. 9.6

We return to our gradient test using values of x slightly below and slightly above $x = 0$.

Using $\dfrac{dy}{dx} = 4x^3$

we substitute for $x = -0.2$ to get

$$\dfrac{dy}{dx} = 4(-0.2)^3 = -0.032.$$

Also for $x = 0.2$ we get

$$\dfrac{dy}{dx} = 4(0.2)^3 = 0.032.$$

Looking at our results in order we have

x	-0.2	0	0.2
$\dfrac{dy}{dx}$	-0.032 $(-)$	0	0.032 $(+)$

As x increases the gradient changes from negative through zero to positive. This means we have a minimum turning point at $x = 0$, confirmed by Fig. 9.6.

■ EXERCISE 9.2 ■

In each question find the nature (type) of the turning points using the i) gradient test, and ii) second differential.

1 $y = 3x^2 + 2x - 6$

2 $y = 7 - 2x - 4x^2$

3 $y = 4x^3 - 3x^2 - 36x$

4 $y = x^3 + x^2 - 1.75x + 0.5$

5 $y = 2 - 6x + x^3$

Maximum and minimum values

This is the third and final stage in our technique. So far we have found that turning points can exist at particular values of x. We also know how to distinguish between maximum and minimum turning points. Now we simply substitute for these values of x (independent variable) into our original equation. This gives us values of y (dependent variable) that are either a maximum or a minimum value. Our substitution will give the y values at A and B in Fig. 9.7. Together with the x values we will have their coordinates.

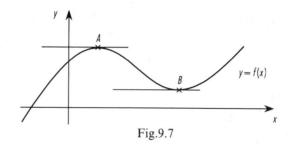

Fig.9.7

■■■■■■ **Example 9.6** ■■■■■■

Find the local maximum and minimum values of
$y = 10 + 30x - 2.25x^2 - x^3$. Also write down the coordinates of these
turning points.

Stage 1   (To find that turning points exist and their values of x.)

Using $y = 10 + 30x - 2.25x^2 - x^3$

we differentiate once,

i.e. $\dfrac{dy}{dx} = 30 - 4.5x - 3x^2.$

At a turning point $\dfrac{dy}{dx} = 0$ so that

$$30 - 4.5x - 3x^2 = 0.$$

This quadratic equation has solutions

$$x = -4, 2.5,$$

i.e. we have turning points where $x = -4$ and $x = 2.5$.

Stage 2   (To find the type (nature) of the turning points.)

Using $\dfrac{dy}{dx} = 30 - 4.5x - 3x^2$

we differentiate again to get

$$\dfrac{d^2y}{dx^2} = -4.5 - 6x.$$

Substituting for $x = -4$,

$$\dfrac{d^2y}{dx^2} = -4.5 - 6(-4) = 19.5, \quad \text{i.e. positive.}$$

This means we have a minimum turning point where $x = -4$.

Also substituting for $x = 2.5$,

$$\dfrac{d^2y}{dx^2} = -4.5 - 6(2.5) = -19.5, \quad \text{i.e. negative.}$$

This means we have a maximum turning point where $x = 2.5$.

Stage 3 (*To find the local maximum and minimum values of* **y**.)

We need to know the values of y at these turning points.

Using $\qquad y = 10 + 30x - 2.25x^2 - x^3$

we substitute for $x = -4$ to get

$$y_{min} = 10 + 30(-4) - 2.25(-4)^2 - (-4)^3 = -82.$$

The local minimum value is -82, occurring at $(-4, -82)$.

Also we substitute for $x = 2.5$ to get

$$y_{max} = 10 + 30(2.5) - 2.25(2.5)^2 - (2.5)^3 = 55.3.$$

The local maximum value is 55.3, occurring at $(2.5, 55.3)$.

▰▰▰ Example 9.7 ▰▰▰

Find the local maximum and minimum values of $y = \dfrac{3x + 3}{x(3 - x)}$.

Using $\qquad y = \dfrac{3x + 3}{x(3 - x)} \quad$ or $\quad \dfrac{3x + 3}{3x - x^2}$

we differentiate once, using the quotient rule,

$\boxed{\dfrac{dy}{dx} = \dfrac{v\dfrac{du}{dx} - u\dfrac{dv}{dx}}{v^2}.}$

i.e. $\qquad \dfrac{dy}{dx} = \dfrac{(3x - x^2)(3) - (3x + 3)(3 - 2x)}{(3x - x^2)^2}$

$$= \dfrac{9x - 3x^2 - 9x + 6x^2 - 9 + 6x}{(3x - x^2)^2}$$

$$= \dfrac{3x^2 + 6x - 9}{(3x - x^2)^2}$$

i.e. $\qquad \dfrac{dy}{dx} = \dfrac{3(x^2 + 2x - 3)}{(3x - x^2)^2}.$

At a turning point $\dfrac{dy}{dx} = 0$ so that

$$\dfrac{3(x^2 + 2x - 3)}{(3x - x^2)^2} = 0$$

i.e. $\qquad 3(x^2 + 2x - 3) = 0$

i.e $\qquad (x + 3)(x - 1) = 0$

$\therefore \qquad x + 3 = 0, \quad x - 1 = 0$

i.e. $\qquad\qquad x = -3, 1,$

> Denominator *cannot* be 0.
>
> Dividing by 3 and factorising.

i.e. we have turning points where $x = -3$ and $x = 1$.

We could differentiate again and test $\dfrac{d^2y}{dx^2}$ at $x=-3$ and $x=1$. However the algebra gets complicated. It is easier to apply the gradient test around $x=-3$ and around $x=1$.

Using $\dfrac{dy}{dx} = \dfrac{3(x^2+2x-3)}{(3x-x^2)^2}$ or $\dfrac{3(x^2+2x-3)}{x^2(3-x)^2}$

we substitute for $x=-3.1$ to get

$$\frac{dy}{dx} = \frac{3((-3.1)^2+2(-3.1)-3)}{(-3.1)^2(3--3.1)^2} = 0.0034.$$

Also for $x=-2.9$ we get

$$\frac{dy}{dx} = \frac{3((-2.9)^2+2(-2.9)-3)}{(-2.9)^2(3--2.9)^2} = -0.0040.$$

Let us look at our results in order

x	-3.1	-3.0	-2.9
$\dfrac{dy}{dx}$	0.0034 (+)	0	-0.0040 (−)

As x increases the gradient changes from positive through zero to negative. This means we have a maximum turning point at $x=-3.0$, confirmed by Fig. 9.8.

For the other turning point, again using $\dfrac{dy}{dx} = \dfrac{3(x^2+2x-3)}{x^2(3-x)^2}$ we substitute for $x=0.9$ to get

$$\frac{dy}{dx} = \frac{3((0.9)^2+2(0.9)-3)}{(0.9)^2(3-0.9)^2} = -0.33.$$

Also for $x=1.1$ we get

$$\frac{dy}{dx} = \frac{3((1.1)^2+2(1.1)-3)}{(1.1)^2(3-1.1)^2} = 0.28.$$

Let us look at these results in order

x	0.9	1.0	1.1
$\dfrac{dy}{dx}$	-0.33 (−)	0	0.28 (+)

As x increases the gradient changes from negative through zero to positive. This means we have a minimum turning point at $x=1.0$, again confirmed by Fig. 9.8.

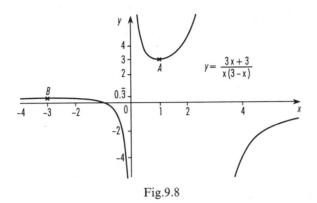

Fig.9.8

Now we need to know the values of y at these turning points.

Using $\quad y = \dfrac{3x+3}{x(3-x)}$

we substitute for $x = -3$ to get

$$y_{max} = \frac{3(-3)+3}{(-3)(3--3)} = \frac{-6}{-18} = 0.\bar{3}.$$

The local maximum value is $0.\bar{3}$, occurring at $(-3, 0.\bar{3})$.

Also we substitute for $x = 1$ to get

$$y_{min} = \frac{3(1)+3}{1(3-1)2} = \frac{6}{2} = 3.$$

The local minimum value is 3, occurring at (1, 3).

We need to look at these turning points on our graph in Fig. 9.8. Notice how the curves show the local minimum point, A, to be above the local maximum point, B. This is an example emphasising the local nature of the turning points. Remember they are not supposed to be absolute maximum or minimum values.

■■■■■ ASSIGNMENT ■■■■■■■■■■■■■■■■

We are going to look for any turning points for our pendulum.

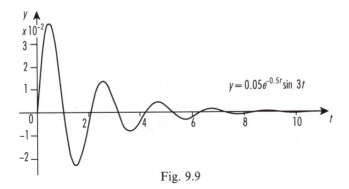

Fig. 9.9

Using $y = 0.05e^{-0.5t} \sin 3t$

we differentiate once, using the product rule,

$$\frac{dy}{dt} = u\frac{dv}{dt} + v\frac{du}{dt}.$$

i.e. $\dfrac{dy}{dt} = (0.05e^{-0.5t})(3\cos 3t) + (\sin 3t)(-0.025e^{-0.5t})$

$$= e^{-0.5t}(0.15\cos 3t - 0.025\sin 3t).$$

At a turning point $\dfrac{dy}{dt} = 0$ so that

$$e^{-0.5t}(0.15\cos 3t - 0.025\sin 3t) = 0$$

i.e. $0.15\cos 3t - 0.025\sin 3t = 0.$

We know that an exponential curve *never* crosses the horizontal axis and so *cannot* be zero.

Then $\dfrac{0.15\cos 3t}{0.025\cos 3t} - \dfrac{0.025\sin 3t}{0.025\cos 3t} = 0$ | Dividing by $0.025\cos 3t$.

i.e. $\dfrac{0.15}{0.025} - \dfrac{\sin 3t}{\cos 3t} = 0$

i.e. $6 - \tan 3t = 0$

i.e. $3t = \tan^{-1}(6)$

∴ $3t = 1.41,\dots$

Remember we use radians because we are differentiating.

Now $3t$ has a cyclical series of solutions as shown in Fig. 9.9. We need to check out the quadrants where tangent is positive. They are the 1st (*) and 3rd (***) quadrants in Fig. 9.10.

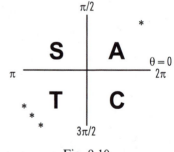

Fig. 9.10

This means that $3t = 1.41,\ \pi + 1.41,\ 2\pi + 1.41,\dots$

∴ $t = 0.47,\ 1.52,\ 2.56,\dots$

These are the early values of t where our graph has turning points. There are others, as you can see from Fig. 9.9.

Our next stage is to find whether each turning point is a maximum or minimum.

Using $\dfrac{dy}{dt} = e^{-0.5t}(0.15\cos 3t - 0.025\sin 3t)$

we differentiate again, using the product rule, to get

$$\frac{d^2y}{dt^2} = (e^{-0.5t})(-0.45\sin 3t - 0.075\cos 3t)$$
$$+ (0.15\cos 3t - 0.025\sin 3t)(-0.5e^{-0.5t})$$
$$= e^{-0.5t}(-0.4375\sin 3t - 0.15\cos 3t).$$

Substituting for $t = 0.47$, | Calculator value is 0.4685...

$$\frac{d^2y}{dt^2} = 0.79(-0.43 - 0.02) = -0.36, \quad \text{i.e. negative.}$$

This means we have a maximum turning point where $t = 0.47$.

Substituting for $t = 1.52$, | Calculator value is 1.5157...

$$\frac{d^2y}{dt^2} = 0.47(0.43 + 0.02) = 0.21, \quad \text{i.e. positive.}$$

This means we have a minimum turning point where $t = 1.52$.

Substituting for t $= 2.56$, | Calculator value is 2.5629...

$$\frac{d^2y}{dt^2} = 0.28(-0.43 - 0.02) = -0.13, \quad \text{i.e. negative.}$$

This means we have a maximum turning point where $t = 2.56$.

We could continue with these calculations for the series of maxima and minima of Fig. 9.9. Notice how their size continues to decrease over time. Instead, we will go on to find the values of the first local maximum and minimum values.

Using $\quad y = 0.05e^{-0.5t}\sin 3t$

we substitute for $t = 0.47$ to get

$$y_{\text{max}} = 0.039.$$

The first local maximum value is 0.039, occurring at (0.47, 0.039).

Also we substitute for $t = 1.52$ to get

$$y_{\text{min}} = -0.023.$$

The first local minimum value is -0.023, occurring at (1.52, -0.023).

▬▬▬ EXERCISE 9.3 ▬▬▬

Find the local maximum and minimum values of y, where appropriate, in each question. Also write down their coordinates, distinguishing between them.

1 $y = 2x^2 - 3x$

2 $y = 4 - 10x - 5x^2$

3 $y = -24x - 3x^2 + x^3$

4 $y = 2 - 7x + 4x^2 + 4x^3$

5 $y = 3x^3 + 4.5x + 1.5$

6 $y = (1.5 + 2.5t)^3 - 30t$

7 $y = \ln(3x^2 + 1)$

8 $y = xe^x$

9 $y = \ln(3\cos\theta)$

10 $y = \dfrac{2\sin\theta}{e^{2\theta}}$

Points of inflexion

A **point of inflexion** is another interesting point that appears on some graphs. It is an exceptional point and so the mathematics needs plenty of care. We have sketched a typical curve in Fig. 9.11.

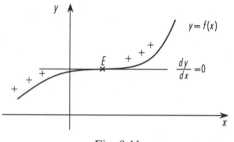

Fig. 9.11

E is the point of inflexion. We can see that close to *E* on both sides the gradient has the same sign. In this particular case the gradient is positive on both sides. (Just as easily in another example the gradient could be negative on both sides.) We start from the left, moving to the right as *x* increases. As this happens the gradient decreases towards *E* and then starts to increase. This means the change in the gradient is negative, becoming zero as it passes through *E* and then positive. Remember that $\frac{d}{dx}\left(\frac{dy}{dx}\right)$, i.e. $\frac{d^2y}{dx^2}$, represents the change in gradient. Mathematically at a **point of inflexion we write that $\frac{d^2y}{dx^2}=0$ and its sign is different on either side. Also the gradient has the same sign on both sides.**

Sometimes $\frac{dy}{dx}=0$ at a point of inflexion, but this is not always true.

━━━━━ **Example 9.8** ━━━━━

The curve $y = 3 + 2x^3 - x^4$ has points of inflexion. Find the coordinates of these points.

Using $\qquad y = 3 + 2x^3 - x^4$

we differentiate once, to get

i.e. $\qquad \frac{dy}{dx} = 6x^2 - 4x^3$

and again to get

$$\frac{d^2y}{dx^2} = 12x - 12x^2.$$

For a point of inflexion $\frac{d^2y}{dx^2}=0$ so that

$$12x - 12x^2 = 0$$

i.e. $\qquad 12x(1 - x) = 0$ $\qquad$ $\boxed{12x \text{ is a common factor.}}$

$\therefore \qquad 12x = 0, \ 1 - x = 0$

i.e. $\qquad x = 0, 1.$

We need to test $\dfrac{d^2y}{dx^2}$ around $x = 0$ and $x = 1$.

Using $\qquad \dfrac{d^2y}{dx^2} = 12x(1 - x)$

we substitute for $x = -0.1$ to get

$$\dfrac{d^2y}{dx^2} = 12(-0.1)(1 - -0.1) = -1.32, \quad \text{i.e. negative.}$$

Also for $x = 0.1$ we get

$$\dfrac{d^2y}{dx^2} = 12(0.1)(1 - 0.1) = 1.08, \quad \text{i.e. positive.}$$

Let us look at our results in order

x	-0.1	0	0.1
$\dfrac{d^2y}{dx^2}$	-1.32 $(-)$	0	1.08 $(+)$

As x increases the second derivative changes sign. This means we have a point of inflexion at $x = 0$, confirmed by Fig. 9.12.

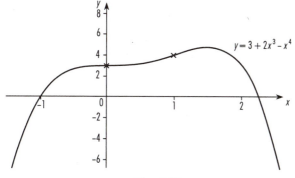

$y = 3 + 2x^3 - x^4$

Fig. 9.12

For the other point, again using $\dfrac{d^2y}{dx^2} = 12x(1 - x)$ we substitute for $x = 0.9$ to get

$$\dfrac{d^2y}{dx^2} = 12(0.9)(1 - 0.9) = 1.08, \quad \text{i.e. positive.}$$

Also for $x = 1.1$ we get

$$\dfrac{d^2y}{dx^2} = 12(1.1)(1 - 1.1) = -1.32, \quad \text{i.e. negative.}$$

Let us look at these results in order

x	0.9	1.0	1.1
$\dfrac{d^2y}{dx^2}$	1.08 (+)	0	−1.32 (−)

As x increases the second derivative changes sign. This means we have a point of inflexion at $x=1$, again confirmed by Fig. 9.12.

We find the coordinates of these points of inflexion by substituting into our original equation.

Using $\quad y = 3 + 2x^3 - x^4$

we substitute for $x=0$ to get $y=3$.

Also for $x=1$ we substitute to get $y=4$.

The coordinates of the points of inflexion are (0, 3) and (1, 4).

As an exercise for yourself you can check that the gradients around both $x=0$ and $x=1$ are positive. This is consistent with our curve in Fig. 9.12.

▰▰▰ Example 9.9 ▰▰▰

Has the curve $y=x^4$ a point of inflexion? If so find the coordinates of the point.

Using $\quad y = x^4$

we differentiate once, to get

i.e. $\quad \dfrac{dy}{dx} = 4x^3$

and again to get

$$\dfrac{d^2y}{dx^2} = 12x^2.$$

For a point of inflexion $\dfrac{d^2y}{dx^2}=0$ so that

$$12x^2 = 0$$

$\therefore \qquad x = 0.$

We need to test $\dfrac{d^2y}{dx^2}$ around $x=0$.

Using $\quad \dfrac{d^2y}{dx^2} = 12x^2$

we substitute for $x=-0.1$ to get

$$\dfrac{d^2y}{dx^2} = 12(-0.1)^2 = 0.12, \quad \text{i.e. positive.}$$

Also for $x = 0.1$ we get

$$\frac{d^2y}{dx^2} = 12(0.1)^2 = 0.12, \quad \text{i.e. positive too.}$$

Immediately we see that the second derivative does not change signs around our value, $x = 0$. This means we do not have a point of inflexion at $x = 0$, confirmed by Fig. 9.13.

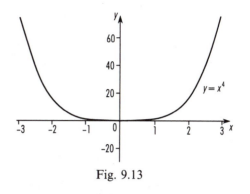

Fig. 9.13

Example 9.10

Find any point(s) of inflexion of the curve $y = \dfrac{27 - x^3}{x}$.

We may re-write our curve as

$$y = 27x^{-1} - x^2$$

and differentiate once, to get

$$\frac{dy}{dx} = -27x^{-2} - 2x$$

and again to get

$$\frac{d^2y}{dx^2} = 54x^{-3} - 2.$$

For a point of inflexion $\dfrac{d^2y}{dx^2} = 0$ so that

$$54x^{-3} - 2 = 0$$

i.e. $$\frac{54}{x^3} = 2$$

i.e. $$x^3 = 27$$

∴ $$x = 3.$$ | Cube root. |

We need to test $\dfrac{d^2y}{dx^2}$ around $x = 3$.

Using $$\frac{d^2y}{dx^2} = \frac{54}{x^3} - 2$$

we substitute for $x = 2.9$ to get

$$\frac{d^2y}{dx^2} = \frac{54}{(2.9)^3} - 2 = 0.21, \quad \text{i.e. positive.}$$

Also for $x = 3.1$ we get

$$\frac{d^2y}{dx^2} = \frac{54}{(3.1)^3} - 2 = -0.19, \quad \text{i.e. negative.}$$

Let us look at our results in order

x	2.9	3.0	3.1
$\dfrac{d^2y}{dx^2}$	0.21 (+)	0	−0.19 (−)

As x increases the second derivative changes sign. This means we have a point of inflexion at $x = 3$, confirmed by Fig. 9.14.

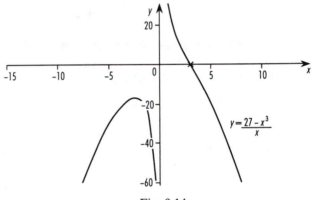

Fig. 9.14

We find the coordinates of this point of inflexion by substituting into our original equation.

Using $\quad y = \dfrac{27 - x^3}{x}$

we substitute for $x = 3$ to get $y = 0$.

As an exercise for yourself you can check that the gradient around $x = 3$ is negative. This is consistent with our curve in Fig. 9.14.

■■■ EXERCISE 9.4 ■■■

Have the following curves any point(s) of inflexion? If so find the coordinates of the points.

1 $y = x^3 + 4$

2 $y = 3 - 2x^4$

3 $y = \dfrac{8}{x} + x^2$

4 $y = (1.5 - x)^3$

5 $y = \dfrac{64 + 3x - x^3}{x}$

Practical applications of maxima and minima

Example 9.11

We have an open cistern with a square base of side x and of height, h. The area of material used is 12 m^2. What dimensions make the volume of this cistern a maximum? Calculate the maximum volume.

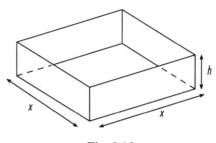

Fig. 9.15

The volume, V, is given by $V = x^2 h$. So far in our differentiation we have only used one independent and one dependent variable. We need to eliminate an independent variable, either x or h. We can do this by using our information about the area of material and substituting for h. The cistern has 4 sides, each of area xh and a square base of area x^2. This means we have

$$\text{Area} = 4xh + x^2 = 12$$

i.e.
$$h = \frac{12 - x^2}{4x}.$$

Then
$$V = x^2 \frac{(12 - x^2)}{4x}$$

i.e.
$$V = 3x - 0.25x^3.$$

We differentiate once to get

i.e.
$$\frac{dV}{dx} = 3 - 0.75x^2.$$

At a turning point $\frac{dV}{dx} = 0$ so that

$$3 - 0.75x^2 = 0$$

i.e.
$$3 = 0.75x^2$$

i.e.
$$x^2 = 4$$

$$\therefore \qquad x = \pm\sqrt{4} = \pm 2.$$

Our practical example involves distances so $x = -2$ has no practical use. We continue our method with just $x = 2$.

Using
$$\frac{dV}{dx} = 3 - 0.75x^2$$

we differentiate again to get

$$\frac{d^2V}{dx^2} = -1.5x.$$

Substituting for $x = 2$ we get

$$\frac{d^2 V}{dx^2} = -1.5(2) = -3, \quad \text{i.e. negative.}$$

This means we have a local maximum for a value of $x = 2$.

Using $\qquad h = \dfrac{12 - x^2}{4x}$

and substituting for $x = 2$ we get

$$h = \frac{12 - 2^2}{4(2)} = 1.$$

Dimensions of 2 m for the base and 1 m for the height give us a maximum volume for the cistern.

Also using $V = x^2 h$ and substituting for $x = 2$ and $h = 1$ we get $V_{\text{max}} = 2^2 \times 1 = 4\,\text{m}^3$.

▰▰▰▰▰▰ **Example 9.12** ▰▰▰▰▰▰

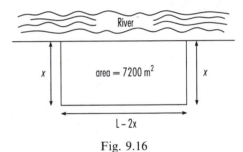

Fig. 9.16

We have a site of area 7200 m² to be developed. It is bounded on one side by a shallow river. Find the minimum length of fencing needed to enclose this rectangular area.

We label one of the sides x and the complete length of fencing L. This means that the other side is $L - 2x$. Linking together our area information for a rectangle we have

$$\text{Area} = (L - 2x)x = 7200$$

i.e. $\qquad L = \dfrac{7200}{x} + 2x \quad \text{or} \quad 7200x^{-1} + 2x.$

We differentiate once to get

i.e. $\qquad \dfrac{dL}{dx} = -7200x^{-2} + 2.$

At a turning point $\dfrac{dL}{dx} = 0$ so that

$$-7200x^{-2} + 2 = 0$$

i.e. $\qquad 2 = \dfrac{7200}{x^2}$

i.e. $\qquad x^2 = 3600$

∴ $\qquad x = \pm\sqrt{3600} = \pm 60.$

Our practical example involves distances so $x = -60$ has no practical use. We continue our method with just $x = 60$.

Using $\qquad \dfrac{dL}{dx} = -7200x^{-2} + 2$

we differentiate again to get

$$\dfrac{d^2L}{dx^2} = 14\,400x^{-3}.$$

Substituting for $x = 60$ we get

$$\dfrac{d^2L}{dx^2} = \dfrac{14\,400}{(60)^3} = 0.0\bar{6}, \quad \text{i.e. positive.}$$

This means we have a local minimum for a value of $x = 60$.

Using $\qquad L = \dfrac{7200}{x} + 2x$

and substituting for $x = 60$ we get

$$L_{\min} = \dfrac{7200}{60} + 2(60) = 240,$$

i.e. we need a minimum of 240 m of fencing.

■ EXERCISE 9.5 ■

1 A rigid beam of length 10 m carries a uniformly distributed load of 10 000 N. The bending moment, M, is related to its distance from one end of the beam, x, according to $M = -10000 + 5000x - 500x^2$. Using differentiation show that the maximum bending moment occurs at the mid-point. What is the value of this maximum bending moment?

2 Sketch the graph of $y = \sin \theta$ in the range $0 \leqslant \theta \leqslant 2\pi$. By differentiating find the coordinates of the local maximum and minimum points. Confirm these with your sketch. Sketch the graph of $y = \cos \theta$ over the same range. Similarly find the turning points for this curve, distinguishing between them. Finally calculate the turning points of the curve $y = \sin \theta - \cos \theta$, again over the same range.

3 A disc is spun from rest. It spins through $y°$ in t seconds according to $y = 100t - 5t^3$. Has this relationship any turning points? If there are any, after what time do they occur and what type are they?

4 One particular machine in an engineering workshop costs £C to lease each week according to the formula $C = 200t - \dfrac{t^3}{30}$. t is the number of hours/week worked by the machine. What is the maximum weekly lease cost?

5   Find the minimum value of y for $y = \frac{1}{2}(e^x + e^{-x})$. Write down the coordinates where this occurs. Attempt to do the same for the curve $y = \frac{1}{2}(e^x - e^{-x})$. This should fail somewhere in the method. Explain why it does fail. Find the point of inflexion for this second curve, $y = \frac{1}{2}(e^x - e^{-x})$.

6   A chemical process plant can produce up to 5 tonnes each week. Its costs, y_1, are related to the tonnage produced, x, by $y_1 = 10 + 3x - x^2$. Its sales income, y_2, is also related to tonnage by $y_2 = 0.2x^3$. Write down a relationship for profit based on tonnage produced. What amount must be produced to make a minimum profit?

7   The diagram shows a submarine signalling cable. The radius of the core is r and the radius of the covering is R. The ratio of these radii is given by $x = \dfrac{r}{R}$. The speed of the signal is v where $v = -x^2 \ln x$. For $x > 0$ find the maximum speed of the signal and the value of x at which this occurs.

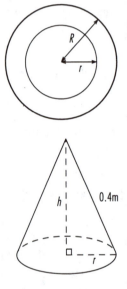

8   The diagram shows a cone of height h, radius r and slant height 0.4 m. Using Pythagoras' theorem write down a relationship linking these dimensions. Make h the subject of this relationship. The volume of a cone, V, is given by $V = \dfrac{1}{3}\pi r^2 h$. Substitute for h and use differentiation to find the maximum and minimum volumes.

9   A rectangular sheet of metal has sides of 2000 mm and 3000 mm. From each corner are cut squares of side x mm as shown in the diagram. The metal is now folded as shown.

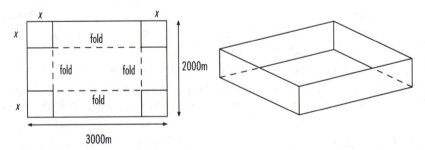

Write down the dimensions of the open rectangular box in terms of x. Similarly write down the volume of this box. Using differentiation find the dimensions that maximise the volume and calculate this maximum volume.

10   A large closed cylindrical metal drum of radius r and height h has a surface area of $24\pi\,\text{m}^2$. Obtain an expression for h in terms of r. If its volume, V, is given by $V = \pi r^2 h$ find the maximum volume and the dimensions of such a drum.

10 Integration

The objectives of this chapter are to:

1 Revise indefinite integration as the reverse process to differentiation.

2 Determine indefinite integrals of functions involving $\cos ax$, $\sin ax$, $\sec^2 ax$, e^{ax} and $\frac{a}{x}$.

3 Evaluate definite integrals of functions involving $\cos ax$, $\sin ax$, $\sec^2 ax$, e^{ax} and $\frac{a}{x}$.

4 Revise $\int_a^b y\, dx$ as the area under the curve between the ordinates $x = a$ and $x = b$.

5 Define and evaluate the mean value of a function over a given range.

6 Define and evaluate the root mean square value of a function over a given range.

Introduction

In Volume 1, Chapter 19, we introduced the idea of integration using some simple functions. Before going on with this chapter you should revise some of those basic ideas from Volume 1. Remember, **integration** is the reverse process of differentiation. You should recall we have **indefinite integration** where we include c, the constant of integration. Also, we have **definite integration** where we avoid c by using **limits of integration**. We have a **lower limit** and an **upper limit** to substitute once we have integrated.

■■■■■ ASSIGNMENT ■■■■■

The Assignment for this chapter involves a sinusoidal current, i, with an amplitude (peak value) of 4 A and a frequency of 50 Hz. From our work on trigonometry we know that

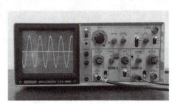

$$\text{frequency} = \frac{1}{\text{time period}} = \frac{\omega}{2\pi}$$

i.e.

$$50 = \frac{\omega}{2\pi}$$

$$\omega = 100\pi.$$

Also, a general sine wave, $y = A \sin \omega x$, has an amplitude of A. Our current, i, varies according to time, t, so our wave is $i = 4 \sin 100\pi t$.

> Substituting i for y,
> t for x,
> 4 for A
> and 100π for ω.

Later we will calculate the mean and root mean square values of the current.

Integration by rule

When we differentiated we saw a pattern of movement of numbers. For integration we have the reverse pattern of movement.

Generally if
$$\frac{dy}{dx} = x^n$$

then
$$y = \frac{x^{n+1}}{n+1} + c \qquad n \neq -1$$

where n is a numerical value.

The important exception to this rule is that $n \neq -1$. We cannot allow the possibility of -1 because this would create a divisor of 0. Remember, we cannot divide by 0 in mathematics.

We say this integration rule as 'increase the power by 1, and divide by the new power'.

Remember, integration is the reverse process of differentiation. You can check for yourself the differentiation of $y = \frac{x^{n+1}}{n+1} + c$ gives the correct answer for $\frac{dy}{dx}$ above.

There is an alternative notation,

$$\int x^n \, dx = \frac{x^{n+1}}{n+1} + c \qquad n \neq -1$$

$\int$ **is the symbol for integration.**

dx means we are going to integrate with respect to x, i.e. x is the variable being processed.

The term(s) between $\int$ and dx is the section to be integrated. At present we have no values to substitute to find c. Where we include a constant of integration we are performing **indefinite integration**.

▰▰▰ Examples 10.1 ▰▰▰

With respect to x integrate   i) $3x^5 + 2x^3 - \dfrac{4}{x^2}$, and   ii) $\dfrac{2x^4 - 3x^3}{x^2} + 6$.

i)   We need to rearrange $\dfrac{4}{x^2}$ before we integrate our expression term by

term. $\dfrac{4}{x^2}$ needs to be rewritten as $4x^{-2}$.

Then $\displaystyle\int (3x^5 + 2x^3 - 4x^{-2})\,dx = \dfrac{3x^{5+1}}{5+1} + \dfrac{2x^{3+1}}{3+1} - \dfrac{4x^{-2+1}}{-2+1} + c$

$$= \dfrac{3x^6}{6} + \dfrac{2x^4}{4} - \dfrac{4x^{-1}}{-1} + c$$

$$= \dfrac{x^6}{2} + \dfrac{x^4}{2} + 4x^{-1} + c.$$

ii)   In this case we need to simplify our expression before attempting to integrate, again term by term. Both $2x^4$ and $3x^3$ may be separated over the common denominator x^2, i.e.

$$\dfrac{2x^4 - 3x^3}{x^2} + 6 = \dfrac{2x^4}{x^2} - \dfrac{3x^3}{x^2} + 6$$

$$= 2x^2 - 3x + 6.$$

Then $\displaystyle\int \left(\dfrac{2x^4 - 3x^3}{x^2} + 6\right) dx = \int (2x^2 - 3x + 6x^0)\,dx$ $\boxed{x^0 = 1.}$

$$= \dfrac{2x^{2+1}}{2+1} - \dfrac{3x^{1+1}}{1+1} + \dfrac{6x^{0+1}}{0+1} + c$$

$$= \dfrac{2x^3}{3} - \dfrac{3x^2}{2} + 6x + c.$$

So far we have concentrated on whole number (integer) powers. We follow exactly the same rule with fractional and decimal powers: i.e. 'increase the power by 1, and divide by the new power'.

▰▰▰ Example 10.2 ▰▰▰

With respect to x, integrate $\dfrac{3\sqrt{x} + 6x^{-1/2}}{4x^2}$.

Before we attempt the integration we need to simplify the algebra. Both $3\sqrt{x}$ and $6x^{-1/2}$ may separated over the common denominator of $4x^2$,

i.e. $\dfrac{3\sqrt{x} + 6x^{-1/2}}{4x^2} = \dfrac{3\sqrt{x}}{4x^2} + \dfrac{6x^{-1/2}}{4x^2}$

$$= \dfrac{3}{4}x^{1/2-2} + \dfrac{6}{4}x^{-1/2-2}$$

$$= 0.75x^{-1.5} + 1.5x^{-2.5}.$$

Notice that we have changed to decimals, which you will probably find easier than fractions. With the algebra simplified now we are able to attempt the integration,

i.e. $\int \left(\dfrac{3\sqrt{x} + 6x^{-1/2}}{4x^2} \right) dx = \int (0.75x^{-1.5} + 1.5x^{-2.5}) \, dx$

$$= \dfrac{0.75x^{-1.5+1}}{-1.5+1} + \dfrac{1.5x^{-2.5+1}}{-2.5+1} + c$$

$$= \dfrac{0.75x^{-0.5}}{-0.5} + \dfrac{1.5x^{-1.5}}{-1.5} + c$$

$$= -1.5x^{-0.5} - x^{-1.5} + c.$$

■ EXERCISE 10.1 ■

Integrate the following expressions.

1 $\int (2x^7 - 4x^6) \, dx$

2 $\int (5t^{-4} + 3t^{-2}) \, dt$

3 $\int \left(\dfrac{5}{x^4} - \dfrac{3}{x^2} \right) dx$

4 $\int \left(\dfrac{3}{2x^4} - 5x^4 \right) dx$

5 $\int \left(\dfrac{3x^4 + 4x^2}{x} \right) dx$

6 $\int \left(\dfrac{2x^2 - x}{\sqrt{x}} \right) dx$

7 $\int 4\sqrt{x} \, dx$

8 $\int 6t^{2/3} \, dt$

9 $\int (\sqrt[4]{x^3} - 2) \, dx$

10 $\int (4t^{0.4} + 2t^{0.2}) \, dt$

11 $\int (6x^{0.5} + 2x^{-0.5}) \, dx$

12 $\int \left(\dfrac{2}{5}x^{-1.5} + \dfrac{1}{5}x^{-0.5} \right) dx$

13 $\int \left(6\sqrt{t} - \dfrac{8}{\sqrt{t}} \right) dt$

14 $\int \left(4\sqrt{x} - \dfrac{1}{6\sqrt{x}} \right) dx$

15 $\int \left(\dfrac{\sqrt{x} + 4x^2}{2x} \right) dx$

Definite integration

So far we have revised our ideas about indefinite integration, involving c but not having a value for it. In contrast, for **definite integration** we get a value for the integral rather than a general expression. This value comes from the numerical substitution of **limits**.

The general rule is

$$\int_a^b x^n \, dx = \left[\dfrac{x^{n+1}}{n+1} \right]_a^b.$$

a is the **lower limit** of x and b is the **upper limit** of x. We expect $a < b$.

We integrate to get some function of x, say $F(x)$. Next we substitute for $x = a$ and $x = b$, and subtract, according to

$$[F(x)]_a^b = F(b) - F(a).$$

In the following examples we show how to substitute for the values in the a and b positions.

▆▆▆▆▆▆ Example 10.3 ▆▆▆▆▆▆

Find the value of $\displaystyle\int_1^3 \left(\frac{3\sqrt{x} + 6x^{-1/2}}{4x^2}\right) dx$.

We know from Example 10.2 that

$$\int \left(3\sqrt{x} + 6x^{-1/2}\right) dx = -1.5x^{-0.5} - x^{-1.5} + c.$$

Remember from Volume 1, Chapter 19, how the cs cancel out during definite integration. Because this always happens we do *not* need to include them.

Thus $\displaystyle\int_1^3 \left(\frac{3\sqrt{x} + 6x^{-1/2}}{4x^2}\right) dx = \left[-1.5x^{-0.5} - x^{-1.5}\right]_1^3$

We substitute for $x = 3$ and for $x = 1$ and subtract to get

$$[-1.5(3)^{-0.5} - (3)^{-1.5}]$$
$$-[-1.5(1)^{-0.5} - (1)^{-1.5}]$$
$$= [-0.866 - 0.192] - [-1.5 - 1]$$
$$= -1.058 + 2.5$$
$$= 1.44.$$

▆▆▆▆▆ EXERCISE 10.2 ▆▆▆▆▆

Find the values of the following definite integrals.

1 $\displaystyle\int_0^1 (12x^5 - 6x^3)\, dx$

2 $\displaystyle\int_{0.5}^{1.5} (3t^{-4} + 2t^{-2})\, dt$

3 $\displaystyle\int_1^2 \left(\frac{15}{x^4} - \frac{3}{x^2}\right) dx$

4 $\displaystyle\int_1^3 \left(\frac{9}{2x^4} - 25x^4\right) dx$

5 $\displaystyle\int_{-1}^1 \left(\frac{30x^4 + 24x^2}{x^2}\right) dx$

6 $\displaystyle\int_4^9 \left(\frac{2x^2 - \sqrt{x}}{3\sqrt{x}}\right) dx$

7 $\displaystyle\int_{0.5}^{1.0} 14\sqrt{x}\, dx$

8 $\displaystyle\int_2^3 16t^{1/3}\, dt$

9 $\displaystyle\int_{0.2}^{0.4} (\sqrt{x^3} + 5)\, dx$

10 $\displaystyle\int_{0.25}^{0.50} (4t^{0.8} + 2t^{0.4})\, dt$

11 $\displaystyle\int_{1.0}^{1.5} (6x^{0.5} + 12x^{-0.5})\, dx$

14 $\displaystyle\int_{5}^{8} \left(6\sqrt{x} - \frac{1}{3\sqrt{x}}\right) dx$

12 $\displaystyle\int_{0.6}^{1.0} \left(\frac{2}{3}x^{1.5} + \frac{1}{6}x^{-0.5}\right) dx$

15 $\displaystyle\int_{1.5}^{2.5} \left(\frac{\sqrt{x} + 3x^2}{4x}\right) dx$

13 $\displaystyle\int_{2}^{4} \left(4\sqrt{t} - \frac{8}{\sqrt{t}}\right) dt$

Integration of trigonometric functions

We looked at some simple examples in Volume 1, Chapter 19. Now we extend them, again remembering that integration is the reverse process of differentiation.

$$\int \cos ax\, dx = \frac{1}{a}\sin ax + c,$$

and $\displaystyle\int \sin ax\, dx = -\frac{1}{a}\cos ax + c$

$$\int \sec^2 ax\, dx = \frac{1}{a}\tan ax + c$$

provided x is in radians and a is a constant.

You need to be careful with the $+/-$ signs. Do *not* confuse differentiation and integration.

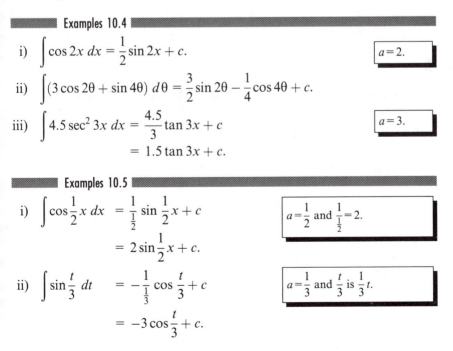

▬▬▬▬ **Examples 10.4** ▬▬▬▬

i) $\displaystyle\int \cos 2x\, dx = \frac{1}{2}\sin 2x + c.$ $a = 2.$

ii) $\displaystyle\int (3\cos 2\theta + \sin 4\theta)\, d\theta = \frac{3}{2}\sin 2\theta - \frac{1}{4}\cos 4\theta + c.$

iii) $\displaystyle\int 4.5\sec^2 3x\, dx = \frac{4.5}{3}\tan 3x + c$ $a = 3.$

$\qquad\qquad\qquad = 1.5\tan 3x + c.$

▬▬▬▬ **Examples 10.5** ▬▬▬▬

i) $\displaystyle\int \cos\frac{1}{2}x\, dx = \frac{1}{\frac{1}{2}}\sin\frac{1}{2}x + c$ $a = \frac{1}{2}$ and $\frac{1}{\frac{1}{2}} = 2.$

$\qquad\qquad\quad = 2\sin\frac{1}{2}x + c.$

ii) $\displaystyle\int \sin\frac{t}{3}\, dt = -\frac{1}{\frac{1}{3}}\cos\frac{t}{3} + c$ $a = \frac{1}{3}$ and $\frac{t}{3}$ is $\frac{1}{3}t.$

$\qquad\qquad\quad = -3\cos\frac{t}{3} + c.$

iii) $\int \dfrac{\sin 4x}{3}\, dx \;=\; \int \dfrac{1}{3}\sin 4x\, dx$

> $a = 4$ and the whole function is divided by 3.

$$= -\dfrac{1}{3} \times \dfrac{1}{4}\cos 4x + c$$

$$= -\dfrac{1}{12}\cos 4x + c.$$

▰▰▰ EXERCISE 10.3 ▰▰▰

Write down and simplify where necessary the following indefinite integrals.

1 $\displaystyle\int \sin 2x\, dx$

2 $\displaystyle\int (4\sin 2x - \cos 2x)\, dx$

3 $\displaystyle\int 3\sec^2 6x\, dx$

4 $\displaystyle\int \sin\dfrac{1}{2}\theta\, d\theta$

5 $\displaystyle\int \left(2\cos\dfrac{t}{2} + \sin\dfrac{t}{4}\right) dt$

6 $\displaystyle\int \left(0.5\sin 2x + \cos\dfrac{x}{4}\right) dx$

7 $\displaystyle\int \left(3 + 4\cos\dfrac{\theta}{2}\right) d\theta$

8 $\displaystyle\int \left(6\sec^2 3x + 2\sin\dfrac{1}{2}x\right) dx$

9 $\displaystyle\int (5 - 10\sec^2 2x)\, dx$

10 $\displaystyle\int \left(\dfrac{\sin 3x}{2} + 3\sin\dfrac{1}{2}x - \cos 2x\right) dx$

Now we combine the integration of trigonometric functions with definite integration. Remember the limits are in radians because the integrals are defined in radians.

▰▰▰ Example 10.6 ▰▰▰

Find the value of $\displaystyle\int_{3\pi/4}^{7\pi/8} \left(4 + \sin\dfrac{2\theta}{3}\right) d\theta$.

Using the standard rules we integrate and then substitute for our limits so that

$$\int_{3\pi/4}^{7\pi/8} \left(4 + \sin\dfrac{2\theta}{3}\right) d\theta = \left[4\theta - \dfrac{1}{\frac{2}{3}}\cos\dfrac{2\theta}{3}\right]_{3\pi/4}^{7\pi/8}$$

> $a = \dfrac{2}{3}$.

$$= \left[4\theta - \dfrac{3}{2}\cos\dfrac{2\theta}{3}\right]_{3\pi/4}^{7\pi/8}$$

> $\dfrac{1}{\frac{2}{3}} = \dfrac{3}{2}$.

$$= \left[4\left(\frac{7\pi}{8}\right) - \frac{3}{2}\cos\frac{2}{3}\left(\frac{7\pi}{8}\right) \right]$$

$$- \left[4\left(\frac{3\pi}{8}\right) - \frac{3}{2}\cos\frac{2}{3}\left(\frac{3\pi}{4}\right) \right]$$

$$= \left[\frac{7\pi}{2} - \frac{3}{2}\cos\frac{7\pi}{12} \right] - \left[3\pi - \frac{3}{2}\cos\frac{\pi}{2} \right].$$

At this stage we can work out each bracket. Alternatively we can gather together similar terms, perhaps removing any common factors to get

$$\frac{7\pi}{2} - 3\pi - \frac{3}{2}\cos\frac{7\pi}{12} - -\frac{3}{2}\cos\frac{\pi}{2}$$

$$= \left(\frac{7}{2} - 3\right)\pi - \frac{3}{2}\left(\cos\frac{7\pi}{12} - \cos\frac{\pi}{2}\right)$$

$$= 1.571 - 1.5(-0.259 - 0)$$

$$= 1.96.$$

■ EXERCISE 10.4 ■

Find the values of the following definite integrals.

1 $\displaystyle\int_{\pi/4}^{\pi/2} 5\cos 3x \, dx$

2 $\displaystyle\int_{\pi/6}^{\pi/3} (\cos 2x - 7\sin 2x) \, dx$

3 $\displaystyle\int_{\pi/12}^{\pi/8} 2\sec^2 2\theta \, d\theta$

4 $\displaystyle\int_{\pi}^{2\pi} 3\sin\frac{x}{4} \, dx$

5 $\displaystyle\int_{\pi/4}^{\pi/3} \left(2 - \sin\frac{3}{2}x\right) dx$

6 $\displaystyle\int_{\pi}^{3\pi/2} \left(5 - \cos\frac{x}{4}\right) dx$

7 $\displaystyle\int_{\pi/24}^{\pi/12} (10 - 5\sec^2 4x) \, dx$

8 $\displaystyle\int_{\pi/6}^{\pi/4} \left(\frac{2\cos t}{3} - \sin\frac{t}{2}\right) dt$

9 $\displaystyle\int_{\pi/3}^{\pi/2} \left(0.5\sin\frac{x}{2} - \cos\frac{x}{3}\right) dx$

10 $\displaystyle\int_{0}^{\pi/2} \left(\frac{2x^2 - 3x\sin 4x}{x}\right) dx$

Integration of exponential functions

Again we extend the ideas of Volume 1, Chapter 19. The integration is as simple as the differentiation, i.e.

$$\int e^{ax} \, dx = \frac{1}{a}e^{ax} + c.$$

▰▰▰▰▰▰ **Examples 10.7** ▰▰▰▰▰▰▰▰▰▰▰▰▰▰▰▰▰▰▰▰▰▰▰▰▰▰▰▰

i) $\int e^{3x} \, dx = \dfrac{1}{3} e^{3x} + c.$

$\boxed{a = 3.}$

ii) $\int 4e^{3x} \, dx = \dfrac{4}{3} e^{3x} + c.$

iii) $\int 14e^{x/2} \, dx = \dfrac{14}{\frac{1}{2}} e^{x/2} + c = 28e^{x/2} + c.$

$\boxed{a = \dfrac{1}{2}.}$

iv) $\int 6e^{1.5x} \, dx = \dfrac{6}{1.5} e^{1.5x} + c = 4e^{1.5x} + c.$

$\boxed{a = 1.5.}$

▰▰▰▰▰▰ **Example 10.8** ▰▰▰▰▰▰▰▰▰▰▰▰▰▰▰▰▰▰▰▰▰▰▰▰▰▰▰▰▰

Find the indefinite integral $\int e^{3x+2} \, dx$.

The power of this exponential contains the addition of $3x$ and 2. Using a law of indices we may rewrite this as $e^{3x} \times e^2$. Now e^2 is a number, a constant, with a calculator value of 7.389... We know we can remove a multiplying constant from the integral so that

$\int e^{3x+2} \, dx = \int e^{3x} \times e^2 \, dx$

$\qquad\qquad = e^2 \times \int e^{3x} \, dx$

$\qquad\qquad = e^2 \times \dfrac{1}{3} e^{3x} + c$

$\boxed{a = 3.}$

$\qquad\qquad = \dfrac{1}{3} e^{3x} \times e^2 + c$

$\qquad\qquad = \dfrac{1}{3} e^{3x+2} + c.$

We see that $+2$ in the power, because it is a pure number, has no effect on the overall integration.

Generally we can write $\int e^{ax+b} \, dx = \dfrac{1}{a} e^{ax+b} + c$ where a and b are constants.

▰▰▰▰▰▰ **Example 10.9** ▰▰▰▰▰▰▰▰▰▰▰▰▰▰▰▰▰▰▰▰▰▰▰▰▰▰▰▰▰

Find the value of $\displaystyle\int_0^{1.5} \left(2e^{3x} - \dfrac{6}{5e^x} \right) dx$.

We rearrange the second exponential to get

$\displaystyle\int_0^{1.5} \left(2e^{3x} - \dfrac{6}{5} e^{-x} \right) dx = \left[\dfrac{2}{3} e^{3x} - \dfrac{6}{5(-1)} e^{-x} \right]_0^{1.5}$

$\qquad\qquad\qquad\qquad\qquad = \left[\dfrac{2}{3} e^{3x} + \dfrac{6}{5} e^{-x} \right]_0^{1.5}$

$$= \left[\frac{2}{3}e^{3\times1.5} + \frac{6}{5}e^{-1.5}\right] - \left[\frac{2}{3}e^{3\times0} + \frac{6}{5}e^{-0}\right]$$

$$= \frac{2}{3}e^{4.5} + \frac{6}{5}e^{-1.5} - \frac{2}{3}e^{0} - \frac{6}{5}e^{0}$$

$$= 60.01 + 0.27 - 0.\overline{66} - 1.20$$

$$= 58.4.$$

EXERCISE 10.5

Write down and simplify where necessary the following indefinite integrals.

1 $\int(e^{2x} + e^{-2x})\,dx$

2 $\int(2e^x - 3e^{-x})\,dx$

3 $\int(e^{x/4} + 2e^{x/2})\,dx$

4 $\int\left(e^{3t} + \frac{4}{e^{-3t}}\right)dt$

5 $\int(2e^{2x+1} + 5e^{1-2x})\,dx$

Find the values of the following definite integrals.

6 $\int_{0}^{1}\left(\frac{e^{4x}}{8} - 4e^{2x}\right)dx$

7 $\int_{-1}^{1}(4e^{0.5x} - 2e^{-1.5x})\,dx$

8 $\int_{-1.5}^{0.5}\left(\frac{e^{2x}}{5} - 5e^{-2x}\right)dx$

9 $\int_{-0.5}^{1.5}\left(2e^t + \frac{3}{e^{2t}}\right)dt$

10 $\int_{1.4}^{2.6}3e^{2x-4}\,dx$

Integration of $\dfrac{a}{x}$

You will remember our general rule for algebraic functions had the exception of $n \neq -1$. Now we will deal with that case where $n = -1$. Remembering the connection between differentiation and integration we have

$$\int\frac{a}{x}\,dx = a\int\frac{1}{x}\,dx = a\ln x + c.$$

You should check back to the differentiation of natural logarithmic functions. This will confirm that integration is the reverse process of differentiation. Notice that we deal only with natural logarithms. Any common logarithms must be converted, changing their base from 10 to e.

Examples 10.10

i) $\int \dfrac{4}{3x} \, dx = \dfrac{4}{3} \ln x + c.$

ii) $\int \left(1 + \dfrac{5}{2x} + \dfrac{1}{x^2}\right) dx = \int \left(1 + \dfrac{5}{2x} + x^{-2}\right) dx$

$$= x + \dfrac{5}{2} \ln x + \dfrac{x^{-1}}{-1} + c$$

or $x + 2.5 \ln x - x^{-1} + c.$

Remember also we can write $x^{-1} = \dfrac{1}{x}.$

Example 10.11

Find the value of $\displaystyle\int_{1.8}^{3.2} \dfrac{3}{4x} \, dx.$

Remember that a logarithmic function is only defined for positive values of x. Both our limits in this example are positive. At this level of mathematics we respect our definition.

Now $\displaystyle\int_{1.8}^{3.2} \dfrac{3}{4x} \, dx = \left[\dfrac{3}{4} \ln x\right]_{1.8}^{3.2}$

> $\dfrac{3}{4}$ or 0.75 is a common factor.

$$= 0.75[\ln 3.2 - \ln 1.8]$$

$$= 0.75[1.163 - 0.588]$$

$$= 0.43.$$

EXERCISE 10.6

Find the values of the following definite integrals.

1 $\displaystyle\int_{2.6}^{3.9} \dfrac{6}{x} \, dx$

2 $\displaystyle\int_{0.50}^{1.75} \dfrac{1}{5x} \, dx$

3 $\displaystyle\int_{2.5}^{4.5} \dfrac{6}{5x} \, dx$

4 $\displaystyle\int_{0.25}^{0.6} \left(\dfrac{10}{x} - \dfrac{1}{2x}\right) dx$

5 $\displaystyle\int_{1}^{2} \left(4 - \dfrac{5}{2x} + \dfrac{3}{x^2}\right) dx$

Area under a curve

Integration has many applications. Finding the area under a curve is one simple case.

Fig. 10.1 shows a general curve. We wish to find the area under the curve bounded by the *x*-axis and the lines $x = a$ and $x = b$. To do this we split the area into a series of thin parallel strips, each of width δx.

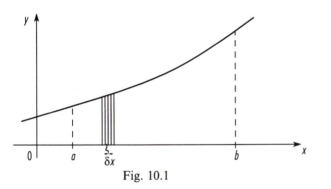

Fig. 10.1

Let us magnify one of those strips. Suppose A is the point (x, y). Remember, a small change in x, δx, causes a small change in y, δy. Because B is a small distance away we know it is the point $(x + \delta x, y + \delta y)$. We approximate this strip to a rectangle. As the strip width gets narrower and narrower (i.e. as $\delta x \to 0$) the accuracy of the approximation improves. This means the extra area above the rectangle becomes very small. Then

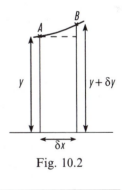

Fig. 10.2

Area of thin strip $\approx$ Area of rectangle

$\approx y\,\delta x$.

$\approx$ is 'equals approximately'.

The area between $x = a$ and $x = b$ is the sum of all such strips, i.e.

$$\text{Area} = \lim_{\delta x \to 0} \sum_{x=a}^{x=b} y\,\delta x$$

In practice we would find it difficult to calculate this sum for the area. We can reach the same result more easily using the integration formula

$$\text{Area} = \int_a^b y\,dx.$$

Integration is the limit of the summation process. We easily apply this formula in the next example.

Example 10.12

Find the area bounded by the curve $y = 1.5e^{2.5x} + 8\cos\dfrac{x}{3}$ and the x-axis between $x = \dfrac{\pi}{6}$ and $x = \dfrac{\pi}{2}$.

We have included a sketch for this curve in Figure 10.3.

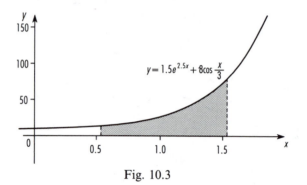

Fig. 10.3

Using our formula

$$\text{Area} = \int_a^b y\, dx$$

we substitute for $a = \dfrac{\pi}{6}$, $b = \dfrac{\pi}{2}$ and y to get

$$\text{Area} = \int_{\pi/6}^{\pi/2} \left(1.5e^{2.5x} + 8\cos\frac{x}{3}\right) dx$$

$$= \left[\frac{1.5}{2.5}e^{2.5x} + \frac{8}{\frac{1}{3}}\sin\frac{x}{3}\right]_{\pi/6}^{\pi/2}$$

$$= \left[0.6e^{2.5x} + 24\sin\frac{x}{3}\right]_{\pi/6}^{\pi/2}$$

$$= \left[0.6e^{2.5\pi/2} + 24\sin\frac{\pi}{2 \times 3}\right] - \left[0.6e^{2.5\pi/6} + 24\sin\frac{\pi}{6 \times 3}\right]$$

$$= 30.452 + 12 - 2.221 - 4.168$$

$$= 36.06 \text{ unit}^2.$$

Example 10.13

Find the area bounded by the curve $y = 10e^{-2x} - 5$, the horizontal axis and the lines $x = 0$ and $x = 1$.

Fig. 10.4 shows the curve partly above and partly below the horizontal axis. Remember from Volume 1, Chapter 19, that an area above the horizontal axis is positive and below is negative. This means we need to split our calculation.

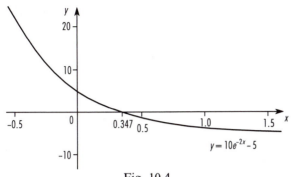

Fig. 10.4

Where the curve cuts the horizontal axis $y = 0$,

i.e. $\quad 10e^{-2x} - 5 = 0$

$\qquad 10e^{-2x} = 5$

$\qquad e^{-2x} = \dfrac{5}{10} = 0.5.$

We take natural logarithms of both sides to get

$\qquad \ln e^{-2x} = \ln 0.5$

i.e. $\qquad -2x \ln e = \ln 0.5$

$\qquad -2x = -0.693 \ldots$

i.e. $\qquad x = 0.347.$

$\boxed{\ln e = 1.}$

Now we use the integration formula for area twice.

The area above the horizontal axis is

$$\text{Area ①} = \int_0^{0.347} (10e^{-2x} - 5) \, dx$$

$$= \left[\frac{10}{-2} e^{-2x} - 5x \right]_0^{0.347}$$

$$= -5 \left[e^{-2x} + x \right]_0^{0.347}$$

$$= -5(0.5 + 0.347 - 1 - 0)$$

$$= 0.77 \text{ unit}^2.$$

The area below the horizontal axis is

$$\text{Area ②} = \int_{0.347}^{1.0} (10e^{-2x} - 5) \, dx$$

$$= -5 \left[e^{-2x} + x \right]_{0.347}^{1.0}$$

$$= -5(0.135 + 1.0 - 0.5 - 0.347)$$

$$= -1.44 \text{ unit}^2,$$

i.e. an area of 1.44 unit² below the horizontal axis.

To find the total area we use the modulus (size) of our answers, i.e.

$$\text{Total area} = |\text{Area} \circled{1}| + |\text{Area} \circled{2}|$$
$$= 0.77 + 1.44$$
$$= 2.21 \text{ unit}^2.$$

▆▆▆▆ EXERCISE 10.7 ▆▆▆▆

1   Find the area bounded by the curve $y = x^3 + e^{2x}$ and the x-axis from $x = 0$ to $x = 1.5$.

2   $y = \dfrac{3x^4 + 2}{x}$. Find the area lying between this curve and the x-axis from $x = 1$ to $x = 2$.

3   From $x = 0.5$ to $x = 1.5$ find the area between the x-axis and the curve $y = \dfrac{4}{x}$.

4   From $\theta = \dfrac{\pi}{4}$ to $\theta = \dfrac{3}{4}\pi$ find the area between the curve

   $y = 2\left(\cos\dfrac{\theta}{3} + \sin\dfrac{\theta}{2}\right)$ and the horizontal axis.

5   Over one cycle on the same pair of axes make a fully labelled sketch of $y = 5\sin 2x$ and $y = 5\cos 2x$. From $x = \dfrac{\pi}{8}$ to $x = \dfrac{\pi}{2}$ find the area between these two curves.

Mean value

In statistics we know how to calculate the **mean (average)** value of a number of items. We add together all our items and divide by that number of items. The mean value is some central measure within a range where there are larger and smaller values. That mean value evens out the variety of large and small values. We can apply the same idea to a continuous curve of y values. We replace the graph of these y values by a horizontal straight line, $\bar{y}$ (Figs. 10.5).

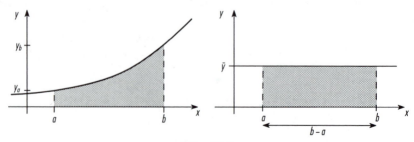

Figs. 10.5

Earlier we saw how integration, applied to an area, is associated with thin strips. The area under the curve is given by $\int_a^b y\,dx$. Our area is *not* a regular shape. Suppose we replace it with a rectangle of equal area over the range $x=a$ to $x=b$. This means the length of one side is $b-a$. Merging together Figs. 10.5 we see the length of the rectangle's other side is $\bar{y}$, lying between y_a and y_b. To achieve this we reposition some of the extra area above $\bar{y}$ to fill in the space below it (Fig. 10.6).

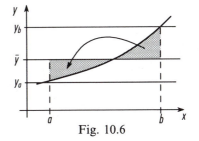

Fig. 10.6

$$\text{Area of rectangle} = \text{Original area}$$

i.e.
$$\bar{y}(b-a) = \int_a^b y\,dx$$

$$\bar{y} = \frac{1}{b-a}\int_a^b y\,dx$$

is the mean value of y between $x=a$ and $x=b$.

We do *not* distinguish between positive and negative values of y. Sometimes it is possible that some negative y values will cancel out some positive ones.

Example 10.14

Find the mean value of $y=10e^{-2x}-5$ from $x=-0.5$ to $x=1.5$.

In Example 10.13 we looked at the area under this curve, splitting it into positive and negative portions. This does *not* apply to a mean value. We simply use our formula

$$\bar{y} = \frac{1}{b-a}\int_a^b y\,dx$$

substituting for $a=-0.5$, $b=1.5$ and $y=10e^{-2x}-5$ to get

$$\bar{y} = \frac{1}{1.5--0.5}\int_{-0.5}^{1.5}(10e^{-2x}-5)\,dx$$

$$= \frac{1}{2}\left[\frac{10}{-2}e^{-2x}-5x\right]_{-0.5}^{1.5}$$

$$= -\frac{5}{2}\left[e^{-2x}+x\right]_{-0.5}^{1.5}$$

$$= -2.5(e^{-3}+1.5-e^1--0.5)$$

i.e.
$$\bar{y} = 1.67 \text{ is the mean value of } y.$$

ASSIGNMENT

We have our equation for current, i, given by $i = 4 \sin 100\pi t$, knowing that it has a frequency of 50 Hz. This tells us that it takes $\dfrac{1}{50}$ th (i.e. 0.02) of a second to complete one cycle.

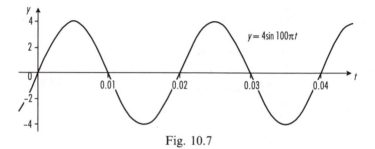

Fig. 10.7

Fig. 10.7 shows some of the cycles of our current. Let us find the mean value, $\bar{i}$, of our current over one cycle. We use our formula with

$$\bar{i} = \frac{1}{0.02 - 0} \int_0^{0.02} 4 \sin 100\pi t \, dt$$

$$= \frac{1}{0.02} \int_0^{0.02} 4 \sin 100\pi t \, dt$$

$$= \frac{1}{0.02} \left[\frac{-4}{100\pi} \cos 100\pi t \right]_0^{0.02}$$

$$= \frac{-4}{2\pi} (\cos 2\pi - \cos 0)$$

$$= -\frac{2}{\pi(1-1)}$$

$$= 0,$$

> $\bar{i}$ for $\bar{y}$,
> 0.02 for a,
> 0 for b and
> t for x.

is the mean value of this current over one cycle. In fact this is true over any cycle for this sinusoidal current. Looking at Fig. 10.7 we expect this result. The positive y values (i.e. above the horizontal axis) are cancelled out by the negative y values (i.e. below the y axis).

EXERCISE 10.8

1   Find the mean value, $\bar{y}$, of $y = 2.5x^4 + e^{-x}$ between $x = 0$ and $x = 1.5$.

2   $y = \dfrac{3x^3 + 2x}{4x^2}$. Find the mean value of y in the range $x = 1.5$ to $x = 3.5$.

3   From $x = 1.25$ to $x = 1.75$ find the mean value of the curve $y = \dfrac{4}{5x}$.

4 From $\theta = 0$ to $\theta = \dfrac{3}{4}\pi$ find the mean value of $y = 2\left(\cos\dfrac{3\theta}{2} + \sin\dfrac{2\theta}{3}\right)$.

5 Over one cycle on the same pair of axes make fully labelled sketches of $y_1 = 3\sin 2x$ and $y_2 = 3\cos 2x$. From $x = 0$ to $x = \dfrac{\pi}{4}$ find the mean value of $y_1 + y_2$.

Root mean square (RMS) value

When we find the mean value sometimes we meet a difficulty. All the negative values can cancel out all the positive values. This happens for sine (and cosine) curves over complete cycles as in the Assignment. We remove this difficulty by **squaring** all the values to make them positive. Then we find the **mean** of all these squares. Finally we compensate for the squares by finding the square **root**. The **root mean square (RMS)** value is the square **root** of the **mean** of the **squares**,

i.e. $$\mathbf{RMS} = \sqrt{\dfrac{1}{b-a}\int_a^b y^2\,dx}.$$

████████ **Example 10.15** ████████████████████████████

Find the RMS value of y for $y = x - x^2$ from $x = -1$ to $x = 1$.

We have drawn the graph of $y = x - x^2$ in Fig. 10.8, noting that some of the curve is below and some is above the horizontal axis.

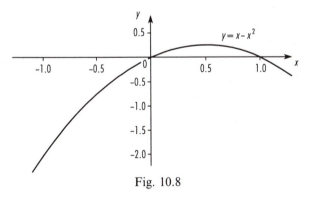

Fig. 10.8

Now using
$$y = x - x^2$$
$$y^2 = (x - x^2)^2 = (x - x^2)(x - x^2) \qquad \boxed{\text{Squaring both sides.}}$$
i.e. $$y^2 = x^2 - 2x^3 + x^4.$$

Substituting into our formula

$$\mathrm{RMS} = \sqrt{\dfrac{1}{b-a}\int_a^b y^2\,dx}$$

we get $\text{RMS} = \sqrt{\dfrac{1}{1--1}\displaystyle\int_{-1}^{1}(x^2-2x^3+x^4)\,dx}$

$= \sqrt{\dfrac{1}{2}\left[\dfrac{x^3}{3}-\dfrac{2x^4}{4}+\dfrac{x^5}{5}\right]_{-1}^{1}}$

$= \sqrt{\dfrac{1}{2}\left[\dfrac{1^3}{3}-\dfrac{1^4}{2}+\dfrac{1^5}{5}\right]-\left[\dfrac{(-1)^3}{3}-\dfrac{(-1)^4}{2}+\dfrac{(-1)^5}{5}\right]}$

$= \sqrt{\dfrac{1}{2}\left\{\dfrac{1}{30}-\left(\dfrac{-31}{30}\right)\right\}}$

$= \sqrt{0.5\overline{3}}$

$= 0.73,$

i.e. the root mean square value of y is 0.73.

■ ASSIGNMENT ■

Remember our current i given by $i=4\sin 100\pi t$ has a mean value of 0. For an alternating current we really need to find the RMS value over one cycle. This is the effective value of the current.

Now using $i = 4\sin 100\pi t$

$i^2 = (4\sin 100\pi t)^2 = 16\sin^2 100\pi t$ | Squaring both sides. |

$= 16\times\dfrac{1}{2}(1-\cos 200\pi t).$

| $\cos 2A = 1 - 2\sin^2 A$ |
| $2\sin^2 A = 1 - \cos 2A$ |
| $\sin^2 A = \frac{1}{2}(1-\cos 2A).$ |

Substituting into our formula

i.e. $\text{RMS} = \sqrt{\dfrac{1}{b-a}\displaystyle\int_{a}^{b}y^2\,dx}$

we get $\text{RMS} = \sqrt{\dfrac{1}{0.02-0}\displaystyle\int_{0}^{0.02}8(1-\cos 200\pi t)\,dt}$

$= \sqrt{50\times 8\displaystyle\int_{0}^{0.02}(1-\cos 200\pi t)\,dt}$

$= \sqrt{400\left[t-\dfrac{1}{200\pi}\sin 200\pi t\right]_{0}^{0.02}}$

$= \sqrt{400\left\{\left[0.02-\dfrac{1}{200\pi}\sin 4\pi\right]-\left[0-\dfrac{1}{200\pi}\sin 0\right]\right\}}$

$= \sqrt{400\times 0.02}$

$$= \sqrt{8}$$
$$= 2.828,$$

i.e. the effective current is 2.828 A.

■■■■■ EXERCISE 10.9 ■■■■■

1   Sketch the curve $y = 3\sin\theta$ over one cycle from $\theta = 0$. Find the RMS value of y from $\theta = 0$ to $\theta = 2\pi$ radians.

2   Given the equation $y = t^2 - 4$ find the RMS value of y from $t = -3$ to $t = 3$.

3   Over one cycle starting at $\theta = 0$ sketch the graph of $y = 4\cos 2\theta$. Find the RMS value of y between $\theta = \dfrac{\pi}{4}$ and $\theta = \dfrac{3\pi}{4}$ radians. HINT: apply $\cos 4\theta = 2\cos^2 2\theta - 1$ and rearrange to make $\cos^2 2\theta$ the subject.

4   In the range $t = 0$ to $t = \dfrac{\pi}{2}$ radians find the RMS value of y if $y = 3\sin 4t$.

5   From $\theta = 0$ to $\theta = \pi$ radians find the RMS value of $(1 + \cos\theta)$.

In our final set of questions we apply our mean and RMS value formulae.

■■■■■ EXERCISE 10.10 ■■■■■

1   $v = 30 - 4t$ relates speed, v (ms^{-1}), to time, t (s). Find the average (mean) speed in the first 10 seconds by integrating v with respect to t.

2   Hooke's Law relates the tension in an elastic spring, T (N), according to $T = \dfrac{\lambda x}{l}$. λ (N) is the coefficient of elasticity, l (m) is the natural length and x (m) is the extension from that natural length. If $\lambda = 30$ N and $l = 0.75$ m show that $T = 40x$. We could plot T vertically against x horizontally.
This means the average tension is given by

$$\text{Average tension} = \frac{1}{b - a}\int_a^b T\,dx.$$

Find the average tension as the spring is stretched from its natural length to 1.125 m.

3   In electrical theory we may plot voltage, V, vertically against current, I (A), horizontally. The mean voltage is given by $\dfrac{1}{b - a}\displaystyle\int_a^b V\,dI$.
For a resistor of $470\,\Omega$ we have $V = 470I$. Find the mean voltage as the current increases from 1×10^{-3} A to 1.08×10^{-3} A.

4 A variable force, F (N), is related to distance, s (m), by $F = 5s^2 + 2s$. The mean force given by

$$\text{Mean force} = \frac{1}{b-a} \int_a^b F \, ds.$$

Calculate the mean force as s increases from 3.50 m to 4.75 m.

5 The pressure, p (Nm^{-2}), and volume, V (m^3) of a gas are related by $pV^{1.5} = k$. If $p = 150 \times 10^3$ Nm^{-2} and $V = 0.5$ m^3, evaluate k.

Rearrange your general formula to make p the subject. The average pressure while expanding the gas from this volume to 0.75 m^3 is given by the integral formula

$$\text{Average pressure} = \frac{1}{0.75 - 0.50} \int_{0.50}^{0.75} p \, dV.$$

Calculate this average pressure.

6 A voltage, V, over time, t, is given by $V = 2 \sin \omega t$. Find the RMS value of this voltage over one cycle, $t = 0$ to $t = \dfrac{2\pi}{\omega}$.

7 Power, P, in an electrical circuit is related to current, I, and resistance, R, according to $P = I^2 R$. We have a resistance of 450 Ω and a current given by $I = 5 \sin \omega t$. Find the mean power over one cycle where

$$\text{mean power} = \frac{1}{\frac{2\pi}{\omega} - 0} \int_0^{2\pi/\omega} P \, dt.$$

8 A sinusoidal current has a peak value of 5 A and a frequency of 50 Hz. By integration find the mean and RMS values over a cycle.

9 A current, i, has a peak value of 50 mA and may be represented by a sine wave of frequency 50 Hz. What is the effective (i.e. RMS) current?

10 Power, P, in an electrical circuit is related to current, I, and resistance, R, according to $P = I^2 R$. We have a resistance of 4 Ω and a current given by $I = \sin \omega t$. Write down an equation for P in terms of $\sin^2 \omega t$ and hence convert it into terms involving $\cos 2\omega t$. To find the RMS value of the power, P, we need to use P^2. Hence write down P in terms involving $\cos^2 2\omega t$. Rewrite it to involve $\cos 2\omega t$ and $\cos 4\omega t$. Find by integration the RMS value of the power in the range $t = 0$ to $t = \dfrac{2\pi}{\omega}$.

Summary of integration rules

y	$\int y\,dx$	
x^n	$\dfrac{x^{n+1}}{n+1} + c$	where n is a number and $n \neq -1$.
$\cos ax$	$\dfrac{1}{a}\sin ax + c$	x in radians.
$\sin ax$	$\dfrac{-1}{a}\cos ax + c$	x in radians.
$\sec^2 ax$	$\dfrac{1}{a}\tan ax + c$	x in radians.
e^{ax}	$\dfrac{1}{a}e^{ax} + c$	
$\dfrac{a}{x}$	$a\ln x + c$	where a is a constant.

11 Numerical Integration

The objectives of this chapter are to:

1 Derive the trapezoidal rule and apply it to numerical integration.
2 Derive the mid-ordinate rule and apply it to numerical integration.
3 Derive Simpson's rule over two intervals and apply it to numerical integration.
4 Generalise Simpson's rule.
5 Apply Simpson's rule to numerical integration.

Introduction

In the previous chapter we looked at some early integration rules. We will look at some more rules after this chapter. They concentrate on integrating functions using exact rules. What happens if we have no rule? Now we know that finding the area under a graph is a simple application of integration. Also in Volume 1, Chapter 16, we looked at three methods for finding irregular shaped areas:

  i) the **mid-ordinate rule**;
 ii) the **trapezoidal rule**; and
iii) **Simpson's rule**.

In this chapter we link together these ideas. It means we can integrate a function between limits by finding the area under its graph. We call this method **numerical integration**. The numerical answer will be approximate, though sometimes differing by very little.

We are going to recall these three rules and apply them to numerical integration. After this chapter we will show you some more integration rules for accurate solutions.

■■■■■ ASSIGNMENT ■■■■■

The Assignment for this chapter looks at a horizontal string tied at both ends. It is plucked at the centre to oscillate about the original horizontal line. The velocity, $V\,\mathrm{ms}^{-1}$, is given in terms of time, t seconds, by $V = 0.1e^{-2t}\cos\dfrac{2t}{3}$.

In Fig. 11.1 we see the velocity plotted against the time. As the wave moves away from the centre both to the left and to the right it is damped. This is due to the exponential part of our velocity equation.

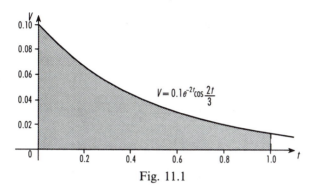

Fig. 11.1

Using each of our three rules we are going to find the distance travelled by the string in the first second. Remember that displacement is the area under a velocity–time graph. As we see in Fig. 11.1 the displacement in the first second is all above the horizontal axis. This tells us the displacement and distance are the same. As a definite integral we can write

$$\text{Distance travelled} = \int_0^1 0.1e^{-2t}\cos\frac{2t}{3}\,dt.$$

Finally we will extend our calculation to find the mean velocity in this first second.

The mid-ordinate rule

We recall from Volume 1 that the ordinate is the vertical value, e.g. in (4, 7) the ordinate is 7. When we apply the mid-ordinate rule usually we split the area into vertical (associated with ordinate) strips. All the strips must be of equal width. Let us look at an irregular area, Fig. 11.2. We split it into strips of equal width, w.

> The method works just as well for horizontal strips.

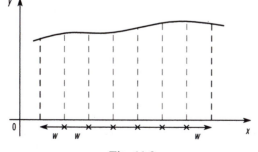

Fig. 11.2

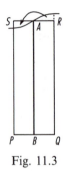

Fig. 11.3

Let us look more closely at one of these strips. Suppose we draw a vertical mid-placed line through AB and then a horizontal line through A. We suggest that the rectangle $PQRS$ is approximately equal in area to the original strip. A small section of the strip lies outside the rectangle. This will approximately fill in the gap where the rectangle is greater than the strip. We repeat this technique for all our strips. Once complete we will have approximated our irregular shape to a series of rectangles. All the rectangles will have the same width. Their heights will be based on the mid-strip heights (i.e. mid-ordinates).

Area = Sum of all rectangular areas

$\quad$ = $(w \times$ mid-ordinate$_1) + (w \times$ mid-ordinate$_2) +$

$\qquad\qquad (w \times$ mid-ordinate$_3) + (w \times$ mid-ordinate$_4) + \ldots$

$\quad$ = $w \times$ (mid-ordinate$_1$ + mid-ordinate$_2$ + mid-ordinate$_3 \ldots$.)

We write this as

Area = Strip width $\times$ Sum of mid-ordinates.

The mid-ordinate rule will work for any area, including standard ones. Generally, the more strips we use the more closely each strip resembles a rectangle. This leads to improved accuracy. In the next example we apply the rule to the area under a natural logarithmic curve.

▨▨▨ Example 11.1 ▨▨▨

We know that $\int \dfrac{a}{x}\, dx = a \ln x + c$ but we do not know how to integrate a natural logarithm.

Now we attempt to integrate such a natural logarithm numerically. We apply the mid-ordinate rule to find $\displaystyle\int_1^4 \ln x \, dx$, Fig. 11.4 reminding you of the graph of $y = \ln x$.

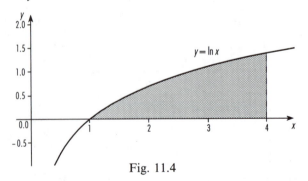

Fig. 11.4

From our lower limit of 1 to our upper limit of 4 we use strip widths of 0.5, i.e. 1.00 to 1.50, 1.50 to 2.00, . . . and 3.50 to 4.00. We need to find the values at the mid-points of these strips, i.e. at $x = 1.25$, 1.75, ... and 3.75, and construct a table.

x	1.25	1.75	2.25	2.75	3.25	3.75
$\ln x$	0.223	0.560	0.811	1.012	1.179	1.322

Using $\displaystyle\int_1^4 \ln x \, dx =$ Strip width $\times$ Sum of ordinates

we get $\displaystyle\int_1^4 \ln x \, dx = 0.50(0.223 + 0.560 + \ldots + 1.322)$

$$= 0.50 \times 5.11$$
$$= 2.55.$$

The calculator value using the memory for the mid-ordinate rule is 2.55285.... The accurate answer by integration is 2.54517..., showing only a small percentage error. Hence the accuracy of our answer is good and could be improved, if necessary, using smaller strip widths.

ASSIGNMENT

For our first look at $V = 0.1e^{-2t}\cos\dfrac{2t}{3}$ we integrate numerically using the mid-ordinate rule. We are going to split the 1 second into 5 equal widths of 0.20 second. They are 0 to 0.20, 0.20 to 0.40, ... and 0.80 to 1.00 with mid-points of 0.10, 0.30, ... and 0.90. Below we show our table of values without the working.

t	0.10	0.30	0.50	0.70	0.90
V	0.082	0.054	0.035	0.022	0.014

Then, using the mid-ordinate rule, we have

$$\int_0^1 0.1e^{-2t}\cos\frac{2t}{3}\,dt = 0.20(0.082 + \ldots + 0.014)$$

$$= 0.20 \times 0.207$$
$$= 0.041,$$

i.e. the distance travelled in the first second is 0.041 metres.

The trapezoidal rule

As the name suggests, this method uses a series of **trapezia** (remember **trapezia** is the plural of **trapezium**) as approximations. We split the area into strips (vertically or horizontally) of equal width as in Fig. 11.5. Each strip is approximated to a trapezium and its area calculated using the formula, Area $= \frac{1}{2}(a + b)h$, in the usual way. The complete area of the

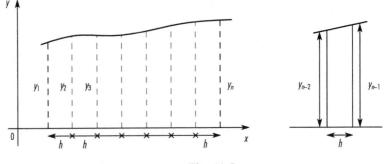

Fig. 11.5

irregular shape is the sum of all these trapezia. You should notice as we develop the formula that most of the y values appear twice. The exceptions are the first and last y values. All the others appear as the right parallel side of one trapezium and the left parallel side of the next trapezium. For example, in the formula you will see $\frac{1}{2}y_3 + \frac{1}{2}y_3$ simplify to y_3. Now we can write

Area = Sum of all trapezia

$$= \tfrac{1}{2}(y_1 + y_2)h + \tfrac{1}{2}(y_2 + y_3)h + \tfrac{1}{2}(y_3 + y_4)h + \ldots$$

$$= h\{\tfrac{1}{2}y_1 + \tfrac{1}{2}y_2 + \tfrac{1}{2}y_2 + \tfrac{1}{2}y_3 + \tfrac{1}{2}y_3 + \ldots + \tfrac{1}{2}y_n\}$$

$$= h\{\tfrac{1}{2}y_1 + \tfrac{1}{2}y_n + y_2 + y_3 + \ldots\}$$

$$= h\{\tfrac{1}{2}(y_1 + y_n) + y_2 + y_3 + \ldots\}.$$

We can think of this formula as

Area = Strip width × {$\frac{1}{2}$(Sum of first and last ordinates)

+ Sum of remaining ordinates}.

When you apply this formula make sure you use each value of y once only in the correct position.

▒▒▒▒ Example 11.2 ▒▒▒▒

Find the value of $\displaystyle\int_0^5 \sqrt{25 - x^2}\,dx$.

Again we could draw a graph of the function. First let us construct a table of values using a strip width of 1.00.

x	0	1.00	2.00	3.00	4.00	5.00
x^2	0	1.00	4.00	9.00	16.00	25.00
$25 - x^2$	25.00	24.00	21.00	16.00	9.00	0
$\sqrt{25 - x^2}$	5.00	4.90	4.58	4.00	3.00	0
	y_1	y_2	y_3	y_4	y_5	y_n

Using $\quad \displaystyle\int_0^5 \sqrt{25 - x^2}\, dx = h\{\tfrac{1}{2}(y_1 + y_n) + y_2 + y_3 + \dots\}$

we get $\quad \displaystyle\int_0^5 \sqrt{25 - x^2}\, dx = 1\{\tfrac{1}{2}(5.00 + 0) + 4.90 + 4.58 + 4.00 + 3.00\}$

$$= 18.98.$$

Let us look more closely at our original function, i.e.

$$y = \sqrt{25 - x^2}.$$

Squaring both sides we get

$$y^2 = 25 - x^2$$

i.e. $\quad x^2 + y^2 = 5^2.$

This is the equation of a circle of radius 5 with centre at the origin. We have just found the area of the shaded quadrant in Fig. 11.6.

$25 = 5^2.$

The accurate answer is 19.63495... (i.e. $\frac{1}{4} \times \pi 5^2$). We can improve the accuracy with smaller strip widths, e.g. 0.5 or 0.25 or ... For example, using a strip width of 0.50 turns our previous underestimate into a better estimate of 19.90.

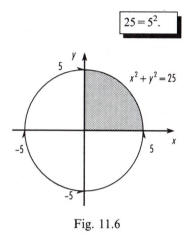

Fig. 11.6

<hr>

■ ASSIGNMENT

For our next look at $V = 0.1e^{-2t}\cos\dfrac{2t}{3}$ we integrate numerically using the trapezoidal rule. Again, we split the 1 second into 5 equal widths of 0.20 second. This gives us 6 ordinates (always one more than the number of strips.) Below we show our table of values, again without the working.

t	0	0.20	0.40	0.60	0.80	1.00
V	0.100	0.066	0.043	0.028	0.017	0.011

Then, using the trapezoidal rule, we have

$$\int_0^1 0.1e^{-2t}\cos\frac{2t}{3}\, dt = 0.20\{\tfrac{1}{2}(0.100 + 0.011)$$
$$+\, 0.066 + 0.043 + 0.028 + 0.017\}$$
$$= 0.20 \times 0.2095$$
$$= 0.042,$$

i.e. the distance travelled in the first second is 0.042 metres.

Notice how close this answer is to the value using the mid-ordinate rule.

Simpson's rule

Again, we split the area into a series of equal width strips. This time it is vital that we have an **even** number of strips, and hence an odd number of ordinates. To help you remember this you can think of an **even** number of wooden fencing panels (i.e. strips) between an odd number of concrete posts (i.e. ordinates).

Let us look at the derivation of Simpson's rule. When we used the mid-ordinate rule we approximated each strip to a rectangle. We used the mid-point value so that part of the area outside the rectangle approximately filled in a missing part inside. The line for the top of the rectangle could be labelled with $y = c$.

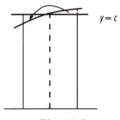

$y = c$

Fig. 11.7

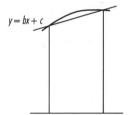

$y = bx + c$

When we used the trapezoidal rule we approximated each strip to a trapezium. The top line making this approximation could be labelled with $y = bx + c$.

Fig. 11.8

For Simpson's rule let us increase the power of x again, this time to x^2. For our approximation to each curve we suggest the quadratic $y = ax^2 + bx + c$. We have drawn this in Fig. 11.9 on either side of the vertical axis. This is simply to ease the algebra for you later. It is *not* necessary for each problem. Now we know how to integrate this quadratic expression, i.e.

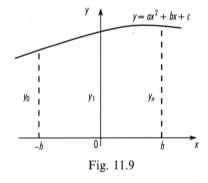

$y = ax^2 + bx + c$

y_0 y_1 y_n

$-h$ 0 h x

Fig. 11.9

$$\int_{-h}^{h} (ax^2 + bx + c)\, dx = \left[\frac{ax^3}{3} + \frac{bx^2}{2} + cx \right]_{-h}^{h}$$

$$= \frac{2}{3} ah^3 + 2ch$$

i.e.   Area $= \dfrac{h}{3}(2ah^2 + 6c)$.

> We could also have removed a factor of 2.

We can rearrange this expression. Let us find the values of y at each important point using $y = ax^2 + bx + c$.

$$\begin{aligned} \text{At } x=-h, \quad & y_0 = ah^2 - bh + c \\ \text{at } x=0, \quad & y_1 = c \\ \text{at } x=h, \quad & y_n = ah^2 + bh + c. \end{aligned}$$

Notice that we can remove the *bh* terms by addition, i.e.

$$y_0 + y_n = 2ah^2 + 2c$$

i.e. $y_0 + y_n - 2c = 2ah^2.$

We substitute for $2ah^2$ in our integrated expression so that

$$\begin{aligned} \text{Area} &= \frac{h}{3}(y_0 + y_n - 2c + 6c) \\ &= \frac{h}{3}(y_0 + y_n + 4c) \\ &= \frac{h}{3}(y_0 + y_n + 4y_1) \end{aligned}$$

Substituting for y_1.

Notice how Simpson's rule is based on two strips of equal width. We could extend the method by splitting each of the strips y_0/y_1 and y_1/y_n into two further strips. We could continue this idea, always with an even number of strips, eventually getting our general formula

$$\text{Area} = \frac{h}{3}\{y_1 + y_n + 4(y_2 + y_4 + y_6 + \ldots) + 2(y_3 + y_5 + \ldots)\}$$

We can write this as

$$\textbf{Area} = \frac{1}{3} \times \textbf{Strip width} \times \left\{ \begin{matrix}\textbf{sum of first and}\\ \textbf{last ordinates}\end{matrix} + 4\begin{pmatrix}\textbf{sum of even}\\ \textbf{ordinates}\end{pmatrix} + 2\begin{pmatrix}\textbf{sum of remaining}\\ \textbf{odd ordinates}\end{pmatrix} \right\}$$

In this formula the *even* of *even ordinates* and the *odd* of *odd ordinates* refer to the subscripts of *y*.

Example 11.3

We know how to separately integrate x^2 and $\cos x$, but *not* $x^2 \cos x$. Now we find the value of $\int_0^{0.9} x^2 \cos x \, dx$.

In Fig. 11.10 we show the graph of $y = x^2 \cos x$ and the area between our limits of 0 and 0.9 radians. The horizontal axis must be labelled in radians because of the cosine.

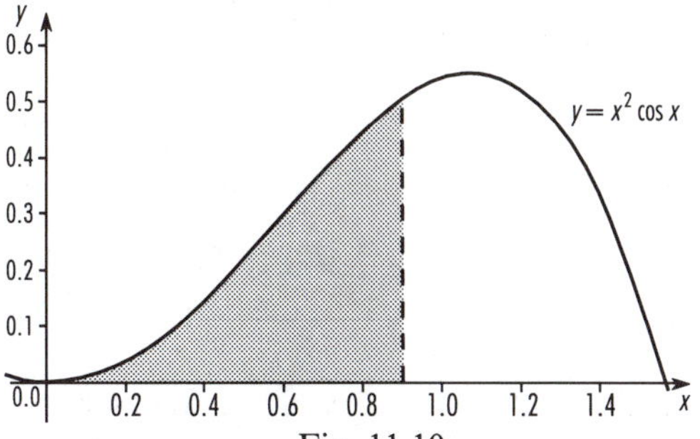

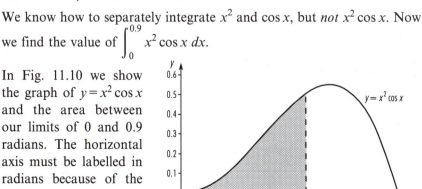

Fig. 11.10

In our table we use six strips of equal width $h = 0.15$.

x	0	0.15	0.30	0.45	0.60	0.75	0.90
x^2	0	0.0225	0.0900	0.2025	0.3600	0.5625	0.8100
$\cos x$	1	0.9888	0.9553	0.9004	0.8253	0.7317	0.6216...
$x^2 \cos x$	0	0.022	0.086	0.182	0.297	0.412	0.504
	y_1	y_2	y_3	y_4	y_5	y_6	y_n

Using Simpson's rule we have

$$\int_0^{0.9} x^2 \cos x \, dx = \frac{1}{3} \times 0.15\{(0 + 0.504) + 4(0.022 + 0.182 + 0.412)$$
$$+ 2(0.086 + 0.297)\}$$
$$= 0.05\{0.504 + 2.464 + 0.766\}$$
$$= 0.187.$$

■■■ ASSIGNMENT ■■■

For our third look at $V = 0.1e^{-2t} \cos \dfrac{2t}{3}$ we integrate numerically using Simpson's rule. We need to split the 1 second into an even number of equal widths. Two strips each of 0.50 second would be too approximate. Four strips each of 0.25 second would be better. For yet greater accuracy we choose 10 strips of width 0.10 second. Below we show our table of values again without the working.

t	0	0.10	0.20	0.30	0.40	0.50
V	0.100	0.082	0.066	0.054	0.043	0.035
	y_1	y_2	y_3	y_4	y_5	y_6
	0.60	0.70	0.80	0.90	1.00	
	0.028	0.022	0.017	0.014	0.011	
	y_7	y_8	y_9	y_{10}	y_n	

Then, using Simpson's rule, we have

$$\int_0^1 0.1e^{-2t} \cos \frac{2t}{3} \, dt = \frac{1}{3} \times 0.10\{(0.100 + 0.011)$$
$$+ 4(0.082 + 0.054 + 0.035 + 0.022 + 0.014)$$
$$+ 2(0.066 + 0.043 + 0.028 + 0.017)\}$$
$$= \frac{1}{3} \times 0.10\{0.111 + 0.828 + 0.308\}$$
$$= 0.042,$$

i.e. the distance travelled in the first second is 0.042 metres.

EXERCISE 11.1

You have seen three different methods for numerical integration,

i.e. i) the mid-ordinate rule;
ii) the trapezoidal rule; and
iii) Simpson's rule.

In this exercise you should practise with more than one method.

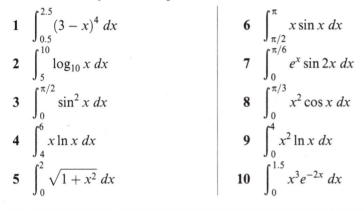

1 $\int_{0.5}^{2.5} (3 - x)^4 \, dx$

2 $\int_{5}^{10} \log_{10} x \, dx$

3 $\int_{0}^{\pi/2} \sin^2 x \, dx$

4 $\int_{4}^{6} x \ln x \, dx$

5 $\int_{0}^{2} \sqrt{1 + x^2} \, dx$

6 $\int_{\pi/2}^{\pi} x \sin x \, dx$

7 $\int_{0}^{\pi/6} e^x \sin 2x \, dx$

8 $\int_{0}^{\pi/3} x^2 \cos x \, dx$

9 $\int_{0}^{4} x^2 \ln x \, dx$

10 $\int_{0}^{1.5} x^3 e^{-2x} \, dx$

Mean value

In the previous chapter we deduced a formula for **mean value**, $\bar{y}$. That formula is

$$\bar{y} = \frac{1}{b - a} \int_{a}^{b} y \, dx$$

where $\bar{y}$ is the mean value betwen $x = a$ and $x = b$.

Because this formula involves integration we can reach a numerical solution using any of our three rules.

Example 11.4

Find the mean value of $y = \ln x$ between $x = 1$ and $x = 4$.

Substituting into our formula we have

$$\bar{y} = \frac{1}{4 - 1} \int_{1}^{4} \ln x \, dx$$

and using our result from Example 11.1 we get

$$\bar{y} = \frac{1}{4 - 1} \times 2.55 = \frac{1}{3} \times 2.55 = 0.85.$$

We do *not* always need to have a curve or an equation. In Examples 11.5 we find the mean values of two sawtooth waves.

![Examples 11.5]

Find the mean value of the following waves

i)

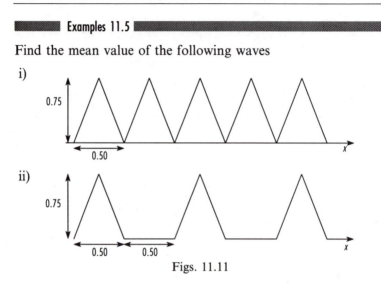

ii)

Figs. 11.11

Notice how both waves use the same isosceles triangle but in slightly different patterns. In each case we need to establish that pattern and the repetitive cycle.

i) Because all the triangles are consecutive each triangle is exactly one cycle. This means we consider just one triangle. Remember that we need a rectangle with a base common to the triangle. Both shapes have the same area, i.e.

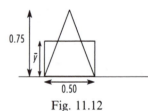

Fig. 11.12

Area of rectangle = Area of triangle

i.e. $\bar{y} \times 0.50 = \dfrac{1}{2} \times 0.50 \times 0.75$

i.e. $\bar{y} = 0.375,$

i.e. the mean value of this complete wave is 0.375.

ii) This time our cycle contains two sections. They are a triangle together with a zero section (i.e. a horizontal line along our axis). Obviously there is no area under our zero section, though we must still consider it as part of the cycle. Similar to our previous calculation we write

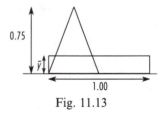

Fig. 11.13

Area of rectangle = Area of triangle + 0

i.e. $\bar{y} \times 1.00 = \dfrac{1}{2} \times 0.50 \times 0.75 + 0$

i.e. $\bar{y} = 0.1875,$

i.e. the mean value of this complete wave is 0.1875.

ASSIGNMENT

This is the final look at our velocity equation in terms of time. We are going to find the mean velocity during the first second of motion. Our formula is

$$\bar{y} = \frac{1}{b-a}\int_a^b y \, dx$$

where $\bar{y}$ is the mean value between $x=a$ and $x=b$.

Substituting for V in place of y and t in place of x we get

$$\text{Mean velocity, } \bar{V} = \frac{1}{1-0}\int_0^1 V \, dt$$

$$= \frac{1}{1-0}\int_0^1 0.1e^{-2t}\cos\frac{2t}{3} \, dt.$$

In our previous looks at our Assignment we have found the value of this integral. Using our last result (Simpson's rule) we get

$$\bar{V} = \frac{1}{1-0}\times\frac{1}{3}\times 0.10\{0.111 + 0.828 + 0.308\}$$

$$= 0.042,$$

i.e. the mean velocity in the first second is 0.042 ms^{-1}.

EXERCISE 11.2

Again in this exercise you should practise the different methods of numerical integration.

1   Find the mean value of $y = x\ln x$ between $x=4$ and $x = 6$.

2   Given $y = \sin^2 x$ between $x=0$ and $x=\frac{\pi}{2}$ radians find the mean value of y. Sketch the curve of $y = \sin^2 x$ and hence work out the mean value over one cycle.

3   Find the mean value of $y = \sqrt{1 + x^2}$ between $x=0$ and $x = 2$.

4   For $y = 3\cos^2 x$ between $x=0$ and $x=\frac{\pi}{2}$ radians find the mean value of y.

5   The diagram shows a fully rectified sine wave. Find the mean value of that wave between $x=0$ and $x=2\pi$ radians.

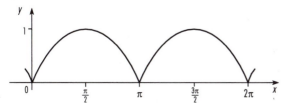

6   For each of the following waveforms find the mean value of y.

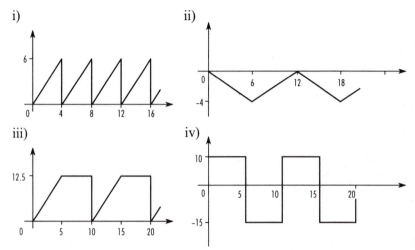

i)

ii)

iii)

iv)

7   The diagram shows a half-rectified sine wave. Find the mean value of that wave between $x = 0$ and $x = 2\pi$ radians.

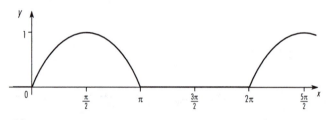

8 $v = \dfrac{10 - t}{30 + t}$ relates speed, v (ms^{-1}), to time, t (s). Find the average (mean) speed in the first 10 seconds by numerically integrating v with repect to t.

9   A variable force, F (N), is related to distance, s (m), by $F = 5s^2 e^{-s}$. The mean force is given by

$$\text{Mean force} = \frac{1}{b - a} \int_a^b F \, ds.$$

Calculate the mean force from $s = 0$ to $s = 3$ m using strip widths of 0.50 m.

10   Power, P, in an electrical circuit is related to current, I, and resistance, R, according to $P = I^2 R$. We have a resistance of $250\,\Omega$ and a current given by $I = \sin \omega t$. Find the mean power over one cycle where

$$\text{mean power} = \frac{1}{\frac{2\pi}{\omega} - 0} \int_0^{2\pi/\omega} P \, dt.$$

Root mean square (RMS) value

Exactly the same formula applies as before,

$$\text{RMS} = \sqrt{\frac{1}{b-a}\int_a^b y^2 \, dx}.$$

Example 11.6

Find the RMS value of y for $y = x(1 - x)$ from $x = -1$ to $x = 1$.

We are going to use Simpson's rule with eight strips, each of width 0.25. Our table of values without the working is

x	−1.00	−0.75	−0.50	−0.25	0	0.25	0.50	0.75	1.00
y^2	4	1.723	0.563	0.098	0	0.035	0.063	0.035	0
	y_1	y_2	y_3	y_4	y_5	y_6	y_7	y_8	y_n

Using our formula

$$\text{RMS} = \sqrt{\frac{1}{b-a}\int_a^b y^2 \, dx}$$

we get $\text{RMS} = \sqrt{\left[\frac{1}{(1--1)} \times \frac{1}{3} \times 0.25\{(4+0) + 4(1.723 + 0.098\right.}$

$$\left. + 0.035 + 0.035) + 2(0.563 + 0 + 0.063)\}\right]}$$

$$= \sqrt{\frac{0.25}{2 \times 3}\{4 + 7.564 + 1.252\}}$$

$$= \sqrt{0.534}$$

$$= 0.73$$

i.e. the root mean square value of y is 0.73.

EXERCISE 11.3

1 Find the RMS value of y if $y = \frac{\sin t}{2}$ over the range $t = 2$ to $t = 4$ radians.

2 A voltage, V, over time, t, is given by $V = 2t \sin t$. Find the RMS value of this voltage from $t = 0$ to $t = 2\pi$ at intervals of $\frac{\pi}{4}$.

3 From $\theta = 0$ to $\theta = 2$ radians find the RMS value of $y = \theta + \sin \theta$.

4 A sinusoidal current has a peak value of 2.5 A and a frequency of 50 Hz. Using the trapezoidal rule with eight strips find the mean and RMS values over a cycle.

5   A current, i, has a peak value of 75 mA and may be represented by a sine wave of frequency 50 Hz. Find the effective (i.e. RMS) current using numerical integration with eight strips.

12 Partial Fractions and Integration

The objectives of this chapter are to:

1 Express compound algebraic fractions

i) $\dfrac{f(x)}{(x-a)(x-b)(x-c)}$, ii) $\dfrac{f(x)}{(x-a)^2(x-b)}$, iii) $\dfrac{f(x)}{(x^2+a)(x-b)}$

in terms of appropriate partial fractions provided $f(x)$ is a polynomial of degree less than 3.

2 Express compound algebraic fractions where the degree of $f(x)$ is *not* less than the denominator.

3 Rewrite the compound algebraic fractions in **1** as partial fractions and then integrate those fractions.

Introduction

This chapter has two sections. In the first section we split compound algebraic fractions into partial fractions. For some students this will be sufficient for their course. In the second section we apply our techniques and then integrate using standard rules.

$f(x)$ is a general way of representing a **polynomial in x**. A polynomial is an added list of terms involving x and constants. Generally the powers of x are positive whole numbers. We look at the exception in Examples 12.1(v). The degree of the polynomial refers to the highest power of x.

▨▨▨▨▨ Examples 12.1 ▨▨▨▨▨

i) $f(x) = x^3 + 5x^2 - 6x + 7$ is a polynomial of degree 3 because the highest power of x is 3.

ii) $f(x) = 4.5 - 2x^3$ is also a polynomial of degree 3. Notice that we do *not* need to have all the powers of x.

iii) $f(x) = x^2 + 7x$ is a polynomial of degree 2 because the highest power of x is 2.

iv) $f(x) = 4 - x$ is a polynomial of degree 1 because we understand x to be x^1.

v) $f(x) = 10$ can also be thought of as a polynomial because $10 = 10 \times 1 = 10x^0$, i.e. $f(x) = 10x^0$.

ASSIGNMENT

Our Assignment for this chapter looks at the overall cost of operating some particular industrial plant. The cost per hour, C_h, is quoted in terms of time, t hours, by

$$C_h = \frac{0.4t^2 + 18t + 120}{(t-15)^2(t+1)}.$$

The units for C_h are thousands of pounds per hour. C_h takes into account all the costs including the lease, installation, maintenance contract, materials, labour and so on.

We show the graph of C_h against t in Fig. 12.1.

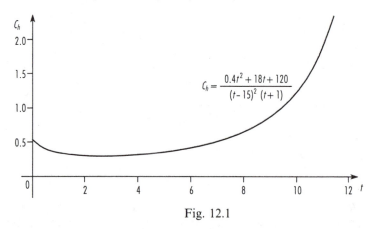

Fig. 12.1

Later we are going to find the total cost of running the plant. However, first we need to rewrite our equation for C_h in terms of partial fractions. Then we will apply some standard integration rules.

Partial fractions – linear factors

Let us remind ourselves how to add simple fractions,

e.g $\dfrac{2}{7} + \dfrac{3}{5} = \dfrac{(2 \times 5) + (3 \times 7)}{7 \times 5} = \dfrac{31}{35}.$

The common denominator, 35, is the lowest common multiple (LCM) of 7 and 5. 35 is common to both (2×5) and (3×7).

$\dfrac{2 \times 5}{35}$ cancels to the original $\dfrac{2}{7}$ and $\dfrac{3 \times 7}{35}$ cancels to the original $\dfrac{3}{5}$.

Exactly the same idea applies to algebraic fractions,

e.g. $\dfrac{2}{x+7} + \dfrac{3}{x-5} = \dfrac{2(x-5) + 3(x+7)}{(x+7)(x-5)} = \dfrac{5x+11}{(x+7)(x-5)}.$

The common denominator, $(x + 7)(x - 5)$, is the LCM of $x + 7$ and $x - 5$. It is usual to leave the common denominator in a factorised (bracketed) form. $\dfrac{5x + 11}{(x + 7)(x - 5)}$ is a compound algebraic fraction which may be split into the partial fractions of $\dfrac{2}{x + 7}$ and $\dfrac{3}{x - 5}$. The addition of these partial fractions is equivalent to the original compound fraction and so we can write

$$\frac{5x + 11}{(x + 7)(x - 5)} \equiv \frac{2}{x + 7} + \frac{3}{x - 5}.$$

> '$\equiv$' means 'equivalent to'.

The method of partial fractions requires us to find the numerators, i.e. in the example above to find the values 2 and 3.

We always need the highest power of x in the denominator to exceed the highest power of x in the numerator. This applies to both the original compound fraction and the resulting partial fractions. Expanding $(x + 7)(x - 5)$ as $x^2 + 2x - 35$ we see this is true originally, i.e. $\dfrac{5x + 11}{x^2 + 2x - 35}$. Also remember we can think of 2 as $2x^0$ to give $\dfrac{2x^0}{x + 7}$ and 3 as $3x^0$ to give $\dfrac{3x^0}{x - 5}$.

Our next series of examples aims to create partial fractions with linear denominators, i.e. the highest power of x is 1. This means the numerators will be pure numbers.

Example 12.2

Express $\dfrac{5x + 11}{(x + 7)(x - 5)}$ in terms of partial fractions.

We concentrate on the factorised denominator noting the rule about powers of x. Its form of two brackets indicates we need two partial fractions, one with a denominator of $x + 7$ and the other with a denominator of $x - 5$. We do *not* know what to put in each numerator. To represent the numbers we use general constants A and B. Our aim is to find A and B. We write

$$\frac{5x + 11}{(x + 7)(x - 5)} \equiv \frac{A}{x + 7} + \frac{B}{x - 5}.$$

On the right-hand side we add these partial fractions. Hence we come back to the original common denominator with

$$\frac{5x + 11}{(x + 7)(x - 5)} \equiv \frac{A(x - 5) + B(x + 7)}{(x + 7)(x - 5)}.$$

The left- and right-hand-side fractions are equivalent. They have the same denominators, hence they must have the same numerators, i.e.

$$5x + 11 \equiv A(x - 5) + B(x + 7).$$ Equating numerators.

The $\equiv$ sign means the left- and right-hand sides are the same for all values of x. This is important. It allows us to choose particular values of x to suit our example. Notice we choose values of x so each bracket in turn has a value of 0. Also, when we change from all values of x to a particular value we change $\equiv$ to $=$.

Let $x - 5 = 0$, i.e. $x = 5$,

so that $\quad 5(5) + 11 = A(0) + B(5 + 7)$

i.e. $\qquad\qquad 36 = 12B$

$\therefore \qquad\qquad\quad 3 = B.$

Also, let $x + 7 = 0$, i.e. $x = -7$,

so that $\quad 5(-7) + 11 = A(-7 - 5) + B(0)$

i.e. $\qquad\qquad -24 = A(-12)$

$\therefore \qquad\qquad\quad 2 = A.$

Now we can write $\quad \dfrac{5x + 11}{(x + 7)(x - 5)} \equiv \dfrac{2}{x + 7} + \dfrac{3}{x - 5}.$

Example 12.3

Express $\dfrac{0.5x + 4}{x^2 + 6x + 8}$ in terms of partial fractions.

Again we concentrate on the denominator, noting the rule about powers of x and that it factorises, i.e. $x^2 + 6x + 8 = (x + 4)(x + 2)$. Then we create partial fractions with each factor as a denominator, i.e.

$$\frac{0.5x + 4}{x^2 + 6x + 8} \equiv \frac{0.5x + 4}{(x + 4)(x + 2)} \equiv \frac{A}{x + 4} + \frac{B}{x + 2}.$$

Again adding the partial fractions we come back to the original common denominator with

$$\frac{0.5x + 4}{x^2 + 6x + 8} \equiv \frac{A(x + 2) + B(x + 4)}{(x + 4)(x + 2)}.$$

The left- and right-hand-side fractions are equivalent. Obviously they have the same denominators. Hence they must have the same numerators, i.e.

$$0.5x + 4 \equiv A(x + 2) + B(x + 4).$$ Equating numerators.

Now we choose particular values of x.

Let $x + 2 = 0$, i.e. $x = -2$,

so that $\quad 0.5(-2) + 4 = A(0) + B(-2 + 4)$

i.e. $\qquad\qquad 3 = 2B$

$\therefore \qquad\qquad 1.5 = B.$

Also, let $x + 4 = 0$, i.e. $x = -4$,

so that $\quad 0.5(-4) + 4 = A(-4 + 2) + B(0)$

i.e. $\qquad\qquad 2 = A(-2)$

$\therefore \qquad\qquad -1 = A.$

Now we can write $\dfrac{0.5x + 4}{x^2 + 6x + 8} \equiv \dfrac{-1}{x + 4} + \dfrac{1.5}{x + 2}.$

Example 12.4

Express $\dfrac{10}{x^2 - 5x}$ in terms of partial fractions.

Again we concentrate on the denominator, noting the rule about powers of x and that it factorises, i.e. $x^2 - 5x = x(x - 5)$. Then we create partial fractions with each factor as a denominator. We can think of x as $(x + 0)$. Then

$$\frac{10}{x^2 - 5x} \equiv \frac{10}{x(x - 5)} \equiv \frac{A}{x} + \frac{B}{x - 5}.$$

Again adding the partial fractions we come back to the original common denominator with

$$\frac{10}{x^2 - 5x} \equiv \frac{A(x - 5) + Bx}{x(x - 5)}.$$

The left- and right-hand-side fractions are equivalent. Obviously they have the same denominators. Hence they must have the same numerators, i.e.

$$10 \equiv A(x - 5) + Bx. \qquad \boxed{\text{Equating numerators.}}$$

Now we choose particular values of x.

Let $x - 5 = 0$, i.e. $x = 5$,

so that $\quad 10 = A(0) + B(5)$

i.e. $\qquad 10 = 5B$

$\therefore \qquad 2 = B.$

Also let $x = 0$,

so that $\quad 10 = A(0 - 5) + B(0)$

i.e. $\qquad 10 = -5A$

$\therefore \qquad -2 = A.$

Now we can write $\dfrac{10}{x^2 - 5x} \equiv \dfrac{-2}{x} + \dfrac{2}{x - 5}.$

We can extend this method to many more than two partial fractions. In Example 12.5 we use three partial fractions as a simple extension.

▬▬▬ Example 12.5 ▬▬▬

Express $\dfrac{30}{(x-1)(x+2)(x-3)}$ in terms of partial fractions.

Again we concentrate on the denominator, noting the rule about powers of x. We create three partial fractions with each factor as a denominator, i.e.

$$\frac{30}{(x-1)(x+2)(x-3)} \equiv \frac{A}{x-1} + \frac{B}{x+2} + \frac{C}{x-3}.$$

Again adding all three partial fractions we come back to the original common denominator with

$$\frac{30}{(x-1)(x+2)(x-3)} \equiv$$
$$\frac{A(x+2)(x-3) + B(x-1)(x-3) + C(x-1)(x+2)}{(x-1)(x+2)(x-3)}.$$

The left- and right-hand-side fractions are equivalent. Obviously they have the same denominators. Hence they must have the same numerators, i.e.

$$30 \equiv A(x+2)(x-3) + B(x-1)(x-3) + C(x-1)(x+2).$$

Now we choose particular values of x.

Let $x + 2 = 0$, i.e. $x = -2$,

so that $30 = A(0)(-5) + B(-3)(-5) + C(-3)(0)$

i.e.　　$30 = \qquad\qquad 15B$

∴　　　$2 = B.$

Next let $x - 3 = 0$, i.e. $x = 3$,

so that $30 = A(5)(0) + B(2)(0) + C(2)(5)$

i.e.　　$30 = \qquad\qquad\qquad 10C$

∴　　　$3 = C.$

Next let $x - 1 = 0$, i.e. $x = 1$,

so that $30 = A(3)(-2) + B(0)(-3) + C(0)(3)$

i.e.　　$30 = -6A$

∴　　　$-5 = A.$

Finally $\dfrac{30}{(x-1)(x+2)(x-3)} \equiv \dfrac{-5}{x-1} + \dfrac{2}{x+2} + \dfrac{3}{x-3}.$

████ **EXERCISE 12.1** ████████████████████████████

Express the following compound algebraic fractions in terms of partial fractions.

1 $\dfrac{6x + 2}{(x - 1)(x + 3)}$

2 $\dfrac{36}{(x + 5)(x - 7)}$

3 $\dfrac{3x}{(x - 4)(x - 1)}$

4 $\dfrac{7x + 5}{x^2 + x - 2}$

5 $\dfrac{2.5x - 3}{x^2 + 3x - 18}$

6 $\dfrac{4.5 + 4x}{x(x + 3)}$

7 $\dfrac{5x + 1}{x^2 - 1}$

8 $\dfrac{11x^2 + 6x - 25}{(x + 3)(x + 1)(x - 4)}$

9 $\dfrac{2x^2 + 21x - 5}{x(x - 1)(x + 5)}$

10 $\dfrac{5x^2 + 9x - 36}{x(x^2 - 9)}$

Partial fractions – repeated roots

This general expression of $\dfrac{f(x)}{(x - a)^2(x - b)}$ has a repeated root at $x - a$.

This is indicated by that bracket being raised to the power 2. There are several ways to express this compound fraction in terms of different fractions. Our method takes into account all possible variations. There are other forms of partial fractions that are not as useful when it comes to integration. Our method makes the integration possible later in the chapter. We remind you that the **common denominators are based on the LCMs**.

▓▓▓▓ **Example 12.6** ▓▓▓▓▓▓▓▓▓▓▓▓▓▓▓▓▓▓▓▓▓▓▓▓▓▓▓▓▓▓▓

Express $\dfrac{5x^2 - 19x + 5}{(x - 1)^2(x + 2)}$ in terms of partial fractions.

Again we concentrate on the denominator, noting the rule about powers of x and that we have a repeated root. We create partial fractions with each factor and the repeated root (factor) as a denominator. Then

$$\frac{5x^2 - 19x + 5}{(x - 1)^2(x + 2)} \equiv \frac{A}{x - 1} + \frac{B}{(x - 1)^2} + \frac{C}{x + 2}.$$

Again adding the partial fractions we come back to the original common denominator with

$$\frac{5x^2 - 19x + 5}{(x - 1)^2(x + 2)} \equiv \frac{A(x - 1)(x + 2) + B(x + 2) + C(x - 1)^2}{(x - 1)^2(x + 2)}.$$

The left- and right-hand-side fractions are equivalent. Obviously they have the same denominators. Hence they must have the same numerators, i.e.

$$5x^2 - 19x + 5 \equiv A(x - 1)(x + 2) + B(x + 2) + C(x - 1)^2.$$

Now we choose particular values of x.

Let $x - 1 = 0$, i.e. $x = 1$,

so that $5(1)^2 - 19(1) + 5 = A(0)(3) + B(3) + C(0)^2$

i.e. $-9 = 3B$

∴ $-3 = B.$

Next let $x + 2 = 0$, i.e. $x = -2$,

so that $5(-2)^2 - 19(-2) + 5 = A(-3)(0) + B(0) + C(-3)^2$

i.e. $63 = 9C$

∴ $7 = C.$

We have equated both brackets to zero, but still need to find the value for A. Remember the numerators are equivalent for all values of x. Now we choose any simple value, e.g. $x = 0$,

so that $5 = A(-1)(2) + B(2) + C(-1)^2$

and substitute our known values of B and C to get

$$5 = -2A + (-3)(2) + (7)(1)$$

i.e. $5 = -2A - 6 + 7$

∴ $-2 = A.$

Now we can write $\dfrac{5x^2 - 19x + 5}{(x - 1)^2(x + 2)} \equiv \dfrac{-2}{x - 1} - \dfrac{3}{(x - 1)^2} + \dfrac{7}{x + 2}.$

■■■■■■ **Example 12.7** ■■■■■■

Express $\dfrac{200}{x^2(10 - x)}$ in terms of partial fractions.

Again we concentrate on the denominator, noting the rule about powers of x and that we have a repeated root. We create partial fractions with each factor and the repeated root (factor) as a denominator. Then

$$\frac{200}{x^2(10 - x)} \equiv \frac{A}{x} + \frac{B}{x^2} + \frac{C}{10 - x}.$$

Again, adding the partial fractions we come back to the original common denominator with

$$\frac{200}{x^2(10 - x)} \equiv \frac{Ax(10 - x) + B(10 - x) + Cx^2}{x^2(10 - x)}.$$

The left- and right-hand-side fractions are equivalent. Obviously they have the same denominators. Hence they must have the same numerators, i.e.

$$200 \equiv Ax(10 - x) + B(10 - x) + Cx^2.$$

Now we choose particular values of x.

Let $10 - x = 0$, i.e. $x = 10$,

so that $\quad 200 = A(10)(0) + B(0) + C(10)^2$

i.e. $\qquad 200 = 100C$

$\therefore \qquad\quad 2 = C.$

Next let $x = 0$,

so that $\quad 200 = A(0)(10) + B(10) + C(0)^2$

i.e. $\qquad 200 = 10B$

$\therefore \qquad\quad 20 = B.$

We have used both possible values of x, but still need to find the value for A. Remember, the numerators are equivalent for all values of x. Now we choose any simple value, e.g. $x = 1$, previously having used the simpler $x = 0$.

Then $\qquad 200 = A(1)(9) + B(9) + C(1)^2.$

We substitute our known values of B and C to get

$$200 = 9A + (20)(9) + (2)(1)$$

i.e. $\qquad 200 = 9A + 180 + 2$

$\therefore \qquad\quad 2 = A.$

Now we can write $\dfrac{200}{x^2(10 - x)} \equiv \dfrac{2}{x} + \dfrac{20}{x^2} + \dfrac{2}{10 - x}.$

■ ASSIGNMENT ■

We are going to express $\dfrac{0.4t^2 + 18t + 120}{(t - 15)^2(t + 1)}$ in terms of partial fractions.

We have a repeated root and create partial fractions. Then

$$\frac{0.4t^2 + 18t + 120}{(t - 15)^2(t + 1)} \equiv \frac{A}{t - 15} + \frac{B}{(t - 15)^2} + \frac{C}{t + 1}.$$

Again adding the partial fractions we come back to the original common denominator with

$$\frac{0.4t^2 + 18t + 120}{(t - 15)^2(t + 1)} \equiv \frac{A(t - 15)(t + 1) + B(t + 1) + C(t - 15)^2}{(t - 15)^2(t + 1)}.$$

The left- and right-hand-side fractions are equivalent. Obviously they have the same denominators. Hence they must have the same numerators, i.e.

$$0.4t^2 + 18t + 120 \equiv A(t - 15)(t + 1) + B(t + 1) + C(t - 15)^2.$$

Now we choose particular values of t.

Let $t - 15 = 0$, i.e. $t = 15$,

so that $0.4(15)^2 + 18(15) + 120 = A(0)(16) + B(16) + C(0)^2$

i.e. $480 = 16B$

$\therefore$ $30 = B.$

Next let $t + 1 = 0$, i.e. $t = -1$,

so that $0.4(-1)^2 + 18(-1) + 120 = A(-16)(0) + B(0) + C(-16)^2$

i.e. $102.4 = 256C$

$\therefore$ $0.4 = C.$

We have equated both brackets to zero, but still need to find the value for A. Remember, the numerators are equivalent for all values of t. Now we choose any simple value, e.g. $t = 0$,

so that $120 = A(-15)(1) + B(1) + C(-15)^2$

and substitute our known values of B and C to get

$$120 = -15A + (30)(1) + (0.4)(225)$$

i.e. $120 = -15A + 30 + 90$

$\therefore$ $0 = A.$

Now we can write $\dfrac{0.4t^2 + 18t + 120}{(t - 15)^2(t + 1)} \equiv \dfrac{30}{(t - 15)^2} + \dfrac{0.4}{t + 1}.$

■■■■■ EXERCISE 12.2 ■■■■■

Express the following compound algebraic fractions in terms of partial fractions.

1 $\dfrac{3x^2 - 10x + 11}{(x + 1)^2(x - 5)}$

2 $\dfrac{-7.5x - 19}{(x - 4)(x + 3)^2}$

3 $\dfrac{\frac{1}{2}(7x^2 - 3x - 8)}{(x - 2)^2 x}$

4 $\dfrac{5x^2 + 15x + 12}{(x + 3)(x + 1)^2}$

5 $\dfrac{6x^2 + 1}{x(2x - 1)^2}$

Partial fractions – quadratic factor

This general expression of $\dfrac{f(x)}{(x^2 + a)(x - b)}$ differs from our previous expression with a different position for the power of 2. $x^2 + a$ is a general quadratic factor. Later we will see quadratics with all three terms, i.e. x^2, x and constant terms. For this method we are *unable* to factorise the quadratic factor. Remember, we need the highest power of x in the denominator to exceed that in the numerator. In the partial fractions we use $Ax + B$ in the numerator. This takes into account all variations, which means it is possible that either A or B may be 0.

Example 12.8

Express $\dfrac{8 - 2x}{(x^2 + 4)(x + 1)}$ in terms of partial fractions.

Again we concentrate on the denominator, noting the rule about powers of x and that $x^2 + 4$ does *not* factorise. We create partial fractions with each factor. Then

$$\frac{8 - 2x}{(x^2 + 4)(x + 1)} \equiv \frac{Ax + B}{x^2 + 4} + \frac{C}{x + 1}.$$

Again adding the partial fractions, we come back to the original common denominator with

$$\frac{8 - 2x}{(x^2 + 4)(x + 1)} \equiv \frac{(Ax + B)(x + 1) + C(x^2 + 4)}{(x^2 + 4)(x + 1)}.$$

The left- and right-hand-side fractions are equivalent. Obviously they have the same denominators. Hence they must have the same numerators, i.e.

$$8 - 2x \equiv (Ax + B)(x + 1) + C(x^2 + 4).$$

Now we choose particular values of x.

Let $x + 1 = 0$, i.e. $x = -1$,

so that $\quad 8 - 2(-1) = (A(-1) + B)(0) + C((-1)^2 + 4)$

i.e. $\qquad\qquad 10 = 5C$

$\therefore \qquad\qquad\quad 2 = C.$

No other bracket, i.e. $(x^2 + 4)$, will simplify to 0 for real values of x. This means we need to choose and substitute simple values of x in the numerator.

Let $x = 0$

so that $\qquad\qquad 8 = (A(0) + B)(1) + C(4)$

and substituting for C we get

$$8 = B + 2(4)$$

∴　　　　　　$0 = B.$

> Obviously $B=0$ in this case.

Again choose a value of x, e.g. $x=1$,

so that　　　$8 - 2(1) = (A(1) + B)(2) + C(5)$

and substituting for B and C we get

$$6 = (A + 0)(2) + 2(5)$$

i.e.　　　　　$6 = 2A + 10$

∴　　　　　　$-2 = A.$

Now we can write　$\dfrac{8 - 2x}{(x^2 + 4)(x + 1)} \equiv \dfrac{-2x}{x^2 + 4} + \dfrac{2}{x + 1}.$

▰▰▰▰ Examples 12.9 ▰▰▰▰

Express in terms of partial fractions

i) $\dfrac{5x + 1}{(x^2 + 4x)(x + 1)}$,

ii) $\dfrac{5x + 1}{(x^2 - 4)(x + 1)}$,

iii) $\dfrac{5x + 1}{(x^2 + 4x + 4)(x + 1)}.$

i)　We concentrate on the denominator, noting that $x^2 + 4x$ factorises, i.e. $x^2 + 4x = x(x + 4)$. This means we have linear factors and so need to use the style

$$\frac{5x + 1}{(x^2 + 4x)(x + 1)} \equiv \frac{5x + 1}{x(x + 4)(x + 1)} \equiv \frac{A}{x} + \frac{B}{x + 4} + \frac{C}{x + 1}.$$

You should complete this as an exercise for yourself, finding $A = \dfrac{1}{4}$, $B = \dfrac{-19}{12}$ and $C = \dfrac{4}{3}.$

ii)　We concentrate on the denominator, noting that $x^2 - 4$ factorises, i.e. $x^2 - 4 = (x + 2)(x - 2)$. This means we have **linear factors** and so need to use the style

$$\frac{5x + 1}{(x^2 - 4)(x + 1)} \equiv \frac{5x + 1}{(x + 2)(x - 2)(x + 1)} \equiv \frac{A}{x + 2} + \frac{B}{x - 2} + \frac{C}{x + 1}.$$

You should complete this as an exercise for yourself, finding $A = \dfrac{-9}{4}$, $B = \dfrac{11}{12}$ and $C = \dfrac{4}{3}.$

iii) We concentrate on the denominator, noting that $x^2 + 4x + 4$ factorises, i.e. $x^2 + 4x + 4 = (x+2)^2$. This means we have a **repeated factor** and so need to use the style

$$\frac{5x+1}{(x^2+4x+4)(x+1)} \equiv \frac{5x+1}{(x+2)^2(x+1)} \equiv \frac{A}{x+2} + \frac{B}{(x+2)^2} + \frac{C}{x+1}.$$

You should complete this as an exercise for yourself, finding $A = 4$, $B = 9$ and $C = -4$.

Example 12.10

Express $\dfrac{3x^2 - 10x - 13}{(x^2 + x + 1)(x - 2)}$ in terms of partial fractions.

Again we concentrate on the denominator, noting the rule about powers of x and that $x^2 + x + 1$ does *not* factorise. We create partial fractions with each factor. Then

$$\frac{3x^2 - 10x - 13}{(x^2 + x + 1)(x - 2)} \equiv \frac{Ax + B}{x^2 + x + 1} + \frac{C}{x - 2}.$$

Again adding the partial fractions we come back to the original common denominator with

$$\frac{3x^2 - 10x - 13}{(x^2 + x + 1)(x - 2)} \equiv \frac{(Ax + B)(x - 2) + C(x^2 + x + 1)}{(x^2 + x + 1)(x - 2)}.$$

The left- and right-hand-side fractions are equivalent. Obviously they have the same denominators. Hence they must have the same numerators, i.e.

$$3x^2 - 10x - 13 \equiv (Ax + B)(x - 2) + C(x^2 + x + 1).$$

Now we choose particular values of x.

Let $x - 2 = 0$, i.e. $x = 2$,

so that $3(2)^2 - 10(2) - 13 = (A(2) + B)(0) + C(2^2 + 2 + 1)$

i.e. $\qquad\qquad -21 = 7C$

$\therefore \qquad\qquad 3 = C.$

No other bracket, i.e. $(x^2 + x + 1)$, will simplify to 0 for real values of x. This means we need to choose and substitute simple values of x in the numerator.

Let $x = 0$

so that $\qquad\qquad -13 = (A(0) + B)(-2) + C(1)$

and substituting for C we get

$$-13 = -2B + (-3)(1)$$

$\therefore \qquad\qquad 5 = B.$

Again, choose a value of x, e.g. $x = 1$,

so that $3(1)^2 - 10(1) - 13 = (A(1) + B)(-1) + C(3)$

and substituting for B and C we get

$$-20 = (A + 5)(-1) + (-3)(3)$$

i.e. $\qquad -20 = -A - 5 - 9$

$\therefore \qquad\qquad 6 = A.$

Now we can write $\dfrac{3x^2 - 10x - 13}{(x^2 + x + 1)(x - 2)} \equiv \dfrac{6x + 5}{x^2 + x + 1} - \dfrac{3}{x - 2}.$

■■■■ EXERCISE 12.3 ■■■■

Express the following compound algebraic fractions in terms of partial fractions.

1 $\dfrac{x^2 - 4x - 3}{(x^2 + 1)(x - 1)}$

2 $\dfrac{6x^2 + 10}{x(x^2 + 2)}$

3 $\dfrac{4x^2 - 50x - 10}{(x + 1)(x^2 + 10)}$

4 $\dfrac{-x - 4}{(x^2 - x + 2)x}$

5 $\dfrac{3x^2 - 8}{(x - 3)(x^2 + 2x + 4)}$

Partial fractions – after division

So far all our examples have had the highest power of x in the denominator exceeding the highest power of x in the numerator. Now we look at cases where this fails. It can fail for all types of compound fractions. When this happens we need to divide out the fraction by long division. This creates a slightly different compound fraction as a remainder. For this part only we apply one of our methods of partial fractions.

We are going to look at two compound fractions in need of long division. Before examining the method let us take a slightly different look. The first is $\dfrac{2x^2 + 9x - 59}{x^2 + 2x - 35}$. We are going to express it in terms of its denominator, $x^2 + 2x - 35$. We can juggle around with the algebra and eventually write

$$2x^2 + 9x - 59 = 2(x^2 + 2x - 35) + 5x + 11.$$

We need the $5x$ and 11 to make up the original $9x$ and -59. For consistency what we do to one side we must do to the other side. Let us divide throughout by $x^2 + 2x - 35$ to get

$$\frac{2x^2 + 9x - 59}{x^2 + 2x - 35} = \frac{2(x^2 + 2x - 35)}{x^2 + 2x - 35} + \frac{5x + 11}{x^2 + 2x - 35}$$

$$= 2 + \frac{5x + 11}{x^2 + 2x - 35}.$$

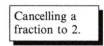

Cancelling a fraction to 2.

The second is $\dfrac{3x^3 + 5x^2 - 102x + 46}{x^2 + 2x - 35}$. We are going to express it in terms of its denominator, $x^2 + 2x - 35$. Again we juggle around with the algebra and eventually write

$$3x^3 + 5x^2 - 102x + 46 = 3x(x^2 + 2x - 35) - 1(x^2 + 2x - 35) + 5x + 11.$$

We need the $5x$ and 11 to make up the original $-102x$ and 46. What we do to one side we must do to the other side. Let us divide throughout by $x^2 + 2x - 35$ to get

$$\frac{3x^3 + 5x^2 - 102x + 46}{x^2 + 2x - 35} = \frac{(3x - 1)(x^2 + 2x - 35)}{x^2 + 2x - 35} + \frac{5x + 11}{x^2 + 2x - 35}$$

$$= 3x - 1 + \frac{5x + 11}{x^2 + 2x - 35}.$$

Cancelling fractions.

In the first case the juggling of the algebra was not too tricky. However, the second case was quite difficult. We have not given you any rules for the algebra. This type of manipulation is not really satisfactory. In Examples 12.11 we look at a proper method.

▨ Examples 12.11 ▨▨▨▨▨▨▨▨▨▨▨▨

Express the following compound fractions in terms of partial fractions

i) $\dfrac{2x^2 + 9x - 59}{x^2 + 2x - 35}$, ii) $\dfrac{3x^3 + 5x^2 - 102x + 46}{x^2 + 2x - 35}$.

In both cases our rule about powers of x fails. This means we need to divide out our fractions before attempting any partial fractions. We attempt the long division in stages. In fact all the terms in the denominator are dividing into all the terms in the numerator.

i) We concentrate on the highest powered terms, x^2 and $2x^2$. For ease we have written them first. x^2 divides into $2x^2$, i.e. $\dfrac{2x^2}{x^2} = 2$. We write this as

$$x^2 + 2x - 35 \overline{)2x^2 + 9x - 59}^{\,2}$$

Now we multiply our first result of 2 by each term of the denominator to get $2x^2 + 4x - 70$, i.e.

$$x^2 + 2x - 35 \overline{)\begin{array}{l} 2x^2 + 9x - 59 \\ 2x^2 + 4x - 70 \end{array}}^{\,2}$$

We subtract to see what remains from the numerator after our first division, i.e.

$$x^2 + 2x - 35 \overline{\smash{\big)}\,2x^2 + 9x - 59} \quad \genfrac{}{}{0pt}{}{2}{}$$
$$2x^2 + 4x - 70$$
$$\overline{ 5x + 11}$$

$(-59) - (-70)$
$= -59 + 70 = 11.$

We are interested in our remainder, $5x + 11$, and our original denominator, $x^2 + 2x - 35$. Again we concentrate on the highest powered terms, x^2 (from the denominator) and $5x$ (the remainder of the numerator). When x^2 divides into $5x$ we get a negative power of x. This is of no use. Hence we take our division no further and write

$$\frac{2x^2 + 9x - 59}{x^2 + 2x - 35} = 2 + \frac{5x + 11}{x^2 + 2x - 35}.$$

We have seen $\dfrac{5x + 11}{x^2 + 2x - 35}$ in Example 12.2 and can use that result to give

$$\frac{2x^2 + 9x - 59}{x^2 + 2x - 35} = 2 + \frac{2}{x + 7} + \frac{3}{x - 5}.$$

ii) We concentrate on the highest powered terms, x^2 and $3x^3$. For ease we have written them first. x^2 divides into $3x^3$, i.e. $\dfrac{3x^3}{x^2} = 3x$. We write this as

$$x^2 + 2x - 35 \overline{\smash{\big)}\,3x^3 + 5x^2 - 102x + 46} \quad \genfrac{}{}{0pt}{}{3x}{}$$

Now we multiply our first result of $3x$ by each term of the denominator to get $3x^3 + 6x^2 - 105x$, i.e.

$$x^2 + 2x - 35 \overline{\smash{\big)}\,3x^3 + 5x^2 - 102x + 46} \quad \genfrac{}{}{0pt}{}{3x}{}$$
$$3x^3 + 6x^2 - 105x$$

We subtract to see what remains from the numerator after our first division, i.e.

$$x^2 + 2x - 35 \overline{\smash{\big)}\,3x^3 + 5x^2 - 102x + 46} \quad \genfrac{}{}{0pt}{}{3x}{}$$
$$3x^3 + 6x^2 - 105x \qquad -$$
$$\overline{ -x^2 + 3x + 46}$$

Again we concentrate on the highest powered terms, x^2 (from the denominator) and $-x^2$ (the remainder of the numerator). When x^2 divides into $-x^2$ we get -1, i.e. $\dfrac{-x^2}{x^2} = -1$. Now we multiply our second result of -1 by each term of the denominator to get $-x^2 - 2x + 35$, i.e.

$$
\begin{array}{r}
3x \\
x^2 + 2x - 35 \overline{\smash{\big)}\ 3x^3 + 5x^2 - 102x + 46} \\
3x^3 + 6x^2 - 105x \qquad - \\
\hline
-x^2 + \quad 3x + 46 \\
-x^2 - \quad 2x + 35
\end{array}
$$

We subtract to see what remains from the numerator after this move, i.e.

$$
\begin{array}{r}
3x \\
x^2 + 2x - 35 \overline{\smash{\big)}\ 3x^3 + 5x^2 - 102x + 46} \qquad - \\
3x^3 + 6x^2 - 105x \\
\hline
-x^2 + \quad 3x + 46 \qquad - \\
-x^2 - \quad 2x + 35 \\
\hline
5x + 11
\end{array}
$$

We have reached this type of situation before. Again we can take our division no further. $5x + 11$ is our remainder and we write

$$
\frac{3x^3 + 5x^2 - 102x + 46}{x^2 + 2x - 35} = 3x - 1 + \frac{5x + 11}{x^2 + 2x - 35}
$$

$$
= 3x - 1 + \frac{2}{x + 7} + \frac{3}{x - 5},
$$

factorising $x^2 + 2x - 35$ as before.

Example 12.12

Express $\dfrac{5x^4 + 2x^3 + 11x^2 + 2x + 4}{x^3 + x}$ in terms of partial fractions.

We concentrate on the highest powered terms, x^3 and $5x^4$. For ease we have written them first. x^3 divides into $5x^4$, i.e. $\dfrac{5x^4}{x^3} = 5x$. We write this as

$$
\begin{array}{r}
5x \\
x^3 + x \overline{\smash{\big)}\ 5x^4 + 2x^3 + 11x^2 + 2x + 4}
\end{array}
$$

Now we multiply our first result of $5x$ by each term of the denominator to get $5x^4 + 5x^2$, i.e.

$$
\begin{array}{r}
5x \\
x^3 + x \overline{\smash{\big)}\ 5x^4 + 2x^3 + 11x^2 + 2x + 4} \\
5x^4 \qquad\quad + \ 5x^2
\end{array}
$$

We subtract to see what remains from the numerator after our first division, i.e.

$$
\begin{array}{r}
5x \\
x^3 + x \overline{\smash{\big)}\ 5x^4 + 2x^3 + 11x^2 + 2x + 4} \qquad - \\
5x^4 \qquad\quad + \ 5x^2 \\
\hline
2x^3 + \quad 6x^2 + 2x + 4
\end{array}
$$

Again we concentrate on the highest powered terms, x^3 (from the denominator) and $2x^3$ (the remainder of the numerator). When x^3 divides into $2x^3$ we get 2, i.e. $\dfrac{2x^3}{x^3} = 2$. Now we multiply our second result of 2 by each term of the denominator to get $2x^3 + 2x$, i.e.

$$
\begin{array}{r}
5x \\
x^3 + x\overline{)\,5x^4 + 2x^3 + 11x^2 + 2x + 4} \\
5x^4 + 5x^2 \\
\hline
2x^3 + 6x^2 + 2x + 4 \\
2x^3 + 2x
\end{array}
$$

We subtract to see what remains from the numerator after this move, i.e.

$$
\begin{array}{r}
5x + 2 \\
x^3 + x\overline{)\,5x^4 + 2x^3 + 11x^2 + 2x + 4} \\
5x^4 + 5x^2 \\
\hline
2x^3 + 6x^2 + 2x + 4 \\
2x^3 + 2x \\
\hline
6x^2 + 4
\end{array}
$$

We have reached this type of situation before. x^3 will only divide into $6x^2$ leaving a negative power of x. We can take our division no further. $6x^2 + 4$ is our remainder so that

$$
\frac{5x^4 + 2x^3 + 11x^2 + 2x + 4}{x^3 + x} = 5x + 2 + \frac{6x^2 + 4}{x^3 + x}
$$

> Factorising the denominator.

$$
= 5x + 2 + \frac{6x^2 + 4}{x(x^2 + 1)}
$$

Notice how we have reduced our original compound fraction. We need only apply the method of partial fractions to $\dfrac{6x^2 + 4}{x(x^2 + 1)}$.

We concentrate on the denominator, noting the rule about powers of x and that $x^2 + 1$ does *not* factorise. We create partial fractions with each factor. Then

$$
\frac{6x^2 + 4}{x(x^2 + 1)} \equiv \frac{A}{x} + \frac{Bx + C}{x^2 + 1}.
$$

Adding the partial fractions we come back to the original common denominator with

$$
\frac{6x^2 + 4}{x(x^2 + 1)} \equiv \frac{A(x^2 + 1) + (Bx + C)x}{x(x^2 + 1)}.
$$

The left- and right-hand-side fractions are equivalent. Obviously they have the same denominators. Hence they must have the same numerators, i.e.

$$
6x^2 + 4 \equiv A(x^2 + 1) + (Bx + C)x.
$$

Now we choose particular values of x.

Let $x = 0$

so that $\qquad 4 = A(0 + 1) + (B(0) + C)0$

i.e. $\qquad 4 = A$.

No other bracket, i.e. $(x^2 + 1)$, will simplify to 0 for real values of x. This means we need to choose and substitute simple values of x in the numerator.

Let $x = 1$

so that $\quad 6(1)^2 + 4 = A(1 + 1) + (B(1) + C)(1)$

and substituting for A we get

$$10 = 4(2) + B + C$$

i.e. $\qquad 2 = B + C. \qquad\qquad ——①$

Again choose a value of x, e.g. $x = -1$,

so that $6(-1)^2 + 4 = A((-1)^2 + 1) + (B(-1) + C)(-1)$

and substituting for A we get

$$10 = 4(2) + B - C$$

i.e. $\qquad 2 = B - C. \qquad\qquad ——②$

Solving equations ① and ② simultaneously we get $B = 2$ and $C = 0$. Then we can write

$$\frac{6x^2 + 4}{x(x^2 + 1)} \equiv \frac{4}{x} + \frac{2x}{x^2 + 1}$$

so our total solution is

$$\frac{5x^4 + 2x^3 + 11x^2 + 2x + 4}{x^3 + x} = 5x + 2 + \frac{4}{x} + \frac{2x}{x^2 + 1}.$$

■ EXERCISE 12.4 ■

In each case divide out the compound fraction. Then apply the method of partial fractions as necessary to create a complete solution.

1 $\dfrac{4x^2 + 3x + 8}{x^2 + x - 6}$

2 $\dfrac{5x^2 - 2}{x^2 - x}$

3 $\dfrac{2x^3 + 4x^2 - x + 1}{x^2 - 1}$

4 $\dfrac{3x^4 + 4x^2 - 24}{(x^2 + 4)x}$

5 $\dfrac{5x^3 - 5x^2 + 8x - 12}{(x - 2)(2 + x^2)}$

Integration

As usual, we look at both indefinite and definite integration. We can apply all our standard integral results. When we use partial fractions remember our aim. It is to get a compound fraction into a form that is easily integrated. You should be able to recognise that integration. For partial fractions within the scope of this book, when we integrate we often get a natural logarithm. In Examples 12.13 we look at some typical fractions that integrate to give a natural logarithm. Your syllabus may include integration by substitution. In that chapter we look in detail at solutions in terms of natural logarithms.

▰▰▰ Examples 12.13 ▰▰▰

i) $\int \dfrac{1}{x}\, dx = \ln x + c.$

ii) $\int \dfrac{4}{x}\, dx = 4\int \dfrac{1}{x}\, dx = 4\ln x + c.$

You can miss out the middle step and go easily to the answer.

iii) $\int \dfrac{4}{x-9}\, dx = 4\int \dfrac{1}{x-9}\, dx = 4\ln(x-9) + c.$

In these first three examples concentrate on the denominator. Differentiate it to get 1. Notice how we have 1 in each numerator. Remember this pattern. We want the differential of the denominator to be in the numerator. If this happens then we integrate to get a natural logarithm of the denominator. We can adjust the numerator with other multipliers but *not* variables.

iv) $\int \dfrac{2x}{x^2+7}\, dx = \ln(x^2+7) + c.$

The differential of the denominator, $x^2 + 7$, is the numerator, $2x$.

v) $\int \dfrac{5x}{3x^2+7}\, dx = \dfrac{5}{6}\int \dfrac{6x}{3x^2+7}\, dx = \dfrac{5}{6}\ln(3x^2+7) + c.$

The differential of the denominator, $3x^2 + 7$, is $6x$. The x in the numerator is good but its coefficient of 5 needs adjusting. If we write $6x$ in the numerator we need to multiply by $\dfrac{5}{6}$. This maintains the correct size of the original integral because $\dfrac{5}{6} \times 6x = 5x$.

Let us apply this technique together with our previous work on partial fractions.

■■■■■ **Examples 12.14** ■■■■■■■■■■■■■■■■■■■■■■■■■■

i) $\int \dfrac{5x+11}{x^2+2x-35}\,dx$. We use our partial fractions result from Example

12.2 remembering that the denominator factorises.

Then $\displaystyle\int \frac{5x+11}{x^2+2x-35}\,dx = \int \frac{2}{x+7}+\frac{3}{x-5}\,dx$

$$= \int 2\times\frac{1}{x+7}+3\times\frac{1}{x-5}\,dx$$

$$= 2\ln(x+7)+3\ln(x-5)+c.$$

ii) $\int \dfrac{2x^2+9x-59}{x^2+2x-35}\,dx$. We use our result from Examples 12.11i) to write

$$\int \frac{2x^2+9x-59}{x^2+2x-35}\,dx = \int 2+\frac{2}{x+7}+\frac{3}{x-5}\,dx$$

$$= 2x+2\ln(x+7)+3\ln(x-5)+c.$$

iii) We find the value of $\displaystyle\int_6^8 \frac{3x^3+5x^2-102x+46}{x^2+2x-35}\,dx$ using our result

from Examples 12.11ii).

$$\int_6^8 \frac{3x^3+5x^2-102x+46}{x^2+2x-35}\,dx$$

$$= \int_6^8 3x-1+\frac{2}{x+7}+\frac{3}{x-5}\,dx$$

$$= \left[\frac{3x^2}{2}-x+2\ln(x+7)+3\ln(x-5)\right]_6^8$$

$$= \left[\frac{3\times 8^2}{2}-8+2\ln 15+3\ln 3\right]-\left[\frac{3\times 6^2}{2}-6+2\ln 13+3\ln 1\right]$$

$$= 96-8+5.416+3.296-54+6-5.130-0$$

$$= 43.58.$$

■■■■■ **EXERCISE 12.5** ■■■■■■■■■■■■■■■■■■■■■■■■■

Find the following integrals, evaluating where necessary. We have seen the compound fractions of Questions 1–10 in earlier Exercises. Integrate using those partial fractions' results.

1 $\displaystyle\int \frac{6x+2}{(x-1)(x+3)}\,dx$

2 $\displaystyle\int \frac{7x+5}{x^2+x-2}\,dx$

3 $\displaystyle\int \frac{200}{x^2(10-x)}\,dx$

4 $\displaystyle\int \frac{x^2-4x-3}{(x^2+1)(x-1)}\,dx$

5 $\displaystyle\int \frac{6x^2 + 10}{x(x^2 + 2)}\, dx$

6 $\displaystyle\int \frac{5x^2 - 2}{x^2 - x}\, dx$

7 $\displaystyle\int_5^{10} \frac{11x^2 + 6x - 25}{(x + 3)(x + 1)(x - 4)}\, dx$

8 $\displaystyle\int_{10}^{20} \frac{5x^2 + 9x - 36}{x(x^2 - 9)}\, dx$

9 $\displaystyle\int_1^5 \frac{100}{x^2(10 - x)}\, dx$

10 $\displaystyle\int_2^4 \frac{3x^4 + 4x^2 - 24}{x(x^2 + 4)}\, dx$

In Questions 11–15 you need to split each compound fraction into partial fractions for yourself. Then you need to integrate your result.

11 $\displaystyle\int_6^9 \frac{2(2x - 3)}{(x + 1)(x - 3)}\, dx$

12 $\displaystyle\int_2^5 \frac{2x^2 + 3x - 1}{(x^2 + 1)(x + 1)}\, dx$

13 $\displaystyle\int_8^{12} \frac{6x^2 + 47x - 168}{x(x + 6)(x - 7)}\, dx$

14 $\displaystyle\int_4^8 \frac{9(4x - 1)}{x^3 - 9x}\, dx$

15 $\displaystyle\int_{2.5}^{7.5} \frac{x^2 + 5x - 1}{x^2 - x}\, dx$

▰▰▰ ASSIGNMENT ▰▰▰

Previously we have applied our method of partial fractions.

Now we can write $\dfrac{0.4t^2 + 18t + 120}{(t - 15)^2(t + 1)} \equiv \dfrac{30}{(t - 15)^2} + \dfrac{0.4}{t + 1}$.

This is our expression for C_h, the hourly cost of running a particular plant, in terms of t hours. By integrating from $t = 0$ to $t = 10$ we can find the total cost per working day for those 10 hours.

$$
\text{Total daily cost} = \int_0^{10} \frac{30}{(t - 15)^2} + \frac{0.4}{t + 1}\, dt
$$

$$
= \int_0^{10} 30(t - 15)^{-2} + \frac{0.4}{t + 1}\, dt
$$

$$
= \left[\frac{30(t - 15)^{-1}}{-1} + 0.4\ln(t + 1) \right]_0^{10}
$$

$$
= \left[\frac{-30}{t - 15} + 0.4\ln(t + 1) \right]_0^{10}
$$

$$
= \left[\frac{-30}{-5} + 0.4\ln 11 \right] - \left[\frac{-30}{-15} + 0.4\ln 1 \right]
$$

$$= 6 + 0.959 - 2 - 0$$
$$= 4.959.$$

Remember our original formula was in terms of thousands of pounds. We interpret our total cost answer of 4.959 as £4.959 $\times 10^3$, i.e. £4959.

Our final Exercise looks at some practical problems. You will find our partial fractions methods and integration useful.

▄▄▄▄▄ EXERCISE 12.6 ▄▄▄▄▄▄▄▄▄▄▄▄▄▄▄▄▄

1　The area under a velocity–time curve represents the distance travelled. The velocity, v ms^{-1}, in terms of time, t s, is given by $v = \dfrac{6t + 16.5}{t^2 + 5t + 6}$. Find the distance travelled in the first 10 seconds.

2　A variable force, F N, is related to displacement, x m, by $F = \dfrac{x^3 + 2}{x^2}$. Integrate F with respect to x to find the mean value of the force between 12 m and 24 m.

3　An acceleration, f ms^{-2}, is related to time, t s, according to $f = \dfrac{3t^2 + 10t + 10}{(t+1)(t+2)^2}$. Find the average acceleration during the first 20 seconds.

4　y is related to x according to $y = \dfrac{x-1}{x-2}$. Divide out the fraction $\dfrac{x-1}{x-2}$ to get $1 + \dfrac{1}{x-2}$. Find the mean value of y between $x = 2.5$ and $x = 5$. Use the usual formula to find the root mean square (RMS) value of y.

5　The formula for kinetic energy is $K = \frac{1}{2}mv^2$ where m (kg) is the mass and v (ms^{-1}) is the velocity. The mass is 0.5 kg and $v = \dfrac{8}{t(t+2)}$ in terms of time t. Find the mean kinetic energy between $t = 5$ and $t = 10$ seconds.

13 Integration by Parts

The objectives of this chapter are to:

1 Derive the integration by parts formula from the product rule for differentiation.
2 Use the integration by parts formula for simple examples of the type $\int xe^{ax}dx$, $\int x \sin ax \, dx$, $\int x \ln x \, dx$.
3 Use the integration by parts formula to find $\int \ln x \, dx$.
4 Use the integration by parts formula twice, e.g. $\int e^{ax} \sin bx \, dx$, $\int x^2 \cos ax \, dx$.

Introduction

We start with the derivation of the formula. Remember integration is the reverse process of differentiation. Our derivation is based on the differentiation of a product. Soon you will see we need to distinguish between differentiation and integration. Do *not* guess standard rules or results. Use the lists of differentials and integrals if you are in any doubt.

Most of our examples are based on the integration of products. This is common, but not exclusive, to integration by parts. In our early examples we apply the method once. Later we apply it twice to an integral. While we stop here the technique may be applied repeatedly.

■■■■■ ASSIGNMENT ■■■■■

The Assignment for this chapter looks again at the horizontal string tied at both ends. It is plucked at the centre to oscillate about the original horizontal line. The velocity, $V \, \text{ms}^{-1}$, is given in

terms of time, t seconds, by $V = 0.1e^{-2t} \cos \dfrac{2t}{3}$.

As the wave moves away from the centre both to the left and to the right is damped. This is due to the exponential part of our velocity equation.

Remember that displacement is the area under a velocity–time graph. We know from Chapter 11 (Numerical Integration) that we can write this as a definite integral

276

Distance travelled $= \int_0^1 0.1e^{-2t} \cos\frac{2t}{3}\, dt$ in the first second.

This time we are going to find the value of this integral using integration by parts.

Derivation of the formula

Suppose we have a product of u and v, i.e. $y = uv$ (or $u \times v$), where u and v are both functions of x. Then the product rule for differentiation is

$$\frac{dy}{dx} = u\frac{dv}{dx} + v\frac{du}{dx}.$$

We can rearrange this into

$$u\frac{dv}{dx} = \frac{d(uv)}{dx} - v\frac{du}{dx}. \qquad \boxed{\text{Substituting for } y = uv.}$$

For each of the three terms we integrate with respect to x, i.e.

$$\int u\frac{dv}{dx}\, dx = \int \frac{d(uv)}{dx}\, dx - \int v\frac{du}{dx}\, dx.$$

Now the middle term, $\int \frac{d(uv)}{dx}\, dx$, is the interesting one. What does it mean? Mathematically we start with uv, representing some combined function of x, and differentiate it to get $\frac{d(uv)}{dx}$. Then we attempt to integrate this answer, i.e. returning to our original uv. This means we can replace $\int \frac{d(uv)}{dx}\, dx$ with our original uv.

The other terms are more of a mixture and *cannot* be simplified like this. Part of each term contains u or v and a differential, $\frac{dv}{dx}$ or $\frac{du}{dx}$. Hence we just write our formula as

$$\int u\frac{dv}{dx}\, dx = uv - \int v\frac{du}{dx}\, dx.$$

In each example we are going to compare the question with the left-hand side of this formula. This means we need to identify a product. We label one part of that product u and differentiate it to get $\frac{du}{dx}$. The general aim is for $\frac{du}{dx}$ to be a simpler function than u. The remaining part of the product is labelled $\frac{dv}{dx}$ and integrated to get v. In this part of the integration custom allows us to omit the constant of integration.

Integration by parts

We look at four different examples and apply our formula once in each case. In the first three examples the product is easily identified. For the fourth example we have to create a simple product.

━━━━━ **Example 13.1** ━━━━━━━━━━━━━━━

Find $\displaystyle\int_0^{0.6} 3xe^{2x}\, dx.$

Our product is $3x$ and e^{2x}. When we differentiate or integrate an exponential that exponential remains together with some multiplier, i.e. it is neither simpler nor more complicated than before. When we differentiate $3x$ we get 3. Now 3 is simpler for us than $3x$ because it is a pure number. This means we are going to differentiate $3x$ and integrate e^{2x}.

If we had attempted the opposite operations the integral of $3x$ would have given us the more complicated $\dfrac{3}{2}x^2$. This endorses our original choice.

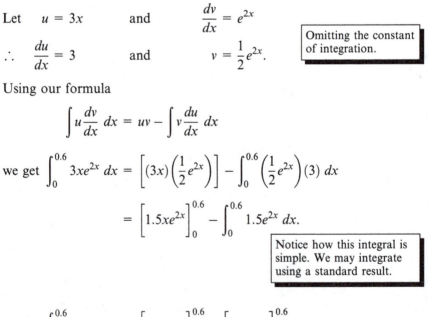

Let $\quad u = 3x \qquad$ and $\qquad \dfrac{dv}{dx} = e^{2x}$

$\therefore \quad \dfrac{du}{dx} = 3 \qquad$ and $\qquad v = \dfrac{1}{2}e^{2x}.$

> Omitting the constant of integration.

Using our formula

$$\int u\frac{dv}{dx}\, dx = uv - \int v\frac{du}{dx}\, dx$$

we get $\displaystyle\int_0^{0.6} 3xe^{2x}\, dx = \left[(3x)\left(\frac{1}{2}e^{2x}\right)\right] - \int_0^{0.6}\left(\frac{1}{2}e^{2x}\right)(3)\, dx$

$$= \left[1.5xe^{2x}\right]_0^{0.6} - \int_0^{0.6} 1.5e^{2x}\, dx.$$

> Notice how this integral is simple. We may integrate using a standard result.

i.e. $\displaystyle\int_0^{0.6} 3xe^{2x}\, dx = \left[1.5xe^{2x}\right]_0^{0.6} - \left[0.75e^{2x}\right]_0^{0.6}$

$$= [1.5(0.6)e^{1.2} - 1.5(0)e^0] - [0.75e^{1.2} - 0.75e^0]$$

$$= 2.988 - 0 - 2.490 + 0.75$$

$$= 1.25.$$

▨▨▨▨▨▨ **Example 13.2** ▨▨▨▨▨▨▨▨▨▨▨▨▨▨▨▨▨▨▨▨▨▨▨▨▨▨▨▨▨▨▨▨

Find $\displaystyle\int_{\pi/2}^{\pi} 2x \sin 4x \, dx$.

Our product is $2x$ and $\sin 4x$. When we differentiate or integrate a sine we get a cosine, i.e. it is neither simpler nor more complicated than before. When we differentiate $2x$ we get 2. Now 2 is simpler for us than $2x$ because it is a pure number. This means we are going to differentiate $2x$ and integrate $\sin 4x$.

Let $\quad u = 2x \quad$ and $\quad \dfrac{dv}{dx} = \sin 4x$.

$\therefore \quad \dfrac{du}{dx} = 2 \quad$ and $\quad v = -\dfrac{1}{4}\cos 4x$

> Omitting the constant of integration.

Using our formula

$$\int u \frac{dv}{dx} \, dx = uv - \int v \frac{du}{dx} \, dx$$

we get

$$\int_{\pi/2}^{\pi} 2x \sin 4x \, dx = \left[(2x)\left(-\frac{1}{4}\cos 4x\right) \right]_{\pi/2}^{\pi} - \int_{\pi/2}^{\pi} \left(-\frac{1}{4}\cos 4x\right)(2) \, dx$$

$$= \left[-0.5x\cos 4x \right]_{\pi/2}^{\pi} + \int_{\pi/2}^{\pi} 0.5 \cos 4x \, dx.$$

> Notice how this integral is simple. We may integrate using a standard result.

i.e. $\displaystyle\int_{\pi/2}^{\pi} 2x \sin 4x \, dx = \left[-0.5x\cos 4x \right]_{\pi/2}^{\pi} + \left[0.125 \sin 4x \right]_{\pi/2}^{\pi}$

$$= \left[-0.5\pi \cos 4\pi - -0.5\frac{\pi}{2}\cos 2\pi \right]$$

$$+ \left[0.125 \sin 4\pi - 0.125 \sin 2\pi \right]$$

$$= -1.571 + 0.785 + 0 - 0$$

$$= -0.79.$$

▨▨▨▨▨▨ **Example 13.3** ▨▨▨▨▨▨▨▨▨▨▨▨▨▨▨▨▨▨▨▨▨▨▨▨▨▨▨▨▨▨▨▨

Find $\displaystyle\int_{1.5}^{3.5} 4x \ln x \, dx$.

Our product is $4x$ and $\ln x$. This time we have a slightly different style for u and $\dfrac{dv}{dx}$. It is necessary because we know only how to differentiate $\ln x$.

Let $\quad u = \ln x \quad$ and $\quad \dfrac{dv}{dx} = 4x$

$\therefore \quad \dfrac{du}{dx} = \dfrac{1}{x} \quad$ and $\quad v = 2x^2$

Using our formula

$$\int u \frac{dv}{dx}\, dx = uv - \int v \frac{du}{dx}\, dx$$

we get $\displaystyle\int_{1.5}^{3.5} 4x \ln x\, dx = \left[(\ln x)(2x^2) \right]_{1.5}^{3.5} - \int_{1.5}^{3.5} (2x^2)\left(\frac{1}{x}\right) dx$

$$= \left[2x^2 \ln x \right]_{1.5}^{3.5} - \int_{1.5}^{3.5} 2x\, dx.$$

> Notice the order so $2x^2$ does *not* become confused with $\ln x$.

i.e. $\displaystyle\int_{1.5}^{3.5} 4x \ln x\, dx = \left[2x^2 \ln x \right]_{1.5}^{3.5} - \left[x^2 \right]_{1.5}^{3.5}$

$$= [2(3.5)^2 \ln 3.5 - 2(1.5)^2 \ln 1.5] - [3.5^2 - 1.5^2]$$

$$= 30.693 - 1.825 - 12.25 + 2.25$$

$$= 18.87.$$

▰▰▰ Example 13.4 ▰▰▰

Find $\displaystyle\int_{1}^{4} \ln x\, dx$.

We have seen this integral before as Example 11.1 in Chapter 11 (Numerical Integration).

There is *no* obvious product and so we need to create one. We think of $\ln x$ as $1 \ln x$. Again we use the slightly varied style of Example 13.3 because we do not know how to integrate $\ln x$.

Let $\quad u = \ln x \quad$ and $\quad \dfrac{dv}{dx} = 1$

$\therefore \quad \dfrac{du}{dx} = \dfrac{1}{x} \quad$ and $\quad v = x$

Using our formula

$$\int u \frac{dv}{dx}\, dx = uv - \int v \frac{du}{dx}\, dx$$

we get $\displaystyle\int_{1}^{4} \ln x\, dx = \left[(\ln x)(x) \right]_{1}^{4} - \int_{1}^{4} (x)\left(\frac{1}{x}\right) dx$

$$= \left[x \ln x \right]_{1}^{4} - \int_{1}^{4} 1\, dx$$

> Again notice the order so x does *not* become confused with $\ln x$.

i.e.
$$\int_1^4 \ln x \, dx = \left[x \ln x \right]_1^4 - \left[x \right]_1^4$$
$$= [4 \ln 4 - 1 \ln 1] - [4 - 1]$$
$$= 5.545 \ldots - 0 - 3$$
$$= 2.54517 \ldots \quad \text{as before.}$$

EXERCISE 13.1

Apply the integration by parts formula to find the values of the following integrals.

1 $\int_{0.5}^{1.5} 2xe^x \, dx$

2 $\int_{\pi/2}^{\pi} x \cos x \, dx$

3 $\int_{-1}^{1} 4xe^{-x} \, dx$

4 $\int_4^6 x \ln x \, dx$

5 $\int_{\pi/6}^{\pi/3} x \cos 2x \, dx$

6 $\int_{\pi/4}^{\pi/2} 4x \sin 2x \, dx$

7 $\int_0^{0.5} xe^{-2x} \, dx$

8 $\int_{-2}^{0} 4xe^{-2x} \, dx$

9 $\int_2^4 x^2 \ln x \, dx$

10 $\int_1^{2.5} 3x^3 \ln 2x \, dx$

Integration by parts – twice

This situation generally occurs when we have x^2 multiplied by either an exponential or a trigonometric function. We use the same technique in both cases. There is another situation which we need to consider later.

Example 13.5

Find $\int_1^3 \frac{1}{2} x^2 e^{-2x} \, dx$.

For easy reference we label this integral as I.

Our product is $\frac{1}{2}x^2$ and e^{-2x}. We return to the pattern of Examples 13.1 and 13.2 for u and $\frac{dv}{dx}$.

Let $u = \frac{1}{2}x^2$ and $\frac{dv}{dx} = e^{-2x}$

$\therefore$ $\frac{du}{dx} = x$ and $v = -\frac{1}{2}e^{-2x}$

Using our formula

$$\int u \frac{dv}{dx} \, dx = uv - \int v \frac{du}{dx} \, dx$$

we get $I = \left[\left(\frac{1}{2}x^2\right)\left(-\frac{1}{2}e^{-2x}\right)\right]_1^3 - \int_1^3 \left(-\frac{1}{2}e^{-2x}\right)(x)\,dx$

$\qquad = \left[-\frac{1}{4}x^2e^{-2x}\right]_1^3 + \int_1^3 \frac{1}{2}xe^{-2x}\,dx.$

Now our last integral is not quite simple enough. Its style is similar to Example 13.1. This means we need to apply integration by parts again to this last integral.

For $\int_1^3 \frac{1}{2}xe^{-2x}\,dx$

let $u = \frac{1}{2}x$ and $\dfrac{dv}{dx} = e^{-2x}$

$\therefore \dfrac{du}{dx} = \dfrac{1}{2}$ and $v = -\dfrac{1}{2}e^{-2x}$

Notice how we have reduced x^2 to $\dfrac{1}{2}$ in two applications of our method. Also notice how the exponential part continues. Using our formula again

$$\int u\frac{dv}{dx}\,dx = uv - \int v\frac{du}{dx}\,dx$$

we get $\int_1^3 \frac{1}{2}xe^{-2x}\,dx = \left[\left(\frac{1}{2}x\right)\left(-\frac{1}{2}e^{-2x}\right)\right]_1^3 - \int_1^3 \left(-\frac{1}{2}e^{-2x}\right)\left(\frac{1}{2}\right)dx$

$\qquad = \left[-\frac{1}{4}xe^{-2x}\right]_1^3 + \int_1^3 \frac{1}{4}e^{-2x}\,dx$

$\qquad = \left[-\frac{1}{4}xe^{-2x}\right]_1^3 + \left[-\frac{1}{8}e^{-2x}\right]_1^3.$

Now we link together both integration by parts, being careful with the $+/-$ signs, i.e.

$I = \int_1^3 \frac{1}{2}x^2e^{-2x}\,dx = \left[-\frac{1}{4}x^2e^{-2x}\right]_1^3 + \left[-\frac{1}{4}xe^{-2x}\right]_1^3 + \left[-\frac{1}{8}e^{-2x}\right]_1^3$

$\qquad = -\frac{1}{4}[3^2e^{-6} - 1^2e^{-2}] - \frac{1}{4}[3e^{-6} - 1e^{-2}] - \frac{1}{8}[e^{-6} - e^{-2}]$

$\qquad = -0.25[0.0223 - 0.1353] - 0.25[0.0074 - 0.1353]$

$\qquad\qquad\qquad\qquad\qquad - 0.125[0.0025 - 0.1353]$

$\qquad = 0.0283 + 0.0320 + 0.0166$

$\qquad = 0.077.$

Example 13.6

Find $\displaystyle\int_0^{0.9} x^2 \cos x \, dx$.

For easy reference we label this integral as *I*. We remind ourselves that the limits of integration are in **radians** because of the trigonometry involved. Our product is x^2 and cos *x*. Again, we use the pattern of Examples 13.1 and 13.2 for *u* and $\dfrac{dv}{dx}$.

Let $\quad u = x^2 \quad$ and $\quad \dfrac{dv}{dx} = \cos x$

$\therefore \quad \dfrac{du}{dx} = 2x \quad$ and $\quad v = \sin x$

Using our formula

$$\int u \frac{dv}{dx} \, dx = uv - \int v \frac{du}{dx} \, dx$$

we get $\qquad I = \left[(x^2)(\sin x) \right]_0^{0.9} - \int_0^{0.9} (\sin x)(2x) \, dx$

$$= \left[x^2 \sin x \right]_0^{0.9} - \int_0^{0.9} 2x \sin x \, dx.$$

Again, our last integral is not quite simple enough. It's style is similar to Example 13.2. This means we need to apply integration by parts again to that last integral.

For $\displaystyle\int_0^{0.9} 2x \sin x \, dx$

let $\quad u = 2x \quad$ and $\quad \dfrac{dv}{dx} = \sin x$

$\therefore \quad \dfrac{du}{dx} = 2 \quad$ and $\quad v = -\cos x$

Notice how we have reduced x^2 to 2 in two applications of our method. Also notice how the trigonometric part alternates between sine and cosine.

Using our formula again

$$\int u \frac{dv}{dx} \, dx = uv - \int v \frac{du}{dx} \, dx$$

we get $\displaystyle\int_0^{0.9} 2x \sin x \, dx = \left[(2x)(-\cos x) \right]_0^{0.9} - \int_0^{0.9} (-\cos x)(2) \, dx$

$$= \left[-2x \cos x \right]_0^{0.9} + \int_0^{0.9} 2 \cos x \, dx$$

$$= \left[-2x \cos x \right]_0^{0.9} + \left[2 \sin x \right]_0^{0.9}.$$

Now we link together both integration by parts being careful with the $+/-$ signs, i.e.

$$I = \int_0^{0.9} x^2 \cos x \, dx = \left[x^2 \sin x \right]_0^{0.9} - \left[-2x \cos x \right]_0^{0.9} - \left[2 \sin x \right]_0^{0.9}$$

$$= \left[x^2 \sin x \right]_0^{0.9} + \left[2x \cos x \right]_0^{0.9} - \left[2 \sin x \right]_0^{0.9}$$

$$= [0.9^2 \sin 0.9 - 0] + [2(0.9) \cos 0.9 - 0]$$
$$- [2 \sin 0.9 - 0]$$

$$= 0.6345 + 1.1189 - 1.5667$$

$$= 0.187$$

which agrees with result of Example 11.3 in Chapter 11, Numerical Integration.

■■■■ EXERCISE 13.2 ■■■■

Apply the integration by parts formula twice to find the values of the following integrals.

1 $\displaystyle\int_0^{\pi/2} x^2 \cos 2x \, dx$

2 $\displaystyle\int_{\pi/3}^{2\pi/3} 2x^2 \sin x \, dx$

3 $\displaystyle\int_{-1}^{1} x^2 e^{-x} \, dx$

4 $\displaystyle\int_0^{1.5} 4x^2 e^{2x} \, dx$

5 $\displaystyle\int_{\pi/4}^{\pi/2} 3x^2 \sin 2x \, dx$

Our final example brings together exponential and trigonometric functions. In Example 13.7 we use $u = e^x$ and $\dfrac{dv}{dx} = \sin x$. You will see the exponential throughout the example with various multipliers. The trigonometric part alternates between sine and cosine.

Alternatively we could have reversed this order for u and $\dfrac{dv}{dx}$. In practice the order is unimportant provided you are consistent.

■■■■ Example 13.7 ■■■■

Find $\displaystyle\int_0^{\pi/2} e^x \sin x \, dx$.

For easy reference later we label this integral I.

Let $u = e^x$ and $\dfrac{dv}{dx} = \sin x$

$\therefore \dfrac{du}{dx} = e^x$ and $v = -\cos x$

Using our formula

$$\int u \frac{dv}{dx} \, dx = uv - \int v \frac{du}{dx} \, dx$$

we get
$$I = \left[(e^x)(-\cos x) \right]_0^{\pi/2} - \int_0^{\pi/2} (-\cos x)(e^x) \, dx$$

$$= \left[-e^x \cos x \right]_0^{\pi/2} + \int_0^{\pi/2} e^x \cos x \, dx.$$

Now our last integral is no simpler than before. We have just replaced one trigonometric function with another one. Let us apply integration by parts again to this last integral and see what happens.

For $\displaystyle\int_0^{\pi/2} e^x \cos x \, dx$

let $\quad u = e^x \quad$ and $\quad \dfrac{dv}{dx} = \cos x$

$\therefore \quad \dfrac{du}{dx} = e^x \quad$ and $\quad v = \sin x$

Using our formula again

$$\int u \frac{dv}{dx} \, dx = uv - \int v \frac{du}{dx} \, dx$$

we get $\displaystyle\int_0^{\pi/2} e^x \cos x \, dx = \left[(e^x)(\sin x) \right]_0^{\pi/2} - \int_0^{\pi/2} (\sin x)(e^x) \, dx$

$$= \left[e^x \sin x \right]_0^{\pi/2} - \int_0^{\pi/2} e^x \sin x \, dx.$$

Notice how this last integral is the one we started with. Let us link together both integration by parts being careful with the $+/-$ signs, i.e.

$$I = \left[-e^x \cos x \right]_0^{\pi/2} + \left[e^x \sin x \right]_0^{\pi/2} - I$$

i.e. $\quad 2I = \left[-e^x \cos x \right]_0^{\pi/2} + \left[e^x \sin x \right]_0^{\pi/2}$

$$= \left[-e^{\pi/2} \cos \frac{\pi}{2} - -e^0 \cos 0 \right] + \left[e^{\pi/2} \sin \frac{\pi}{2} - e^0 \sin 0 \right]$$

$$= 0 + 1 + 4.8105 - 0$$

$$= 5.81.$$

EXERCISE 13.3

Apply the integration by parts formula twice to find the values of the following integrals.

1 $\displaystyle\int_0^{\pi/6} e^x \sin 2x \, dx$

2 $\displaystyle\int_0^{\pi} e^{-x} \cos 2x \, dx$

3 $\displaystyle\int_{\pi/6}^{\pi/3} e^{-3x} \cos x \, dx$

4 $\displaystyle\int_{\pi/2}^{\pi} 2e^{2x} \sin \tfrac{1}{2}x \, dx$

5 $\displaystyle\int_{\pi/4}^{\pi} 2e^{2x} \cos 4x \, dx$

Now we are in a position to look at our Assignment. Then there is a final exercise for you to apply the methods of this chapter.

ASSIGNMENT

We are going to find the distance travelled in the first second. We know from early in this chapter that

$$\text{Distance travelled} = \int_0^1 0.1e^{-2t} \cos \frac{2t}{3} \, dt$$

$$= 0.1I \qquad\qquad \text{where } I = \int_0^1 e^{-2t} \cos \frac{2t}{3} \, dt.$$

Integrating by parts we use the standard techniques.

Let $u = e^{-2t}$ and $\dfrac{dv}{dt} = \cos \dfrac{2t}{3}$

$\therefore$ $\dfrac{du}{dt} = -2e^{-2t}$ and $v = \dfrac{3}{2} \sin \dfrac{2t}{3}$

Using our formula

$$\int u \frac{dv}{dt} \, dt = uv - \int v \frac{du}{dt} \, dt$$

we get $I = \left[(e^{-2t}) \left(\dfrac{3}{2} \sin \dfrac{2t}{3} \right) \right]_0^1 - \int_0^1 \left(\dfrac{3}{2} \sin \dfrac{2t}{3} \right) (-2e^{-2t}) \, dt$

$= \left[\dfrac{3}{2} e^{-2t} \sin \dfrac{2t}{3} \right]_0^1 + \int_0^1 3e^{-2t} \sin \dfrac{2t}{3} \, dt$

> Simplifying signs and fractions.

Now our last integral is no simpler than before. We have just replaced one trigonometric function with another one. Let us apply integration by parts again to this last integral and see what happens.

For $\int_0^1 3e^{-2t} \sin \frac{2t}{3} \, dt$

let $\quad u = 3e^{-2t} \quad$ and $\quad \dfrac{dv}{dt} = \sin \dfrac{2t}{3}$

$\therefore \quad \dfrac{du}{dt} = -6e^{-2t} \quad$ and $\quad v = -\dfrac{3}{2} \cos \dfrac{2t}{3}$

Using our formula again

$$\int u \frac{dv}{dt} \, dt = uv - \int v \frac{du}{dt} \, dt$$

we get $\displaystyle\int_0^1 3e^{-2t} \sin \frac{2t}{3} \, dt = \left[(3e^{-2t})\left(-\frac{3}{2} \cos \frac{2t}{3} \right) \right]_0^1$

$$- \int_0^1 \left(-\frac{3}{2} \cos \frac{2t}{3} \right)(-6e^{-2t}) \, dt$$

$$= \left[-\frac{9}{2} e^{-2t} \cos \frac{2t}{3} \right]_0^1 - 9\int_0^1 e^{-2t} \cos \frac{2t}{3} \, dt.$$

> Simplifying signs and fractions.

Notice how this last integral is the one we started with, I. Let us link together both integration by parts, being careful with the $+/-$ signs, i.e.

$$I = \left[\frac{3}{2} e^{-2t} \sin \frac{2t}{3} \right]_0^1 + \left[-\frac{9}{2} e^{-2t} \cos \frac{2t}{3} \right]_0^1 - 9I$$

i.e. $\quad 10I = \left[\frac{3}{2} e^{-2t} \sin \frac{2t}{3} \right]_0^1 + \left[-\frac{9}{2} e^{-2t} \cos \frac{2t}{3} \right]_0^1$

> Limits are radians.

$$= \left[\frac{3}{2} e^{-2} \sin \frac{2}{3} - \frac{3}{2} e^0 \sin 0 \right] + \left[-\frac{9}{2} e^{-2} \cos \frac{2}{3} - -\frac{9}{2} e^0 \cos 0 \right]$$

$$= 0.1255 - 0 - 0.4786 + 4.5$$

i.e. $\quad I = \dfrac{1}{10} \times 4.1469$

$$= 0.41.$$

Remember that the distance travelled is $0.1I = 0.041$ m when substituting for I.

EXERCISE 13.4

1 Find the mean value of $y = x^2 \ln x$ between $x = 5$ and $x = 10$.

2 The work done by a force, F, is given by the integral, $\displaystyle\int_a^b F \, ds$. If $F = (10 - s)e^{0.2s}$, where s is the displacement, find the work done over the first 10 metres.

3   A voltage, V, over time, t, is given by $V = 2t \sin t$. Find the mean value of this voltage from $t = 0$ to $t = 6$ seconds.

4   A variable force, F (N), is related to distance, s (m), by $F = 5s^2 e^{-s}$. The mean force given by

$$\text{Mean force} = \frac{1}{b-a} \int_a^b F \, ds.$$

Calculate the mean force from $s = 0$ to $s = 3$ m.

5   Find where the curve $y = (9 - x^2) \ln x$ cuts the horizontal axis (i.e. where $y = 0$). (Remember that a logarithm is defined only for positive values of x.) Use these values as your limits of integration. Find the area bounded by the curve above the horizontal axis.

14 Integration by Substitution

The objectives of this chapter are to:

1   Use a given algebraic substitution to reduce an indefinite integral to a standard format, e.g. $\displaystyle\int \frac{x}{1 + x^2}\, dx, \quad \int \frac{x}{\sqrt{a^2 + x^2}}\, dx.$

2   Use a given trigonometric substitution to reduce an indefinite integral to a standard format, e.g. $\displaystyle\int \sqrt{a^2 - x^2}\, dx, \quad \int \frac{1}{x^2\sqrt{1 + x^2}}\, dx, \quad \int \frac{1}{a\cos\theta + b\sin\theta}\, d\theta.$

3   Work out new limits of integration for the substitution into a definite integral.

4   Use trigonometric identities to reduce an indefinite integral to a standard format.

Introduction

For both types of integration by substitution we give the appropriate substitution. In each case the basic method is the same. Next we look at definite integration. The limits in each example refer to the original variable. When we change the variable we must change the limits. This maintains the consistency. Our final type of integration is more of a replacement than a substitution. We use trigonometric identities to simplify the function before attempting the integration.

ASSIGNMENT

The Assignment for this chapter returns to the ellipse we introduced in Volume 1, Chapter 16. Remember, the general equation of an ellipse is $\dfrac{x^2}{a^2} + \dfrac{y^2}{b^2} = 1$. It crosses the axes at $(-a, 0)$, $(a, 0)$, $(0, -b)$ and $(0, b)$. In that chapter we applied the formula for area, πab. Since then we have become expert at finding the area under a curve using integration. In this Assignment we are going to integrate to find the area of the ellipse $\dfrac{x^2}{9} + \dfrac{y^2}{4} = 1$. Then we are going to derive the general area formula for an ellipse.

Algebraic substitution

We have a variety of types to demonstrate in our examples. They all involve some function of x, $f(x)$, and its derivative, $f'(x)$. Remember we are attempting to integrate yet the method involves a derivative. You must be able to distinguish between differentiation and integration. Do *not* guess standard rules or results. Use the lists of differentials and integrals if you are in any doubt.

The original variable must be replaced completely by the new variable. Later we will see this means we need to change the limits of integration too.

▨▨▨▨ **Example 14.1** ▨▨▨▨▨▨▨▨▨▨▨▨▨▨

Find $\int (3x^2 - 2)(2x^3 - 4x + 7)^5 \, dx$.

Our integral looks to have two functions. We concentrate on the more complicated one, i.e. the bracket raised to the power 5. We substitute u for this part of the integral and then differentiate.

Let $\qquad u = 2x^3 - 4x + 7$

$\therefore \qquad \dfrac{du}{dx} = 6x^2 - 4$

i.e. $\qquad du = (6x^2 - 4) \, dx$ | Multiplying through by dx.

$\qquad\qquad = 2(3x^2 - 2) \, dx$ | Removing a common factor.

i.e. $\quad \dfrac{1}{2} \, du = (3x^2 - 2) \, dx$.

All these adjustments after the differentiation are important. You will see the first bracket, $(3x^2 - 2)$, and dx are linked. We can automatically substitute for them with $\dfrac{1}{2} du$.

We rewrite our original integral in stages, i.e.

$$\int (3x^2 - 2)(2x^3 - 4x + 7)^5 \, dx = \int (2x^3 - 4x + 7)^5 (3x^2 - 2) \, dx$$

$$= \int u^5 \times \frac{1}{2} \, du \qquad \boxed{\text{Substituting.}}$$

$$= \frac{1}{2} \int u^5 \, du$$

$$= \frac{1}{2} \times \frac{u^6}{6} + c$$

$$= \frac{1}{12} (2x^3 - 4x + 7)^6 + c.$$

Our original integral was in terms of x. This means we need to substitute back to replace u. Then for consistency our final answer is also in terms of x.

▰▰▰▰▰ **Example 14.2** ▰▰▰▰▰

Find $\displaystyle\int \frac{9x^2 - 6}{(9 + 4x - 2x^3)^4}\, dx$.

Again our integral looks to have two functions. We concentrate on the more complicated one, i.e. the bracket raised to the power 4. It does not matter that it is in the denominator. We substitute u for this part of the integral and then differentiate.

Let $\qquad u = 9 + 4x - 2x^3$

$\therefore \qquad \dfrac{du}{dx} = 4 - 6x^2$

i.e. $\qquad du = (4 - 6x^2)\, dx$ | Multiplying through by dx.

$\qquad\qquad = -2(3x^2 - 2)\, dx$ | Removing a common factor.

i.e. $\qquad -\dfrac{1}{2}\, du = (3x^2 - 2)\, dx.$

Again, all these adjustments after the differentiation are important. They are similar to the previous example, differing only by a minus sign.

We rewrite our original integral in stages. The complete bracket in the denominator is raised to the power 4. This means the power 4 becomes -4 as we reposition the denominator into the numerator position, i.e.

$$\int \frac{9x^2 - 6}{(9 + 4x - 2x^3)^4}\, dx = \int 3(3x^2 - 2)(9 + 4x - 2x^3)^{-4}\, dx$$

| Factorising the numerator.

$$= 3\int (9 + 4x - 2x^3)^{-4}(3x^2 - 2)\, dx$$

$$= 3\int u^{-4} \times \frac{1}{-2}\, du$$ | Substituting.

$$= -\frac{3}{2}\int u^{-4}\, du$$

$$= -\frac{3}{2} \times \frac{u^{-3}}{(-3)} + c$$

$$= \frac{1}{2}(9 + 4x - 2x^3)^{-3} + c.$$ | Cancelling -3s.

Again we substitute to replace u with our original variable, x.

████████ **Example 14.3** ████████

Find $\int \dfrac{3x}{1+x^2}\, dx$.

Our integral looks to have two functions. We concentrate on the denominator, $1 + x^2$. Its first derivative is $2x$, which is close to $3x$, i.e. $3x = 1.5 \times 2x$. We can make this numerical adjustment, but *cannot* easily adjust any algebra. These facts lead us to our important function $1 + x^2$. We substitute u for this part of the integral and then differentiate.

Let $\qquad u = 1 + x^2$

$\therefore \qquad \dfrac{du}{dx} = 2x$

i.e. $\qquad du = 2x\,dx$ | Multiplying through by dx.

i.e. $\qquad 1.5\,du = 1.5 \times 2x\,dx$ | Multiplying through by 1.5.

$\qquad\qquad 1.5\,du = 3x\,dx.$

All these adjustments after the differentiation are important. We substitute to get

$$\int \frac{3x}{1+x^2}\, dx = \int \frac{1.5}{u}\, du \qquad \boxed{\text{Substituting.}}$$

$$= 1.5 \int \frac{1}{u}\, du$$

$$= 1.5 \ln u + c$$

$$= 1.5 \ln(1 + x^2) + c. \qquad \boxed{\text{Substituting } u = 1 + x^2.}$$

As usual, for consistency our final answer is in terms of x as in the original question.

Example 14.3 is a typical one of the general type

$$\int \frac{ax}{b+x^2}\, dx = \frac{a}{2} \ln(b + x^2) + c \quad \text{where } a, b \text{ and } c \text{ are constants.}$$

████████ **Example 14.4** ████████

Find $\int \dfrac{-5x}{\sqrt{4+x^2}}\, dx$.

Once again our integral looks to have two functions. We concentrate on the more complicated one, i.e. the square root in the denominator. We substitute u for part of the denominator and then differentiate.

Let $\qquad u = 4 + x^2$ | Excluding the square root.

$\therefore \qquad \dfrac{du}{dx} = 2x$

i.e. $du = 2x \, dx$

Multiplying through by dx.

i.e. $\dfrac{1}{2} du = x \, dx$.

Now with these adjustments we substitute to get

$$\int \frac{-5x}{\sqrt{4 + x^2}} \, dx = -5 \int \frac{x}{\sqrt{4 + x^2}} \, dx$$

$$= -5 \int \frac{1/2}{\sqrt{u}} \, du$$

$$= -2.5 \int u^{-0.5} \, du$$

$$= -2.5 \times \frac{u^{0.5}}{0.5} + c$$

$$= -5u^{0.5} + c$$

$$= -5(4 + x^2)^{0.5} + c$$

$$\text{or } -5\sqrt{(4 + x^2)} + c.$$

Substituting.

$-5 \times \dfrac{1}{2} = 2.5$.

$\dfrac{1}{\sqrt{u}} = \dfrac{1}{u^{0.5}} = u^{-0.5}$

Example 14.5

Find $\int \sin \theta \cos^2 \theta \, d\theta$.

At first glance this is the integration of a trigonometric function. We can include it within our section on Algebraic Substitution. This is because we are going to use an algebraic substitution for $\cos \theta$. Once again our integral looks to have two functions. We concentrate on the more complicated one, i.e. the cosine raised to the power 2. We think of this as $\cos^2 \theta = (\cos \theta)^2$ and substitute partially for u.

Let $u = \cos \theta$

$\therefore$ $\dfrac{du}{d\theta} = -\sin \theta$

i.e. $du = -\sin \theta \, d\theta$

i.e. $-du = \sin \theta \, d\theta$.

Multiplying through by $d\theta$.
Multiplying through by -1.

Now with these adjustments we substitute to get

$$\int \sin \theta \cos^2 \theta \, d\theta = \int \sin \theta (\cos \theta)^2 \, d\theta$$

$$= \int (\cos \theta)^2 \sin \theta \, d\theta$$

$$= -\int u^2 \, du$$

Substituting.

$$= \frac{-u^3}{3} + c$$

$$= \frac{-\cos^3 \theta}{3} + c \text{ with our final answer in terms of } \theta.$$

We may generalise Example 14.5 to the standard results

$$\int \sin\theta \cos^n\theta \, d\theta = \frac{-\cos^{n+1}\theta}{n+1} + c$$

and $\quad \int \cos\theta \sin^n\theta \, d\theta = \frac{\sin^{n+1}\theta}{n+1} + c.$

In the first Exercise we give you the correct substitutions for each integral.

▮▮▮ EXERCISE 14.1 ▮▮▮▮▮▮▮▮

1 $\int \dfrac{x}{x^2+1} \, dx$ $\qquad\qquad$ let $u = x^2 + 1$

2 $\int x(x^2+5)^2 \, dx$ $\qquad\qquad$ let $u = x^2 + 5$

3 $\int \dfrac{6x}{x^2+6} \, dx$ $\qquad\qquad$ let $u = x^2 + 6$

4 $\int \dfrac{4x}{(2x^2+5)^3} \, dx$ $\qquad\qquad$ let $u = 2x^2 + 5$

5 $\int (x - x^3)(1 + 2x^2 - x^4)^3 \, dx$ $\quad$ let $u = 1 + 2x^2 - x^4$

6 $\int \dfrac{t}{\sqrt{t^2-7}} \, dt$ $\qquad\qquad$ let $u = t^2 - 7$

7 $\int \dfrac{x}{2(2-x^2)} \, dx$ $\qquad\qquad$ let $u = 2 - x^2$

8 $\int 12x\sqrt{4 + 3x^2} \, dx$ $\qquad\qquad$ let $u = 4 + 3x^2$

9 $\int \cos\theta \, (1 + \sin\theta)^2 \, d\theta$ $\qquad$ let $u = 1 + \sin\theta$

10 $\int (e^t - e^{-t})(e^t + e^{-t})^3 \, dt$ $\qquad$ let $u = e^t + e^{-t}$

Definite integration

So far we have learned how to change from one variable to another. For indefinite integration our final answer was in terms of the original variable. It included a constant of integration. Remember for definite integration we do *not* need a constant of integration.

For definite integration, when we change the variable we need to change the limits as well. The original limits refer to the original variable. This means we need new limits consistent with our new variable.

▮▮▮ Example 14.6 ▮▮▮▮▮▮▮▮

Find $\displaystyle\int_0^1 \dfrac{-5x}{\sqrt{4 + x^2}} \, dx.$

We have seen a similar indefinite integral in Example 14.4. We are going to apply that result to this definite integral. There we used the substitution

$u = 4 + x^2$. Now we need to use it further to change our limits of integration.

Using $\quad u = 4 + x^2$

for $x = 1$, $\quad u_U = 4 + 1^2 = 5$,

for $x = 0$, $\quad u_L = 4 + 0^2 = 4$.

> u_U is the new upper limit.
> u_L is the new lower limit.

Hence $\quad \displaystyle\int_0^1 \frac{-5x}{\sqrt{4 + x^2}}\, dx = -5\int_4^5 \frac{1/2}{\sqrt{u}}\, du$

> Substituting.

$$= -5\left[u^{0.5} \right]_4^5$$

> Using our integration from Example 14.4.

$$= -5[\sqrt{5} - \sqrt{4}]$$

$$= -1.18.$$

Example 14.7

Find $\displaystyle\int_0^{\pi/2} \sin\theta(2 - \cos\theta)^3\, d\theta$.

As usual we concentrate on the more complicated part of our integral, i.e. $(2 - \cos\theta)^3$. We substitute u for the part within the bracket, i.e.

let $\quad u = 2 - \cos\theta$

$\therefore \quad \dfrac{du}{d\theta} = \sin\theta$

i.e. $\quad du = \sin\theta\, d\theta$.

> Multiplying through by $d\theta$.

Also we need to change our limits using

$$u = 2 - \cos\theta.$$

For $\theta = \dfrac{\pi}{2}$, $\quad u_U = 2 - \cos\dfrac{\pi}{2} = 2$,

for $\theta = 0$, $\quad u_L = 2 - \cos 0 = 1$.

> u_U is the new upper limit.
> u_L is the new lower limit.

Now we replace our functions in terms of θ, $d\theta$ and our original limits to get

$$\int_0^{\pi/2} \sin\theta(2 - \cos\theta)^3\, d\theta = \int_0^{\pi/2} (2 - \cos\theta)^3 \sin\theta\, d\theta$$

$$= \int_1^2 u^3\, du$$

> Substituting.

$$= \left[\frac{u^4}{4} \right]_1^2$$

$$= \frac{2^4}{4} - \frac{1^4}{4}$$

$$= 3.75.$$

Notice our earlier consistency changing the limits with the variable. This leads directly to a numerical answer.

◼◼◼◼ EXERCISE 14.2 ◼◼◼◼◼◼◼

Find the values of the following integrals using the suggested substitutions.

1 $\displaystyle\int_1^4 \frac{2x}{3+x^2}\, dx$ let $u = 3 + x^2$

2 $\displaystyle\int_0^2 5x\sqrt{3x^2 + 1}\, dx$ let $u = 3x^2 + 1$

3 $\displaystyle\int_4^5 \frac{y}{2y^2 + 1}\, dy$ let $u = 2y^2 + 1$

4 $\displaystyle\int_5^{7.5} \frac{3x}{\sqrt{x^2 - 4}}\, dx$ let $u = x^2 - 4$

5 $\displaystyle\int_1^2 \frac{3x}{(5 - 3x^2)^2}\, dx$ let $u = 5 - 3x^2$

6 $\displaystyle\int_0^1 5(x^4 + 3x^2)(x^5 + 5x^3 - 7)^2\, dx$ let $u = x^5 + 5x^3 - 7$

7 $\displaystyle\int_{0.5}^{1.0} \frac{4e^{2x}}{\sqrt{1 + e^{2x}}}\, dx$ let $u = 1 + e^{2x}$

8 $\displaystyle\int_1^{1.5} \frac{t}{\sqrt{7 - t^2}}\, dt$ let $u = 7 - t^2$

9 $\displaystyle\int_0^{0.5} 2te^{3t^2 - 1}\, dt$ let $u = 3t^2 - 1$

10 $\displaystyle\int_{\pi/8}^{\pi/4} (1 + \sin 2\theta)^2 \cos 2\theta\, d\theta$ let $u = 1 + \sin 2\theta$

Trigonometric substitution

In this section our style of substitution differs slightly from the algebraic substitution. We tend to introduce a new trigonometric function into the integration. There is a function but *not* its derivative as we had before. Again we differentiate the substitution and rearrange the terms. The same rules apply to the changing of limits for consistency. You may find the following trigonometric identities from Chapter 4 useful

　i)　$\sin^2\theta + \cos^2\theta = 1,$

　ii)　$\sec^2\theta = 1 + \tan^2\theta,$

　iii)　$\operatorname{cosec}^2\theta = 1 + \cot^2\theta,$

　iv)　$\cos 2\theta = 2\cos^2\theta - 1,$ and

　v)　$\cos 2\theta = 1 - 2\sin^2\theta.$

We will need to re-arrange these identities when we apply them.

▰▰▰▰ Example 14.8 ▰▰▰▰

Find i) $\int_1^2 \sqrt{4 - x^2}\, dx$, and ii) $\int_0^{1.5} \dfrac{1}{\sqrt{4 - x^2}}\, dx$.

The approach to both examples is similar. Much of the early work on the substitution for x and hence dx is the same. Initially our choice of substitution may seem strange. We need the multiplier of 2 to use later as a common factor.

Let $\qquad x = 2 \sin \theta$

$\therefore \qquad \dfrac{dx}{d\theta} = 2 \cos \theta$

> Multiplying through by $d\theta$.

i.e. $\qquad dx = 2 \cos \theta$

Also we need to change our limits. This is where the solutions separate, though the technique continues similarly.

i) Using $\qquad x = 2 \sin \theta$

for $x = 2$, $\qquad 2 = 2 \sin \theta_U$

> θ_U is the new upper limit.

i.e. $\qquad 1 = \sin \theta_U$

so that $\qquad \theta_U = \dfrac{\pi}{2}$.

For $x = 1$, $\qquad 1 = 2 \sin \theta_L$

> θ_L is the new lower limit.

i.e. $\qquad 0.5 = \sin \theta_L$

so that $\qquad \theta_L = \dfrac{\pi}{6}$.

Now we replace our function in terms of x, dx and our original limits to get

$$\int_1^2 \sqrt{4 - x^2}\, dx = \int_{\pi/6}^{\pi/2} \sqrt{4 - 4\sin^2 \theta}\, 2 \cos \theta\, d\theta$$

> $x = 2 \sin \theta$
> $\therefore\ x^2 = (2 \sin \theta)^2$
> $\qquad = 4 \sin^2 \theta.$

$$= \int_{\pi/6}^{\pi/2} \sqrt{4(1 - \sin^2 \theta)}\, 2 \cos \theta\, d\theta$$

$$= \int_{\pi/6}^{\pi/2} \sqrt{4 \cos^2 \theta}\,.\, 2 \cos \theta\, d\theta$$

> $\cos^2 \theta + \sin^2 \theta = 1$
> $\therefore\ \cos^2 \theta = 1 - \sin^2 \theta.$

$$= \int_{\pi/6}^{\pi/2} 2 \cos \theta\,.\, 2 \cos \theta\, d\theta$$

$$= 4 \int_{\pi/6}^{\pi/2} \cos^2 \theta\, d\theta$$

> $2 \cos^2 \theta - 1 = \cos 2\theta$
> $2 \cos^2 \theta = 1 + \cos 2\theta$
> $\cos^2 \theta = \dfrac{1 + \cos 2\theta}{2}.$

$$= 4 \times \frac{1}{2} \int_{\pi/6}^{\pi/2} (1 + \cos 2\theta)\, d\theta$$

$$= 2 \left[\theta + \frac{1}{2} \sin 2\theta \right]_{\pi/6}^{\pi/2}$$

$$= 2\left(\left[\frac{\pi}{2} + \frac{1}{2}\sin 2\frac{\pi}{2}\right] - \left[\frac{\pi}{6} + \frac{1}{2}\sin 2\frac{\pi}{6}\right]\right)$$

$$= 2(1.571 + 0 - 0.524 - 0.433)$$

$$= 1.23.$$

ii) Using $\quad\quad\quad\quad x = 2\sin\theta$

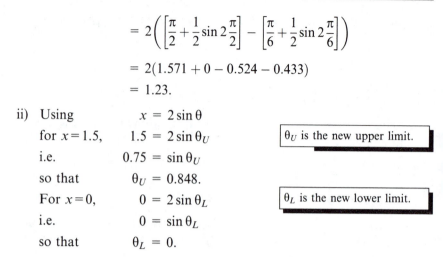

for $x = 1.5$, $\quad\quad 1.5 = 2\sin\theta_U$ $\quad\quad$ θ_U is the new upper limit.

i.e. $\quad\quad\quad\quad 0.75 = \sin\theta_U$

so that $\quad\quad\quad \theta_U = 0.848.$

For $x = 0$, $\quad\quad\quad 0 = 2\sin\theta_L$ $\quad\quad$ θ_L is the new lower limit.

i.e. $\quad\quad\quad\quad\quad 0 = \sin\theta_L$

so that $\quad\quad\quad\quad \theta_L = 0.$

Now we replace our function in terms of x, dx and our original limits to get

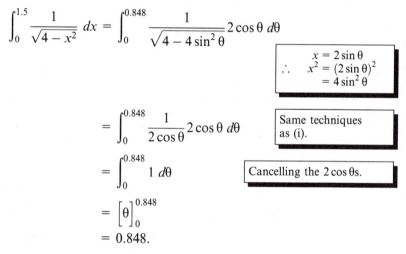

$$\int_0^{1.5} \frac{1}{\sqrt{4 - x^2}}\, dx = \int_0^{0.848} \frac{1}{\sqrt{4 - 4\sin^2\theta}} 2\cos\theta\, d\theta$$

$$\begin{aligned} x &= 2\sin\theta \\ \therefore\quad x^2 &= (2\sin\theta)^2 \\ &= 4\sin^2\theta \end{aligned}$$

$$= \int_0^{0.848} \frac{1}{2\cos\theta} 2\cos\theta\, d\theta$$

Same techniques as (i).

$$= \int_0^{0.848} 1\, d\theta$$

Cancelling the $2\cos\theta$s.

$$= \Big[\theta\Big]_0^{0.848}$$

$$= 0.848.$$

Again, notice our earlier consistency in changing the limits with the variable leads directly to a numerical answer.

▬▬▬▬ **Example 14.9** ▬▬▬▬

Find $\displaystyle\int_0^1 \frac{x^3}{\sqrt{1 - x^2}}\, dx.$

We substitute for x and hence dx, once again using trigonometry. For variety this time we use t instead of θ.

Let $\quad\quad\quad\quad x = \sin t$

$$\frac{dx}{dt} = \cos t$$

i.e. $\quad\quad\quad\quad dx = \cos t\, dt.$ $\quad\quad$ Multiplying through by dt.

Also we need to change our limits using

$$x = \sin t.$$

For $x = 1$, $\qquad 1 = \sin t_U$

t_U is the new upper limit.

so that $\qquad t_U = \dfrac{\pi}{2}$

For $x = 0$, $\qquad 0 = \sin t_L$

t_L is the new lower limit.

so that $\qquad t_L = 0.$

Now we replace our function in terms of x, dx and our original limits to get

$$\int_0^1 \frac{x^3}{\sqrt{1 - x^2}} \, dx = \int_0^{\pi/2} \frac{\sin^3 t}{\sqrt{1 - \sin^2 t}} \cos t \, dt.$$

$$= \int_0^{\pi/2} \frac{\sin^3 t \cos t}{\cos t} \, dt$$

$$\sqrt{1 - \sin^2 t} = \sqrt{\cos^2 t}$$
$$= \cos t.$$

$$= \int_0^{\pi/2} \sin^3 t \, dt$$

Cancelling $\cos t$s.

The next step is not readily obvious. We rewrite $\sin^3 t$ and use a basic trigonometric identity, i.e.

$$\sin^3 t = \sin t \times \sin^2 t$$
$$= \sin t(1 - \cos^2 t)$$
$$= \sin t - \sin t \cos^2 t.$$

Now our integral becomes

$$\int_0^{\pi/2} (\sin t - \sin t \cos^2 t) \, dt$$

$$= \left[-\cos t + \frac{\cos^3 t}{3} \right]_0^{\pi/2}$$

Using our result from Example 14.5.

$$= \left[-\cos \frac{\pi}{2} + \frac{1}{3} \cos^3 \frac{\pi}{2} \right] - \left[-\cos 0 + \frac{1}{3} \cos^3 0 \right]$$

$$= \frac{2}{3}.$$

Example 14.10

Find $\displaystyle\int_1^{\sqrt{3}} \frac{1}{x^2\sqrt{1 + x^2}} \, dx.$

We substitute for x^2 and hence dx. Again our choice is influenced by a trigonometric identity we mentioned earlier.

Let $\qquad x = \tan \theta$

$\therefore \qquad \dfrac{dx}{d\theta} = \sec^2 \theta$

i.e. $\qquad dx = \sec^2 \theta \, d\theta.$

Multiplying through by $d\theta$.

Also we need to change our limits using

$$x = \tan\theta.$$

For $x = \sqrt{3}$, $\sqrt{3} = \tan\theta_U$ θ_U is the new upper limit.

i.e $\theta_U = \dfrac{\pi}{3}.$

For $x = 1$, $1 = \tan\theta_L$ θ_L is the new lower limit.

i.e. $\theta_L = \dfrac{\pi}{4}.$

Now we replace our functions in terms of x, dx and our original limits to get

$$\int_1^{\sqrt{3}} \frac{1}{x^2\sqrt{1+x^2}}\, dx = \int_{\pi/4}^{\pi/3} \frac{1}{\tan^2\theta\sqrt{1+\tan^2\theta}}\sec^2\theta\, d\theta$$ Substituting.

$$= \int_{\pi/4}^{\pi/3} \frac{\sec^2\theta}{\tan^2\theta.\sec\theta}\, d\theta$$ $1 + \tan^2\theta = \sec^2\theta$
 $\therefore\ \sqrt{1+\tan^2\theta} = \sqrt{\sec^2\theta}$
 $= \sec\theta.$

$$= \int_{\pi/4}^{\pi/3} \frac{\sec\theta}{\tan^2\theta}\, d\theta.$$

Now we are not used to dealing with $\sec\theta$ and variations of $\tan\theta$. Remember $\sec\theta = \dfrac{1}{\cos\theta}$ and $\tan\theta = \dfrac{\sin\theta}{\cos\theta}$. Then

$$\frac{\sec\theta}{\tan^2\theta} = \frac{\dfrac{1}{\cos\theta}}{\dfrac{\sin^2\theta}{\cos^2\theta}} = \frac{1}{\cos\theta} \times \frac{\cos^2\theta}{\sin^2\theta} = \frac{\cos\theta}{\sin^2\theta}.$$ $\div$ changes to $\times$, <br> invert fraction.
 Cancelling $\cos\theta$.

This still leaves us with a mixture of trigonometric functions,

$$\int_{\pi/4}^{\pi/3} \frac{\cos\theta}{\sin^2\theta}\, d\theta.$$

The integral requires another substitution if it is not immediately obvious to you.

Let $t = \sin\theta$

$\therefore$ $\dfrac{dt}{d\theta} = \cos\theta$

i.e. $dt = \cos\theta\, d\theta.$ Multiplying through by $d\theta$.

Again we need to change our limits using

$$t = \sin\theta.$$

For $\theta = \dfrac{\pi}{3}$, $t_U = \sin\dfrac{\pi}{3} = 0.866.$ t_U is the new upper limit.

For $\theta = \dfrac{\pi}{4}$, $t_L = \sin\dfrac{\pi}{4} = 0.707.$ t_L is the new lower limit.

Now we replace our functions in terms of θ, $d\theta$ and our original limits to get

$$\int_{\pi/4}^{\pi/3} \frac{\cos \theta}{\sin^2 \theta} \, d\theta = \int_{0.707}^{0.866} \frac{1}{t^2} \, dt$$

Substituting again.

$$= \int_{0.707}^{0.866} t^{-2} \, dt$$

$$= \left[-\frac{1}{t} \right]_{0.707}^{0.866}$$

$$= \frac{-1}{0.866} - \frac{-1}{0.707}$$

$$= 0.260.$$

EXERCISE 14.3

Find the values of the following integrals using the suggested substitutions.

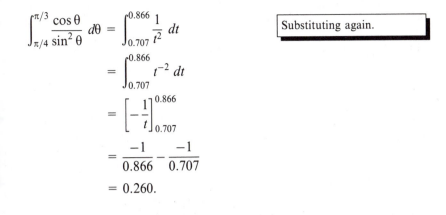

1	$\int_0^1 \frac{2}{1+x^2} \, dx$	$x = \tan \theta$	**6**	$\int_{0.5}^1 \sqrt{9 - x^2} \, dx$	$x = 3 \sin \theta$
2	$\int_0^{0.5} \frac{1}{\sqrt{1 - t^2}} \, dt$	$t = \cos \phi$	**7**	$\int_{0.25}^{0.50} \frac{1}{\sqrt{t}\sqrt{1 - t}} \, dt$	$t = \sin^2 \theta$
3	$\int_{-\sqrt{3}}^{\sqrt{3}} \frac{4}{9 + x^2} \, dx$	$x = 3 \tan \theta$	**8**	$\int_{0.5}^1 \frac{1}{2x^2 \sqrt{4 + x^2}} \, dx$	$x = 2 \tan \theta$
4	$\int_0^1 \sqrt{1 - x^2} \, dx$	$x = \sin \theta$	**9**	$\int_1^2 \frac{1}{t^2 \sqrt{1 + 4t^2}} \, dt$	$t = \frac{1}{2} \tan \theta$
5	$\int_{0.25}^{0.75} \frac{1}{\sqrt{t}} \sin \sqrt{t} \, dt$	$\theta = \sqrt{t}$	**10**	$\int_0^{\sqrt{2}} \frac{2x}{\sqrt{4 - x^2}} \, dx$	$x = 2 \sin t$

ASSIGNMENT

We are going to find the area of a particular ellipse, $\dfrac{x^2}{9} + \dfrac{y^2}{4} = 1$.
Afterwards we will generalise this to any ellipse and deduce the formula πab.

First we rearrange the equation to express y in terms of x.

$$\frac{x^2}{9} + \frac{y^2}{4} = 1$$

becomes

$$\frac{y^2}{4} = 1 - \frac{x^2}{9}$$

$$y^2 = 4\left(1 - \frac{x^2}{9}\right)$$

$$\therefore \qquad y = 2\left(1 - \frac{x^2}{9}\right)^{1/2}.$$

Positive square root of both sides.

In Fig. 14.1 we remind ourselves of the shape of this ellipse and where it crosses the axes. Our equation for y uses the positive square root. This is the portion of the curve above the x-axis. If we use the negative square root we will find that negative area below the x-axis.

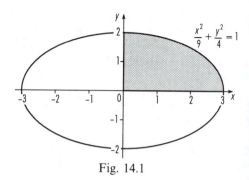

$$\frac{x^2}{9} + \frac{y^2}{4} = 1$$

Fig. 14.1

Each axis is a line of symmetry so the area in each quadrant is the same. This means we can find the area in any quadrant and multiply by 4 to find the total area. The area (shaded) in the first quadrant is the easiest to calculate. Using the area integral,

$$\text{Area} = \int_a^b y \, dx$$

we get $\text{Area} = \int_0^3 2\left(1 - \frac{x^2}{9}\right)^{1/2} dx$

> Limits from $x=0$ to $x=3$ for the first quadrant.

We substitute for x and hence dx, using trigonometry.

Let $x = 3 \sin t$

$\therefore$ $\dfrac{dx}{dt} = 3 \cos t$

i.e. $dx = 3 \cos t \, dt$.

> Multiplying through by dt.

Also we need to change our limits using

$$x = 3 \sin t.$$

For $x = 3$, $3 = 3 \sin t_U$

> t_U is the new upper limit.

so that $t_U = \dfrac{\pi}{2}$.

For $x = 0$, $0 = \sin t_L$

> t_L is the new lower limit.

so that $t_L = 0$.

Now we replace our function in terms of x, dx and our original limits to get

$$\int_0^3 2\left(1 - \frac{x^2}{9}\right)^{1/2} dx = 2 \int_0^{\pi/2} \left(1 - \frac{(3 \sin t)^2}{9}\right)^{1/2} 3 \cos t \, dt$$

$$= 2 \int_0^{\pi/2} \left(1 - \frac{9 \sin^2 t}{9}\right)^{1/2} 3 \cos t \, dt$$

$$= 2 \int_0^{\pi/2} (1 - \sin^2 t)^{1/2} \, 3 \cos t \, dt$$

$$(1 - \sin^2 t)^{1/2} = \sqrt{\cos^2 t}$$
$$= \cos t.$$

$$= 2 \int_0^{\pi/2} \cos t . 3 \cos t \, dt$$

$$= 6 \int_0^{\pi/2} \cos^2 t \, dt$$

$$2 \cos^2 t - 1 = \cos 2t$$
$$\cos^2 t = \frac{1 + \cos 2t}{2}.$$

$$= 6 \times \frac{1}{2} \int_0^{\pi/2} (1 + \cos 2t) \, dt$$

$$= 3 \left[t + \frac{1}{2} \sin 2t \right]_0^{\pi/2}$$

$$= 3 \left(\left[\frac{\pi}{2} + \frac{1}{2} \sin 2 \frac{\pi}{2} \right] \left[0 - \frac{1}{2} \sin 0 \right] \right)$$

$$= \frac{3}{2} \pi.$$

This means the total area is $4 \times \frac{3}{2} \pi = 6\pi \, \text{unit}^2$. Our ellipse has $a = 3$ and $b = 2$. This area answer agrees with the formula, $\pi a b = \pi(3)(2) = 6\pi \, \text{unit}^2$. Now we have worked through this example we can generalise it to deduce our general formula $\pi a b$. This time we use $\dfrac{x^2}{a^2} + \dfrac{y^2}{b^2} = 1$ and again rearrange the equation to express y in terms of x, i.e.

$$\frac{x^2}{a^2} + \frac{y^2}{b^2} = 1$$

becomes $\quad y = b \left(1 - \dfrac{x^2}{a^2} \right)^{1/2}.$

> Positive square root of both sides.

Using the area integral we get

$$\text{Area} = \int_0^a b \left(1 - \frac{x^2}{a^2} \right)^{1/2} dx$$

> Limits from $x = 0$ to $x = a$ for the first quadrant.

We substitute for x and hence dx, once again using trigonometry.

Let $\qquad x = a \sin t$

$\therefore \qquad \dfrac{dx}{dt} = a \cos t$

i.e. $\qquad dx = a \cos t \, dt.$

> Multiplying through by dt.

Also we need to change our limits using

$$x = a \sin t.$$

For $x = a$, $a = a \sin t_U$

> t_U is the new upper limit.

so that $t_U = \dfrac{\pi}{2}.$

For $x = 0$, $0 = \sin t_L$

> t_L is the new lower limit.

so that $t_L = 0.$

Now we replace our function in terms of x, dx and our original limits to get

$$\int_0^a b\left(1 - \frac{x^2}{a^2}\right)^{1/2} dx = b\int_0^{\pi/2} \left(1 - \frac{(a \sin t)^2}{a^2}\right)^{1/2} a \cos t \, dt$$

We use the same trigonometric identities as before for $\sin^2 t$ and $\cos^2 t$ and integrate. Eventually we get

$$\frac{ab}{2}\left[t + \frac{1}{2}\sin 2t\right]_0^{\pi/2}$$

$$= \frac{ab}{2}\left(\left[\frac{\pi}{2} + \frac{1}{2}\sin 2\frac{\pi}{2}\right] - \left[0 - \frac{1}{2}\sin 0\right]\right)$$

$$= \frac{1}{4}ab\pi.$$

This means the total area is $4 \times \dfrac{1}{4}ab\pi = \pi ab$ unit^2, so agreeing with our formula from Volume 1, Chapter 16.

Using the *t* substitution

In this section we use the t substitution from Chapter 4. You will remember this is based on half angles. It is possible to rewrite our double angle formulae and keep our patterns, i.e.

$$\sin A = 2\sin\frac{A}{2}\cos\frac{A}{2},$$

> Replacing A with $\dfrac{A}{2}$ and $2A$ with A.

$$\cos A = \cos^2\frac{A}{2} - \sin^2\frac{A}{2} \text{ and others,}$$

$$\tan A = \frac{2\tan\dfrac{A}{2}}{1 - \tan^2\dfrac{A}{2}}.$$

However, there are alternative forms using $t = \tan\dfrac{A}{2}$. We can think of $t = \dfrac{t}{1}$ and draw the right-angled triangle shown in Fig. 14.2. We find the length of the hypotenuse using Pythagoras' theorem.

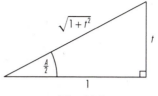

Fig. 14.2

Then $\sin\dfrac{A}{2} = \dfrac{t}{\sqrt{1+t^2}}$ and $\cos\dfrac{A}{2} = \dfrac{1}{\sqrt{1+t^2}}$. We substitute these into our above identities to get

$$\sin A = 2 \times \dfrac{t}{\sqrt{1+t^2}} \times \dfrac{1}{\sqrt{1+t^2}}$$

i.e. $\qquad \mathbf{\sin A} = \dfrac{2t}{1+t^2}.$

Also $\qquad \cos A = \left(\dfrac{1}{\sqrt{1+t^2}}\right)^2 - \left(\dfrac{t}{\sqrt{1+t^2}}\right)^2$

i.e. $\qquad \mathbf{\cos A} = \dfrac{1-t^2}{1+t^2}.$

Finally $\quad \mathbf{\tan A} = \dfrac{2t}{1-t^2}.$

> Simple substitution in the formula for $\tan\dfrac{A}{2}$.

▨▨▨▨ Example 14.11 ▨▨▨▨

Find $\displaystyle\int \dfrac{1}{\cos\theta}\, d\theta.$

We substitute for θ and hence $d\theta$ using the t substitution.

Let $\qquad t = \tan\dfrac{\theta}{2}$ or $\tan\dfrac{1}{2}\theta$

$\therefore \qquad \dfrac{dt}{d\theta} = \dfrac{1}{2}\sec^2\dfrac{\theta}{2} = \dfrac{1}{2}\left(1 + \tan^2\dfrac{\theta}{2}\right)$

> Substituting $t = \tan\dfrac{\theta}{2}$.

$\qquad\qquad \dfrac{dt}{d\theta} = \dfrac{1}{2}(1 + t^2)$

i.e. $\qquad \dfrac{2}{1+t^2}\, dt = d\theta.$

Now we replace our function in terms of θ and $d\theta$ to get

$$\int \dfrac{1}{\cos\theta}\, d\theta = \int \dfrac{1}{\dfrac{1-t^2}{1+t^2}} \cdot \dfrac{2}{1+t^2}\, dt$$

> $\cos\theta = \dfrac{1-t^2}{1+t^2}.$

$$= \int \dfrac{1+t^2}{1-t^2} \cdot \dfrac{2}{1+t^2}\, dt$$

> Inverting the first fraction.

$$= \int \dfrac{2}{1-t^2}\, dt$$

> Cancelling the $(1+t^2)$s.

$$= \int \left(\frac{1}{1+t} + \frac{1}{1-t} \right) dt \qquad \boxed{\text{Using partial fractions.}}$$

$$= \int \left(\frac{1}{1+t} - \frac{-1}{1-t} \right) dt$$

$$= \ln(1+t) - \ln(1-t) + c$$

$$= \ln \left(\frac{1+t}{1-t} \right) + c \qquad \boxed{\text{Law of logarithms.}}$$

$$= \ln \left(\frac{1 + \tan \dfrac{\theta}{2}}{1 - \tan \dfrac{\theta}{2}} \right) + c.$$

You need to be aware that there are other forms of this answer. Remember that $\tan \dfrac{\pi}{4} = 1$. We can write

$$\frac{1 + \tan \dfrac{\theta}{2}}{1 - \tan \dfrac{\theta}{2}} = \frac{1 + \tan \dfrac{\theta}{2}}{1 - 1. \tan \dfrac{\theta}{2}}$$

$$= \frac{\tan \dfrac{\pi}{4} + \tan \dfrac{\theta}{2}}{1 - \tan \dfrac{\pi}{4} . \tan \dfrac{\theta}{2}}$$

$$\boxed{\begin{aligned} \tan(A+B) \\ = \frac{\tan A + \tan B}{1 - \tan A . \tan B} \\ \text{with } A = \frac{\pi}{4} \text{ and } B = \frac{\theta}{2}. \end{aligned}}$$

This gives us the alternative answer of

$$\int \frac{1}{\cos \theta} \, d\theta = \ln \left(\tan \left(\frac{\pi}{4} + \frac{\theta}{2} \right) \right) + c.$$

It is possible that you may meet yet another version,

$$\int \frac{1}{\cos \theta} \, d\theta = \ln \left(\frac{1 + \sin \theta}{\cos \theta} \right) + c.$$

You can check this is an equivalent form using trigonometric identities.

Example 14.12

Find $\displaystyle \int_{\pi/4}^{\pi/2} \frac{1}{\sin \theta + \cos \theta} \, d\theta$.

We know from Chapter 4 that we can rewrite this as
$\sin \theta + \cos \theta \equiv R \sin(\theta + \alpha)$.
Using the usual method $R = \sqrt{2}$ and $\alpha = \dfrac{\pi}{4}$. We substitute for $(\theta + \alpha)$ and hence $d\theta$ using the t substitution.

Let $\qquad t = \tan\left(\dfrac{\theta + \alpha}{2}\right)$

$\therefore \qquad \dfrac{dt}{d\theta} = \dfrac{1}{2}\sec^2\left(\dfrac{\theta + \alpha}{2}\right) = \dfrac{1}{2}\left(1 + \tan^2\left(\dfrac{\theta + \alpha}{2}\right)\right)$

$\qquad \dfrac{dt}{d\theta} = \dfrac{1}{2}(1 + t^2)$

> Substituting $t = \tan\left(\dfrac{\theta + \alpha}{2}\right)$.

i.e. $\quad \dfrac{2}{1 + t^2}\,dt = d\theta.$

Also we need to change our limits using

$$t = \tan\left(\dfrac{\theta + \alpha}{2}\right).$$

For $\theta = \dfrac{\pi}{2}$, $\quad t_U = \tan\left(\dfrac{\dfrac{\pi}{2} + \dfrac{\pi}{4}}{2}\right) = 2.414,$

> t_U is the new lower limit.

for $\theta = \dfrac{\pi}{4}$, $\quad t_L = \tan\left(\dfrac{\dfrac{\pi}{4} + \dfrac{\pi}{4}}{2}\right) = 1$

> t_L is the new lower limit.

Now we replace our functions in terms of $(\theta + \alpha)$, $d\theta$ and our original limits to get

$$\int_{\pi/4}^{\pi/2} \dfrac{1}{\sin\theta + \cos\theta}\,d\theta = \int_{\pi/4}^{\pi/2} \dfrac{1}{R\sin(\theta + \alpha)}\,d\theta$$

> $R = \sqrt{2},\ \alpha = \dfrac{\pi}{4}.$

$$= \dfrac{1}{\sqrt{2}} \int_{1}^{2.414} \dfrac{1}{\dfrac{2t}{1 + t^2}} \cdot \dfrac{2}{1 + t^2}\,dt$$

> $\sin(\theta + \alpha) = \dfrac{2t}{1 + t^2}.$

$$= \dfrac{1}{\sqrt{2}} \int_{1}^{2.414} \dfrac{1 + t^2}{2t} \cdot \dfrac{2}{1 + t^2}\,dt$$

$$= \dfrac{1}{\sqrt{2}} \int_{1}^{2.414} \dfrac{1}{t}\,dt$$

$$= \dfrac{1}{\sqrt{2}} \Big[\ln t\Big]_{1}^{2.414}$$

> Cancelling $(1 + t^2)$s and 2s.

$$= \dfrac{1}{\sqrt{2}}(\ln 2.414 - \ln 1)$$

$$= 0.623.$$

████ **EXERCISE 14.4** ████

Find the values of the following integrals using the suggested substitutions.

1 $\displaystyle\int_{\pi/4}^{\pi/2} \frac{1}{\sin\theta}\, d\theta$ $t = \tan\dfrac{\theta}{2}$

2 $\displaystyle\int_{0}^{\pi/2} \frac{1}{(1+\cos\theta)}\, d\theta$ $t = \tan\dfrac{\theta}{2}$

3 $\displaystyle\int_{0}^{\pi/3} \frac{1}{1-\sin\theta}\, d\theta$ $t = \tan\dfrac{\theta}{2}$

4 $\displaystyle\int_{\pi/6}^{\pi/4} \frac{1}{4\sin\theta + 3\cos\theta}\, d\theta$ $t = \tan\left(\dfrac{\theta+\alpha}{2}\right)$

5 $\displaystyle\int_{\pi/2}^{3\pi/4} \frac{1}{6\sin\theta - 8\cos\theta}\, d\theta$ $t = \tan\left(\dfrac{\theta+\alpha}{2}\right)$

Trigonometric replacement

We have used plenty of trigonometry in this chapter already. In this section we look at replacement using trigonometric identities rather than substitution. We gather together some of our previous integration work in the following selection of examples.

████ **Example 14.13** ████

Find $\int \sin^2\theta\, d\theta$.

We have attempted this type of integral before.

$$\int \sin^2\theta\, d\theta = \frac{1}{2}\int(1 - \cos 2\theta)\, d\theta$$

> $\cos 2\theta = 1 - 2\sin^2\theta$
> i.e. $2\sin^2\theta = 1 - \cos 2\theta$.

$$= \frac{1}{2}\left(\theta - \frac{1}{2}\sin 2\theta\right) + c.$$

████ **Example 14.14** ████

Find $\displaystyle\int_{\pi/6}^{\pi/3} \cos^2 3\theta\, d\theta$.

Again we use a trigonometric identity involving cosine and the double angle. In the general formula $\cos 2A = 2\cos^2 A - 1$ we replace A with 3θ.

$$\int_{\pi/6}^{\pi/3} \cos^2 3\theta\, d\theta = \frac{1}{2}\int_{\pi/6}^{\pi/3}(1 + \cos 6\theta)\, d\theta$$

$$= \frac{1}{2}\left[\theta + \frac{1}{6}\sin 6\theta\right]_{\pi/6}^{\pi/3}$$

> $\cos 6\theta = 2\cos^2 3\theta - 1$
> i.e. $2\cos^2 3\theta = 1 + \cos 6\theta$.

$$= \frac{1}{2}\left(\left[\frac{\pi}{3}+\frac{1}{6}\sin 6\frac{\pi}{3}\right]-\left[\frac{\pi}{6}+\frac{1}{6}\sin 6\frac{\pi}{6}\right]\right)$$

$$= 0.5(1.047 + 0 - 0.524 - 0)$$

$$= 0.262.$$

Example 14.15

Find $\displaystyle\int_{0.15}^{0.30} \tan^2 5\theta.$

We use the identity $\sec^2 A = 1 + \tan^2 A$, replacing A with 5θ. Then we write

$$\int_{0.15}^{0.30} \tan^2 5\theta \, d\theta = \int_{0.15}^{0.30} (\sec^2 5\theta - 1) \, d\theta$$

$$= \left[\frac{1}{5}\tan 5\theta - \theta\right]_{0.15}^{0.30}$$

$$= \left[\frac{1}{5}\tan 1.50 - 0.30\right] - \left[\frac{1}{5}\tan 0.75 - 0.15\right]$$

$$= 2.820 - 0.30 - 0.186 + 0.15$$

$$= 2.48.$$

Example 14.16

Find $\int \sin^2 \theta \cos^3 \theta \, d\theta.$

This technique applies where we have a product of sines and cosines. The order does not matter. One trigonometric function is to an odd power and the other is to an even power. We rearrange the trigonometric function to the odd power.

Think of
$$\cos^3 \theta = \cos \theta \times \cos^2 \theta$$
$$= \cos \theta \times (1 - \sin^2 \theta)$$
$$= \cos \theta - \cos \theta \sin^2 \theta.$$

Then $\quad \sin^2 \theta \cos^3 \theta = \sin^2 \theta (\cos \theta - \cos \theta \sin^2 \theta)$

i.e. $\quad \sin^2 \theta \cos^3 \theta = \cos \theta \sin^2 \theta - \cos \theta \sin^4 \theta.$

Hence $\quad \int \sin^2 \theta \cos^3 \theta \, d\theta = \int(\cos \theta \sin^2 \theta - \cos \theta \sin^4 \theta) \, d\theta.$

We have two sections to this integral separated by a minus sign. Each is a function involving $\sin \theta$ to a power and the differential of $\sin \theta$, i.e. $\cos \theta$.

Hence $\int \sin^2 \theta \cos^3 \theta \, d\theta = \dfrac{1}{3}\sin^3 \theta - \dfrac{1}{5}\sin^5 \theta + c$, in the same style as Example 14.5.

████ **Example 14.17** ████

Find $\int \sin^4 \theta \, d\theta$.

This technique applies to either sine or cosine raised to an even power.

$$\sin^4 \theta = \sin^2 \theta \times \sin^2 \theta$$

$$= \frac{1}{2}(1 - \cos 2\theta) \times \frac{1}{2}(1 - \cos 2\theta)$$

| $\cos 2\theta = 1 - 2\sin^2 \theta$ |
| $2\sin^2 \theta = 1 - \cos 2\theta.$ |

$$= \frac{1}{4}(1 - 2\cos 2\theta + \cos^2 2\theta)$$

$$= \frac{1}{4} - \frac{1}{2}\cos 2\theta + \frac{1}{4}\cos^2 2\theta.$$

We need to apply the double angle formula again; this time to $\cos^2 2\theta$.

Using $\cos 2A = 2\cos^2 A - 1$

i.e. $2\cos^2 A = 1 + \cos 2A$

$\therefore$ $\cos^2 2\theta = \dfrac{1 + \cos 4\theta}{2}.$
| Replacing A with 2θ. |

We substitute into our expression for $\sin^4 \theta$ to get

$$\sin^4 \theta = \frac{1}{4} - \frac{1}{2}\cos 2\theta + \frac{1}{4}\left(\frac{1 + \cos 4\theta}{2}\right)$$

$$= \frac{1}{4} - \frac{1}{2}\cos 2\theta + \frac{1}{8} + \frac{1}{8}\cos 4\theta.$$

$\therefore$ $$\int \sin^4 \theta \, d\theta = \int \left(\frac{3}{8} - \frac{1}{2}\cos 2\theta + \frac{1}{8}\cos 4\theta\right) d\theta$$

$$= \frac{3}{8}\theta - \frac{1}{4}\sin 2\theta + \frac{1}{32}\sin 4\theta + c.$$

████ **Example 14.18** ████

Find $\int \cos 3\theta \sin 2\theta \, d\theta$.

We recall from Chapter 4 the trigonometric identity

$$2\cos\left(\frac{x + y}{2}\right)\sin\left(\frac{x - y}{2}\right) = \sin x - \sin y.$$

Comparing the left-hand side of this identity to our integral gives us

$$\frac{x + y}{2} = 3\theta$$

and

$$\frac{x - y}{2} = 2\theta.$$

We solve these equations simultaneously to get $x = 5\theta$ and $y = \theta$. Then our integral becomes

$$\int \cos 3\theta \sin 2\theta \, d\theta = \frac{1}{2} \int (\sin 5\theta - \sin \theta) \, d\theta$$

$$= \frac{1}{2} \left(-\frac{1}{5} \cos 5\theta + \cos \theta \right) + c$$

$$= \frac{1}{2} \cos \theta - \frac{1}{10} \cos 5\theta + c.$$

We can apply this technique to products of sines and to products of cosines. Remember there is a series of four formulae in Chapter 4.
Our final exercise contains practical questions. It allows you to apply the integration techniques introduced throughout this chapter.

EXERCISE 14.5

Find the following integrals using the necessary trigonometric identities.

1 $\int \cos^2 \theta \, d\theta$

2 $\int \cos t \tan t \, dt$

3 $\int \tan^2 2\theta \, d\theta$

4 $\int \sin^3 x \cos^2 x \, dx$

5 $\int \sin 3\phi \cos 2\phi \, d\phi$

6 $\int \sin 4\theta \sin 2\theta \, d\theta$

7 $\int_{\pi/6}^{\pi/2} \cos 2\theta \cos \theta \, d\theta$

8 $\int_{0}^{0.10} 4 \cos^4 \theta \, d\theta$

9 $\int_{\pi/6}^{\pi/2} 6 \cos^2 3t \, dt$

10 $\int_{-\pi/6}^{\pi/6} \cos \theta \cos(-\theta) \, d\theta$

11 $\int_{\pi/6}^{\pi/3} \sin \theta \cos \theta \, d\theta$

12 $\int_{\pi/4}^{\pi} \frac{1}{1 + \cos 2\theta} \, d\theta$

13 $\int_{0}^{0.25} 2 \sin^2 5t \, dt$

14 $\int_{0}^{\pi/8} \sin^2 x \cos^2 x \, dx$

HINT: $\sin x \cos x = \frac{1}{2} \sin 2x$

15 $\int_{\pi/3}^{2\pi/3} \frac{1}{1 + \tan^2 t} \, dt$

EXERCISE 14.6

1 Sketch the ellipse $\frac{x^2}{4} + \frac{y^2}{16} = 1$. Using the substitution $x = 2 \sin t$ find the area in the first, positive, quadrant. Hence calculate the area within the ellipse.

2 Find the value of the integral $\int_{0}^{5} \sqrt{25 - x^2} \, dx$ using the substitution $x = 5 \sin \theta$. Leave your answer as a multiple of π. Explain why this is a quarter of the area of a circle of radius 5 units.

3   The velocity, $v\,\mathrm{ms}^{-1}$, of a unit mass is $2\sin t$ in terms of time t seconds. Find the average kinetic energy, $K.E.$, between $t=0$ and $t=\dfrac{2}{3}\pi$ seconds. $K.E.=\dfrac{1}{2}mv^2$.

4   Water is flowing at a rate of r gallons/hour where $r = 10^3\sqrt{16-4t^2}$. This continues for 1 hour after the valve has been opened. Then the flow settles to a steady state. Find the total volume of water through the valve in that first hour, using Volume $=\displaystyle\int_0^1 r\,dt$.
   HINT: Use the substitution $t=2\sin\phi$.

5   We can plot the reciprocal of velocity, $\dfrac{1}{v}$, against displacement, x. The area under the graph represents time, t, i.e. Time $=\displaystyle\int \dfrac{1}{v}\,dx$. In terms of the displacement, x m, we have $v=3\cos x\;\mathrm{ms}^{-1}$. Find the time taken (seconds) to cover the first 0.5 m.

6   The general equation for a circle with centre $(0, 0)$ and radius r is $x^2+y^2=r^2$. Using integration deduce the area of the circle is πr^2. A suggested substitution during the integration is $x=r\cos\theta$.

7   On one set of axes sketch the circle $x^2+y^2=1$ and the ellipse $\dfrac{x^2}{4}+y^2=1$. Using integration find the area between them. Check your answer using the standard formulae πr^2 and πab respectively.

8   Equipment maintenance costs are to include consumable items. The costs rise with fair wear and tear according to age. The total cost is given by the value of the integral, $\int C\,dt$. C is the rate of change of cost with respect to time, t years. $C=kte^{0.01t^2+1}$ where k is a constant. Use the substitution $u=0.01t^2+1$ during the integration to find the maintenance costs for the 1st and 5th years. Leave both answers as multiples of k.

9   An engineering company manufactures and installs packaging plant. Its total profit is linked to its sales and costs. Find the total profit on sales of $(x=)$ 75 by finding the value of the integral $\displaystyle\int_0^{75} \dfrac{10^4}{\sqrt{25-0.0025x^2}}\,dx$. Use the substitution $x=100\sin t$.

10   With the power cut off an engine slows. Its acceleration, f, is given by $f=-\dfrac{(9+v^2)}{300}\;\mathrm{ms}^{-2}$ in terms of the velocity, $v\,\mathrm{ms}^{-1}$. The time taken to come to rest from $14\,\mathrm{ms}^{-1}$ is given by the integral
$$\text{Time taken} = \int_{14}^{0} \dfrac{1}{f}\,dv.$$
Using the substitution $v=3\tan\theta$ find the value of this integral.

15 Introduction to Differential Equations

The objectives of this chapter are to:

1   Define the terms differential equation, order and degree of a differential equation.

2   For a simple function determine and sketch a family of curves given the derivative.

3   Determine a particular curve of the family by specifying a point on it.

4   Distinguish between general and particular solutions.

5   Solve a differential equation of the type $\dfrac{dy}{dx} = f(x)$ for given boundary conditions.

6   Differentiate $y = Ae^{kx}$ and verify by substitution that it satisfies $\dfrac{dy}{dx} = ky$.

7   Solve differential equations of the form $\dfrac{dy}{dx} = f(y)$.

8   Solve differential equations of the form $\dfrac{dy}{dx} = f_1(x)f_2(y)$ by separating the variables and integrating.

9   Derive equations of the form $\dfrac{dy}{dx} = ky$ arising in engineering and science.

10   Generally formulate differential equations.

Introduction

This chapter is a simple introduction to the vast topic of differential equations. There are many other solution techniques that are far beyond us at this stage. We introduce some basic terms and look at a range of useful methods for solving differential equations.

The term 'differential equations' is a little strange in some respects. We solve a differential equation by integration. A **differential equation** involves **differentials (derivatives)** within an **equation**. Our aim, as with all equations, is to find a solution, i.e. to solve the differential equation. The solution is itself an equation, but without any differential. We will have two variables, generally using x and y. x is the independent variable and y is the dependent variable.

████ ASSIGNMENT ████

This Assignment looks at a chemical solution in a tank. We solve three types of differential equations. The first time we look at the mathematics we are filling the tank. Then we dilute the chemical concentration by flushing with water. Next we amend the flushing routine with different flow rates. In the later sections we look at formulating the differential equations we have previously solved.

Order and degree

Differential equations can have **order** and **degree**. The order refers to the **type of differential** within the equation. A first order differential equation has $\dfrac{dy}{dx}$ (the first derivative) as the highest derivative. A second order differential equation has $\dfrac{d^2y}{dx^2}$ (the second derivative) as the highest derivative. It may or may *not* include $\dfrac{dy}{dx}$ as well.

In shortened form we have alternative notation. We use y' (y dashed) for $\dfrac{dy}{dx}$ and y'' (y double dashed) for $\dfrac{d^2y}{dx^2}$. Then we can extend this to y''' and beyond. When we use $\dot{y}$ (y dot) and $\ddot{y}$ (y double dot) we understand time, t, to be the independent variable. We use $\dot{y}$ for $\dfrac{dy}{dt}$ and $\ddot{y}$ for $\dfrac{d^2y}{dt^2}$. Again, we can extend these ideas to further derivatives.

████ Examples 15.1 ████

i) $2\dfrac{dy}{dx} + y = 0$ is of order 1.

ii) $x\dfrac{d^2y}{dx^2} + 3\dfrac{dy}{dx} = 1$ is of order 2.

iii) $2\dfrac{d^3y}{dx^3} = x + y$ is of order 3.

The **degree** of a differential equation refers to the power of the **highest derivative**.

████ Examples 15.2 ████

i) $3\left(\dfrac{dy}{dx}\right)^2 = 2x - e^y$ is of degree 2.

ii) $2\dfrac{dy}{dx} + y = 0$ is of degree 1.

iii) $x\left(\dfrac{d^2y}{dx^2}\right)^4 + 3\left(\dfrac{dy}{dx}\right)^6 = \cos^7 x$ is of degree 4.

Remember, we concentrate on the derivative of highest order. In this example it is raised to the power 4.

In this introductory chapter we solve differential equations of order and degree 1.

■ EXERCISE 15.1 ■

For each differential equation write down the
   i) independent variable,
   ii) dependent variable,
   iii) order, and
   iv) degree.

1 $\dfrac{dy}{dx} = 7 - x$

2 $\dfrac{d^2y}{dx^2} + 3x\dfrac{dy}{dx} - 4y = 0$

3 $t\left(\dfrac{dy}{dt}\right)^2 + \dfrac{dy}{dt} = \cos t$

4 $\left(\dfrac{d^2y}{dx^2}\right)^3 + \left(\dfrac{dy}{dx}\right)^2 - 6y = e^x$

5 $\dfrac{d^3y}{dx^3} - 4\left(\dfrac{dy}{dx}\right)^2 = \dfrac{1}{x}$

6 $\dfrac{d^3y}{dx^3} = \left(\dfrac{d^2y}{dx^2}\right)^2$

7 $L\dfrac{d^2Q}{dt^2} + R\dfrac{dQ}{dt} + \dfrac{Q}{C} = 0$

8 $\dfrac{d^3x}{dt^3} + 6\left(\dfrac{d^2x}{dt^2}\right)^3 - \dfrac{dx}{dt} = e^{2t}$

9 $\left(\dfrac{dy}{dx}\right)^2 + 2x\dfrac{dy}{dx} = e^y$

10 $(4 + \cos\theta)\dfrac{dy}{d\theta} = -y\sin\theta$

Families of curves

We look at some specific cases to demonstrate families of curves. Suppose we have $\dfrac{dy}{dx} = 4$, i.e. the gradient is 4. The gradient of the graph is 4 all the way along it. From Volume 1, Chapter 2, we know that straight lines have the same gradient all the way along them. The equation of a straight line,

$$y = mx + c$$

becomes $y = 4x + c$ in our case.

This is the equation of a straight line of gradient 4 intercepting the vertical axis at the point $(0, c)$.

Alternatively $\dfrac{dy}{dx} = 4$ integrates directly to give

$y = 4x + c$ where c is the constant of integration. Both approaches attach the same meaning to c using alternative words. At this stage we do *not* know the value of c. $y = 4x + c$ is the **general solution** of our first order differential equation $\dfrac{dy}{dx} = 4$. We can sketch a family of parallel straight lines all with gradient 4. While they all have the same gradient they have different intercepts (cs) on the vertical axis. In Fig. 15.1 we sketch a selection of these lines.

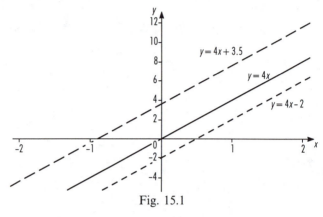

Fig. 15.1

Example 15.3

Sketch the family of curves linked by the differential equation $\dfrac{dy}{dx} - 2x + 3 = 0$.

We rewrite our differential equation as $\dfrac{dy}{dx} = 2x - 3$.

Remember our work at the end of Volume 1 and earlier in Volume 2. We know we need a parabola $y = x^2 - 3x + c$ to have this gradient. Again, c is the constant of integration/intercept on the vertical axis. In Fig. 15.2 we sketch a selection of curves with this gradient. They all have different values for c.

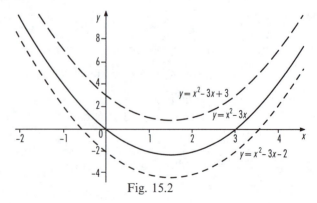

Fig. 15.2

We recall our early integration work. We used the rule 'increase the power by 1 and divide by the new power'. Notice how this has happened; also notice the graphical interpretation. For $\dfrac{dy}{dx} = 4$ to give $y = 4x + c$ we have a constant gradient. This gives a family of straight lines equally inclined to the horizontal axis. For $\dfrac{dy}{dx} = 2x - 3$ to give $y = x^2 - 3x + c$ we have a linear (highest power of x is 1) gradient. This gives a family of curves (parabolas) based on quadratic (highest power of x is 2) functions. We can extend this. The highest power of x in the curve is always 1 greater than the highest power of x in the gradient. By way of example, for a curve of $y = cubic\ function$ we start with $\dfrac{dy}{dx} = quadratic\ function$.

We are able to choose just one curve from the family. We do this by specifying a particular point through which the curve must pass. When we specify numerical values we are stating **boundary conditions**. They allow us to find the value(s) of any constant, including c. Then we replace those constant(s) in the general solution with the calculated values. This converts the general solution into a particular solution. Different boundary conditions will give us different particular solutions. Overall, our pattern of solution is

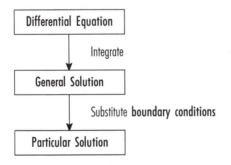

███ **Example 15.4** ███████████████████████████████████

A family of curves is linked by the differential equation $\dfrac{dy}{dx} + 2e^x = 0$. Find the particular curve passing through the point (1, 3).

We rewrite our differential equation as

$$\frac{dy}{dx} = -2e^x.$$

The type of curve with this gradient is an exponential curve,

$$y = -2e^x + c.$$

| General solution. Check this is true by differentiation. |

We substitute our boundary conditions (1, 3) to give

$$3 = -2e^1 + c$$

i.e. $c = 8.44$.

Substituting back for $c = 8.44$ into our general solution for y we get

$$y = -2e^x + 8.44.$$

| Particular solution. |

▬▬▬ EXERCISE 15.2 ▬▬▬

1   Sketch the graphical family based on the differential equation $\dfrac{dy}{dx} = -2$. What type of graph has this gradient?
Write down the general equation of the graph giving y as a function of x. Find the particular solution of the graph passing through the point $(2, 5)$.

2   Sketch the graphical family based on the differential equation $\dfrac{dy}{dx} = 4x$. Write down the general equation of the graph giving y as a function of x. Find the particular solution of the graph passing through the point $(0, -1)$.

3   Rearrange $\dfrac{dy}{dx} + x - 4 = 0$ to make $\dfrac{dy}{dx}$ the subject. Sketch the graphical family based on this differential equation. Write down the general equation of the graph with y as a function of x. Find the particular solution of the graph passing through the point $(1, \frac{1}{2})$.

4   Sketch the graphical family based on the differential equation $\dfrac{dy}{d\theta} - \cos\theta = 0$. What type of graph has this gradient? Write down the general equation of the graph with y as a function of θ. Find the particular solutions where the graph passes through the points
i) $\left(\dfrac{\pi}{2}, 1\right)$ and ii) $\left(\dfrac{\pi}{2}, 2\right)$.

5   Sketch the graphical family based on the differential equation $\dfrac{dy}{dx} - 3x^2 = 0$. Find the particular solution where the graph passes through the point $(2, 8)$.

Equations of the type $\dfrac{dy}{dx} = f(x)$

Here we look at the same type of differential equations that we have sketched graphs for. Now our method of solution is based on integration rather than by recognition of the gradient type.

████ **Example 15.5** ████████████████████████████████████

Find the general solution of the differential equation $\dfrac{dy}{dx} - 2x + 3 = 0$.

Given the boundary conditions that $y = -4$ where $x = 1$ find the particular solution.

We rewrite our differential equation as

$$\frac{dy}{dx} = 2x - 3$$

and integrate directly

i.e. $\qquad y = \int (2x - 3)\, dx$

so that $\qquad y = x^2 - 3x + c.$ | General solution. |

Substituting $y = -4$ and $x = 1$ we get | Boundary conditions. |

$$-4 = 1^2 - 3(1) + c$$

$\therefore \qquad\qquad c = -2.$

Finally, substituting this value of c into the general solution we get

$$y = x^2 - 3x - 2.$$ | Particular solution. |

Now look back at Fig. 15.2. You can see this is one of our sketched curves.

We say/write that the solution $y = f(x)$ satisfies the differential equation. Using Example 15.5 we say/write that $y = x^2 - 3x - 2$ satisfies the differential equation $\dfrac{dy}{dx} = 2x - 3$. We can check that this is true. Simply by differentiating $y = x^2 - 3x - 2$ we get $\dfrac{dy}{dx} = 2x - 3$.

████ **Example 15.6** ████████████████████████████████████

Find the particular solution of the differential equation $\dfrac{dy}{dx} = 4 - 6x + 3x^3$ given that $y = 1$ at $x = -2$.

We start with $\qquad \dfrac{dy}{dx} = 4 - 6x + 3x^3$

and integrate directly

i.e. $\qquad\qquad y = \int (4 - 6x + 3x^3)\, dx$

so that $\qquad\qquad y = 4x - 3x^2 + \dfrac{3}{4}x^4 + c.$ | General solution. |

Substituting $y = 1$ and $x = -2$ we get | Boundary conditions. |

$$1 = 4(-2) - 3(-2)^2 + \frac{3}{4}(-2)^4 + c$$

i.e. $\qquad\qquad 1 = -8 - 3(4) + \dfrac{3}{4}(16) + c$

so that $\qquad\qquad c = 9.$

Finally, substituting for $c = 9$ into the general solution we get

$$y = 9 + 4x - 3x^2 + \frac{3}{4}x^4.$$

or $\qquad y = \frac{3}{4}x^4 - 3x^2 + 4x + 9.$ ___Particular solution.___

As usual in algebra we write a function in either ascending or descending powers of x.

▓▓▓▓ Example 15.7 ▓▓▓▓

Find the particular solution of the differential equation $\dfrac{dy}{d\theta} + 2\sin\theta = 0$ given that $y = 3$ at $\theta = 0$.

We rewrite our differential equation in the form

$$\frac{dy}{d\theta} = -2\sin\theta$$

and integrate directly

i.e. $\qquad y = \int -2\sin\theta \, d\theta$

so that $\qquad y = 2\cos\theta + c.$ ___General solution.___

Substituting $y = 3$ and $\theta = 0$ we get ___Boundary conditions.___

$\qquad 3 = 2\cos 0 + c$ ___$\cos 0 = 1$.___

so that $\qquad c = 1.$

Finally substituting for $c = 1$ into the general solution we get

$$y = 2\cos\theta + 1$$ ___Particular solution.___

▓▓▓ ASSIGNMENT ▓▓▓

Remember, we are looking at a tank and a chemical in solution. First of all we are going to fill the tank. The filling process is in two stages. Towards the end of this Chapter we will consider the formulation of the Mathematics. For this section we look at the first 5 minutes. The volume of solution in the tank is given by V (litres) and the time by t (minutes). The rate of increase of volume is $\dfrac{dV}{dt}$, measured in litres/minute. We are given the differential equation $\dfrac{dV}{dt} = 8t$. Starting with an empty tank what is the volume after 5 minutes?

Using our differential equation

$$\frac{dV}{dt} = 8t$$

we integrate directly

i.e. $\qquad V = \int 8t \, dt$

so that $\qquad V = 4t^2 + c.$ ___General solution.___

Substituting $V = 0$ and $t = 0$ we get

$$0 = 0 + c$$

so that $c = 0$.

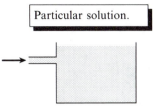

Boundary conditions.

Finally, substituting for $c = 0$ into the general solution we get

$$V = 4t^2.$$

Particular solution.

Now we wish to find the volume, V, after 5 minutes, i.e. we substitute $t = 5$ into our particular solution to get

$$V = 4(5^2) = 100 \text{ litres.}$$

EXERCISE 15.3

In each case, by direct integration find the particular solution of the differential equation.

1 $\dfrac{dy}{dx} = 3x + 1$ given the boundary conditions $y = 2$ where $x = 0$.

2 $\dfrac{dy}{dx} + x^2 - 4x = 2$ given the boundary conditions $y = 3$ where $x = 1$.

3 $\dfrac{dy}{dx} + \dfrac{4}{x^2} = 2x$ given the boundary conditions $y = 3.5$ where $x = 2$.

4 $x\dfrac{dy}{dx} + 4x - \dfrac{1}{x} = 0$ given the boundary conditions $y = 2$ where $x = 1$.

(**HINT**: Rearrange the differential equation to make $\dfrac{dy}{dx}$ the subject.)

5 $\dfrac{dy}{dt} + 2\cos t - 3\sin t = 1$ given that $y = 5$ initially.

Equations of the type $\dfrac{dy}{dx} = ky$

This differential equation type tells us the gradient, $\dfrac{dy}{dx}$, is some multiple of the original function, y. We have met this situation before with exponential functions. Let us look at an example,

$$y = 4e^{3x}.$$

Now $\qquad \dfrac{dy}{dx} = 4 \times 3e^{3x} = 3 \times 4e^{3x}$

i.e. $\qquad \dfrac{dy}{dx} = 3y.$

Substituting for $y = 4e^{3x}$.

This form is similar to our general form, $\dfrac{dy}{dx} = ky$, with $k = 3$.

More generally, if $\quad y = Ae^{kx}$

then $\qquad\qquad\qquad \dfrac{dy}{dx} = Ake^{kx} = k(Ae^{kx})$

i.e. $\qquad\qquad\qquad \dfrac{dy}{dx} = ky,$ | Substituting for $y = Ae^{kx}$.

i.e. $y = Ae^{kx}$ satisfies the differential equation $\dfrac{dy}{dx} = ky$.

▰▰▰ Examples 15.8 ▰▰▰

Solve the differential equation $\dfrac{dy}{dx} - 2y = 0$ with the boundary conditions

i) $y = 8$ at $x = 0$,
ii) $y = 14$ at $x = 1.5$.

We rewrite our differential equation in the form

$$\dfrac{dy}{dx} = 2y.$$

By recognition $k = 2$. This means our general solution is

$\qquad y = Ae^{2x}$ where A is a constant.

i) We substitute our boundary condition, $y = 8$, $x = 0$, to give

$\qquad\qquad 8 = Ae^{2(0)} = Ae^0$

i.e. $\qquad 8 = A.$ | $e^0 = 1.$

This gives us the particular solution $y = 8e^{2x}$.

ii) We substitute our boundary condition, $y = 14$, $x = 1.5$, to give

$\qquad\qquad 14 = Ae^{2(1.5)} = Ae^3$

i.e. $\qquad \dfrac{14}{e^3} = A$

$\qquad\qquad 0.70 = A.$

This gives us the particular solution $y = 0.70e^{2x}$.

▰▰▰ EXERCISE 15.4 ▰▰▰

In each case, by recognition, find the particular solution of the differential equation.

1 $\dfrac{dy}{dx} = 2y$ given the boundary condition $y = 3$ where $x = 0$.

2 $\dfrac{dy}{dt} = -4y$ given the boundary condition $y = 3$ when $t = 1$.

3 $\dfrac{dy}{dx} = \dfrac{y}{2}$ given the boundary condition $y = 6$ where $x = -2$.

4 $\dfrac{dQ}{dt} = 3Q$ given the boundary condition $Q = 0.5$ initially.

5 $\dfrac{2}{3}\dfrac{dy}{dx} + \dfrac{5}{8}y = y$ given the boundary condition $y = 2$ where $x = 1$.

(**HINT**: Rearrange the differential equation to make $\dfrac{dy}{dx}$ the subject.)

Solution by separating the variables

So far we have looked at two types of first order differential equations. They are $\dfrac{dy}{dx} = f(x)$ and $\dfrac{dy}{dx} = f(y)$. We look again at their solutions to introduce this new technique. More importantly, we later apply the method to solve our third type of differential equation, $\dfrac{dy}{dx} = f_1(x)f_2(y)$. We will see the essential feature is to separate the variables x and y.

First we look at the type $\dfrac{dy}{dx} = f(x)$ reusing Example 15.5.

Example 15.9

Find the particular solution of the differential equation $\dfrac{dy}{dx} - 2x + 3 = 0$ given that $y = -4$ at $x = 1$.

We rewrite our differential equation in the form

$$\frac{dy}{dx} = 2x - 3.$$

Remember, this is an equation. We need to be consistent throughout the equation. The operation we apply to the left-hand side we must also apply to the right-hand side. We **integrate** throughout with respect to **x**. $\int$ means we are going to **integrate** and **dx** means we are going to integrate **with respect to x**, i.e.

$$\int \frac{dy}{dx}\, dx = \int (2x - 3)\, dx.$$

It can be shown that $\dfrac{dy}{dx}dx$ can be replaced by dy to leave

$$\int dy = \int (2x - 3)\, dx. \qquad \boxed{\text{We understand } \int dy = \int 1\, dy.}$$

Having separated the variables, y and x appear exclusively on either side of the $=$ sign. We need to be more precise with our integration. We need to include the given limits of integration. For the lower limits we use the boundary condition $y = -4$ where $x = 1$. For the upper limits we use y and x. This is because we need a particular solution with y in terms of x. We write

$$\int_{-4}^{y} 1 \, dy = \int_{1}^{x} (2x - 3) \, dx$$

so $\qquad \left[y \right]_{-4}^{y} = \left[x^2 - 3x \right]_{1}^{x}$ $\qquad$ Definite integration.

$$[y] - [-4] = [x^2 - 3x] - [1^2 - 3(1)]$$

i.e. $\qquad y + 4 = x^2 - 3x - 1 + 3$

i.e. $\qquad y = x^2 - 3x - 2,$ as before.

Our next example is the type $\dfrac{dy}{dx} = f(y)$, reusing Examples 15.8.

▩▩▩▩ Example 15.10 ▩▩▩▩▩▩▩▩▩▩▩▩▩▩▩▩▩▩▩▩▩▩▩▩

Solve the differential equation $\dfrac{dy}{dx} - 2y = 0$ with the boundary conditions

i) $y = 8$ at $x = 0$,

ii) $y = 14$ at $x = 1.5$.

We rewrite our differential equation in the form

$$\frac{dy}{dx} = 2y.$$

We need to separate the variables x and y. Before we do that notice the position of dy. This indicates we need y on that same side, i.e.

$$\frac{1}{y} \frac{dy}{dx} = 2. \qquad \boxed{\text{Dividing through by } y.}$$

We integrate with respect to x, i.e.

$$\int \frac{1}{y} \frac{dy}{dx} \, dx = \int 2 \, dx.$$

The dxs appear to cancel and we have the variables exclusively on separate sides of the $=$ sign. Also, we include the limits of integration.

i) For the lower limits we use the boundary condition $y = 8$ where $x = 0$. For the upper limits we use y and x. This is because we need a particular solution with y in terms of x. Now we write

$$\int_{8}^{y} \frac{1}{y} \, dy = \int_{0}^{x} 2 \, dx$$

so $\qquad \left[\ln y \right]_{8}^{y} = \left[2x \right]_{0}^{x}$

$$\ln y - \ln 8 = 2x - 0$$

$$\ln\left(\frac{y}{8}\right) = 2x \qquad \boxed{\text{Law of logarithms.}}$$

$\therefore \qquad \dfrac{y}{8} = e^{2x} \qquad \boxed{\text{Definition of logarithms.}}$

i.e. $\qquad y = 8e^{2x},$ as before.

ii) Suppose we had the other boundary conditions. The theory follows similarly until we reach

$$\left[\ln y\right]_{14}^{y} = \left[2x\right]_{1.5}^{x}$$

$$\ln y - \ln 14 = 2x - 2(1.5)$$

$$\ln\left(\frac{y}{14}\right) = 2x - 3 \qquad \boxed{\text{Law of logarithms.}}$$

$$\therefore \qquad \frac{y}{14} = e^{2x-3} \qquad \boxed{\text{Definition of logarithms.}}$$

i.e. $y = 14e^{2x-3}$, as before.

There are other versions of this equation. We can rewrite the exponential,

$$e^{2x-3} = \frac{e^{2x}}{e^3}. \qquad \boxed{\begin{array}{l}\text{Subtraction of powers is} \\ \text{division of bases.}\end{array}}$$

$$y = \frac{14e^{2x}}{e^3} = 0.70\, e^{2x} \quad \text{as before.} \qquad \boxed{\dfrac{14}{e^3} = 0.70.}$$

We can generalise this technique using

$$\frac{dy}{dx} = ky,$$

i.e. $\qquad \dfrac{1}{y}\dfrac{dy}{dx} = k.$ $\qquad \boxed{\text{Dividing through by } y.}$

We integrate with respect to x so that

$$\int \frac{1}{y}\frac{dy}{dx}\, dx = \int k\, dx$$

i.e. $\qquad \displaystyle\int \frac{1}{y}\, dy = \int k\, dx$ $\qquad \boxed{dxs \text{ appear to cancel.}}$

$$\therefore \qquad \ln y = kx + c$$

i.e. $\qquad y = e^{kx+c}$

or $\qquad y = Ae^{kx}.$

$$\boxed{\begin{array}{l}\text{Definition of logarithms} \\ e^{kx+c} = e^{kx} \times e^c \\ \qquad\quad = Ae^{kx} \\ \text{where } A = e^c.\end{array}}$$

Notice we have included the constant of integration, c, on one side only. Alternatively, we could have included c_1 on the left and c_2 on the right. Then we could have gathered them together as $c = c_2 - c_1$.

◼ ASSIGNMENT

For this second look at our Assignment we start with the tank containing 500 litres of chemical solution. Our aim is to reduce the concentration of the chemical to 50 litres. We do this by flushing with water. How long will it take? We are given the rate of change of chemical solution with respect to time, i.e. $\dfrac{dx}{dt} = -\dfrac{40x}{500}.$

We divide through by x to link it with dx and integrate with respect to x. Our lower limits are $x = 500$ and $t = 0$. Our upper limits are $x = 50$ and t. We write

$$\int_{500}^{50} \frac{1}{x} \frac{dx}{dt} \, dt = \int \frac{-40}{500} \, dt$$

i.e. $$\int_{500}^{50} \frac{1}{x} \, dx = \int_{0}^{t} -0.08 \, dt$$

> dts appear to cancel.

$$\left[\ln x \right]_{500}^{50} = \left[-0.08t \right]_{0}^{t}$$

$$\ln 50 - \ln 500 = -0.08t + 0$$

i.e $$\ln \left(\frac{50}{500} \right) = -0.08t$$

$$-\frac{1}{0.08} \ln 0.1 = t$$

$$28.78 = t,$$

i.e. it takes 28.78 minutes to reduce the concentration.

We come to our final, most important type, of differential equation, $\frac{dy}{dx} = f_1(x) f_2(y)$. This format shows the functions of x and y to be multiplied. Alternatively, they may be divided. Our method deals with both variations.

Examples 15.11

Find the particular solution of the following differential equations

i) $\frac{dy}{dx} = y(2x + 3)$ given that $y = 2$ at $x = 0$,

ii) $\frac{dy}{dx} - \frac{(x + 3)}{y} = 0$ given that $y = 0$ at $x = 3$,

iii) $(x + 3)\frac{dy}{dx} - y = 0$ given that $y = 1$ at $x = 1$.

The position of dy influences the separation of the variables in all cases.

i) For $\frac{dy}{dx} = y(2x + 3)$

we divide through by y to link it with dy, i.e.

$$\frac{1}{y} \frac{dy}{dx} = 2x + 3.$$

We integrate with respect to x, i.e.

$$\int \frac{1}{y}\frac{dy}{dx}\,dx = \int (2x+3)\,dx$$

and include the boundary condition to give

$$\int_2^y \frac{1}{y}\,dy = \int_0^x (2x+3)\,dx \qquad \boxed{dx\text{s appear to cancel.}}$$

so $\Big[\ln y\Big]_2^y = \Big[x^2 + 3x\Big]_0^x$

$\ln y - \ln 2 = x^2 + 3x - 0$

$$\ln\left(\frac{y}{2}\right) = x^2 + 3x$$

$\therefore \qquad \dfrac{y}{2} = e^{x^2+3x}$

i.e. $\qquad y = 2e^{x^2+3x}.$

ii) We make $\dfrac{dy}{dx}$ the subject of our differential equation, i.e.

$$\frac{dy}{dx} = \frac{x+3}{y}.$$

This time we multiply through by y to link it with dy,

i.e. $\quad y\dfrac{dy}{dx} = x + 3.$

We integrate with respect to x, i.e.

$$\int y\frac{dy}{dx}\,dx = \int (x+3)\,dx \qquad \boxed{dx\text{s appear to cancel.}}$$

and include the boundary conditions to give

$$\int_0^y y\,dy = \int_3^x (x+3)\,dx$$

so $\left[\dfrac{y^2}{2}\right]_0^y = \left[\dfrac{x^2}{2} + 3x\right]_3^x$

$$\frac{y^2}{2} - 0 = \left[\frac{x^2}{2} + 3x\right] - \left[\frac{3^2}{2} + 3(3)\right]$$

i.e. $\quad y^2 = x^2 + 6x - 27 \qquad \boxed{\text{Multiplying through by 2.}}$

or $\quad y = \pm\sqrt{x^2 + 6x - 27}.$

iii) In stages we rewrite our differential equation, i.e.

$$(x+3)\frac{dy}{dx} - y = 0$$

becomes $\qquad (x+3)\dfrac{dy}{dx} = y.$

Again, the position of dy influences our next steps. We divide by y to link it with dy. Also we divide by $(x + 3)$ to separate it from dy, i.e.

$$\frac{1}{y}\frac{dy}{dx} = \frac{1}{x + 3}.$$

We integrate with respect to x, i.e

$$\int \frac{1}{y}\frac{dy}{dx}\,dx = \int \frac{1}{x + 3}\,dx$$

and include the boundary conditions to give

$$\int_1^y \frac{1}{y}\,dy = \int_1^x \frac{1}{x + 3}\,dx \qquad \boxed{dxs \text{ appear to cancel.}}$$

so

$$\left[\ln y\right]_1^y = \left[\ln(x + 3)\right]_1^x$$

$$\ln y - \ln 1 = \ln(x + 3) - \ln 4$$

i.e.

$$\ln\left(\frac{y}{1}\right) = \ln\left(\frac{x + 3}{4}\right)$$

i.e.

$$y = \frac{x + 3}{4}. \qquad \boxed{\text{Anti-logs of both sides.}}$$

■■■■■ ASSIGNMENT ■■■■■

For this third look at our Assignment we again start with the tank containing 500 litres of chemical solution. Once more our aim is to reduce the chemical concentration by flushing with water. This time we are interested in the concentration after 20 minutes. The inflow and outflow are different from last time. $\dfrac{dy}{dt}$ is the rate of change of chemical solution with respect to time and we are given

$$\frac{dy}{dt} = -\frac{60y}{500 - 20t}.$$

We divide through by y to link it with dy and integrate with repect to y. Our lower limits are $y = 500$ and $t = 0$. Our upper limits are y and $t = 20$. We write

$$\int_{500}^y \frac{1}{y}\frac{dy}{dt}\,dt = \int_0^{20} \frac{-60}{500 - 20t}\,dt$$

i.e.

$$\int_{500}^y \frac{1}{y}\,dt = 3\int_0^{20} \frac{-20}{500 - 20t}\,dt \qquad \boxed{dts \text{ appear to cancel.}}$$

Notice we adjust the right-hand side by removing a factor of 3. This creates the correct form so we integrate to a natural logarithm,

$$\left[\ln y\right]_{500}^y = 3\left[\ln(500 - 20t)\right]_0^{20}$$

so $\ln y - \ln 500 = 3[\ln 100 - \ln 500]$

i.e. $\ln y = 3\ln 100 - 2\ln 500$

$\ln y = 1.386$

∴ $y = e^{1.386} = 4,$

Definition of logarithms.

i.e. after 20 minutes there are 4 litres of the original chemical solution.

■ EXERCISE 15.5 ■

In each case separate the variables to find the particular solution of the differential equation.

1 $\dfrac{dy}{dx} = \dfrac{x}{y^2}$ given the boundary condition $y=0$ where $x=4$.

2 $\dfrac{dy}{dx} = \dfrac{x}{y+1}$ given the boundary condition $y=1$ where $x=1$.

3 $\dfrac{dy}{dx} = (\cos x)(\sec y)$ given the boundary condition $y=\dfrac{3\pi}{2}$ where $x=\dfrac{\pi}{2}$.

4 $\dfrac{dy}{dx} = \dfrac{y}{x}$ given the boundary condition $y=4$ where $x=1$.

5 $\dfrac{dy}{dx} = y\sin x$ given the boundary condition $y=1$ where $x=0$.

6 $(x^2+1)\dfrac{dy}{dx} = \dfrac{2x}{y}$ given the boundary condition $y=4$ where $x=0$.

7 $\dfrac{dy}{dx} = \dfrac{x-1}{y+1}$ given the boundary condition $y=0$ where $x=0$.

8 $\dfrac{dy}{dx} = \dfrac{y+1}{x-1}$ given the boundary condition $y=7$ where $x=9$.

9 $xy\dfrac{dy}{dx} = x+1$ given the boundary condition $y=2$ where $x=1$.

10 $\dfrac{dy}{dx} = 2e^{x-y}$ given the boundary condition $y=1=x$.

Forming differential equations of the type $\dfrac{dy}{dx} = f(x)$

We know how to solve this type of differential equation. In this section we translate a problem into a mathematical form, $\dfrac{dy}{dx} = f(x)$. Also we look at a practical type of problem. The pattern of letters will be the same, though not always x and y.

Example 15.12

A rigid beam is 10 m long. Its bending moment, M, is related to the distance from one end, x. $\dfrac{dM}{dx}$ is proportional to $(5 - x)$. We need to find M in terms of x given some boundary conditions; $M = -10000$ N at the original end and $M = 2500$ N at the centre. Finally we find the bending moment at the other end of the beam.

We start with '$\dfrac{dM}{dx}$ is proportional to $(5 - x)$' which we can write as

$$\frac{dM}{dx} \propto (5 - x)$$

i.e.

$$\frac{dM}{dx} = k(5 - x).$$

> $\propto$ represents 'is proportional to'.
>
> k is a constant of proportionality.

We integrate directly with respect to x,

i.e.

$$M = \int k(5 - x)\, dx$$

so that

$$M = k\left(5x - \frac{x^2}{2}\right) + c.$$

> General solution.

We have two sets of boundary conditions. They are $M = -10000$ N where $x = 0$ m and $M = 2500$ N where $x = 5$ m. We substitute the first conditions into our general solution to get

$$-10\,000 = k(0 - 0) + c$$

i.e. $-10\,000 = c.$

Replacing c in our general solution we get

$$M = k\left(5x - \frac{x^2}{2}\right) - 10\,000.$$

We substitute the second conditions into our amended solution to get

$$2500 = k\left(5(5) - \frac{5^2}{2}\right) - 10\,000$$

i.e. $12\,500 = k(25 - 12.5)$

$$\frac{12500}{12.5} = k$$

$$1000 = k.$$

Replacing k we get a particular solution,

$$M = 1000\left(5x - \frac{x^2}{2}\right) - 10000.$$

Our final task is to find the bending moment at the other end of the beam, i.e. where $x = 10$ m. Substituting, we get

$$M = 1000\left(5(10) - \frac{10^2}{2}\right) - 10000$$

i.e. $M = -10000$ N.

ASSIGNMENT

Let us return to the original filling of the tank with chemical in solution. It is in two stages. Now we are able to formulate the Mathematics of these stages. The volume of solution in the tank is given by V (litres) and the time by t (minutes). The rate of increase of volume is $\dfrac{dV}{dt}$, measured in litres/minute. In this problem it is proportional to time until it reaches 40 litres/minute and the volume is 100 litres. Then it remains at this rate to complete the filling. We start with '$\dfrac{dV}{dt}$ is proportional to t'

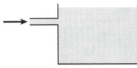

which we can write as

$$\frac{dV}{dt} \propto t$$

i.e.
$$\frac{dV}{dt} = kt.$$

> $\propto$ represents 'is proportional to'.
>
> k is a constant of proportionality.

We integrate directly with respect to t,

i.e.
$$V = \int kt \, dt$$

so that
$$V = \frac{kt^2}{2} + c.$$

> General solution.

When we start filling the tank it is empty, i.e. $V=0$ at $t=0$. We substitute these boundary conditions into our general solution to get

$$0 = 0 + c$$

i.e.
$$0 = c.$$

This simplifies our general solution to

$$V = \frac{kt^2}{2}.$$

Now we use our other pieces of information,

i.e.
$$\frac{dV}{dt} = 40 \quad \text{or} \quad kt = 40$$

and
$$V = 100 \quad \text{or} \quad \frac{kt^2}{2} = 100.$$

We rewrite
$$\frac{kt^2}{2} = 100$$

as
$$\frac{(kt)(t)}{2} = 100$$

and substitute to get

$$\frac{40t}{2} = 100$$

i.e.
$$t = 5 \quad \text{and} \quad k = 8.$$

> Using $kt=40$.

Substituting for k we have the differential equation $\dfrac{dV}{dt} = 8t$ with a particular solution $V = 4t^2$.

In Fig. 15.3 we sketch $\dfrac{dV}{dt}$ against t. We see the flow rate increasing until it reaches 40 litres/minute and then remaining constant. The area under this graph is the volume in the tank.

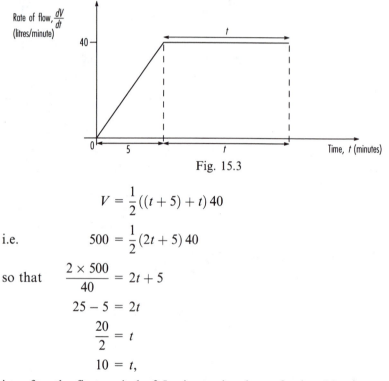

Fig. 15.3

$$V = \frac{1}{2}((t + 5) + t)\,40$$

i.e.
$$500 = \frac{1}{2}(2t + 5)\,40$$

so that
$$\frac{2 \times 500}{40} = 2t + 5$$

$$25 - 5 = 2t$$

$$\frac{20}{2} = t$$

$$10 = t,$$

i.e. after the first period of 5 minutes it takes a further 10 minutes to fill the tank.

■ EXERCISE 15.6 ■

1   The gradient of a curve is 4 times the independent variable. Write this as a differential equation. The curve passes through the point (1, 5). Find the particular equation of this curve.

2   An object of mass 4.9 kg is thrown vertically upwards. It is subject only to the acceleration due to gravity, 9.8 ms^{-2}. Apply Newton's law, $F = ma$, to create a differential equation. Solve this differential equation for the velocity, v, in terms of time, t. You are given the initial velocity to be 25 ms^{-1}.

3 The temperature, $T\,°\text{K}$, between two faces of material is related to the thickness, x m, by

$$\frac{dT}{dx} = \frac{-Q}{kA}.$$

Q is the rate of heat energy transfer, (W),
k is the coefficient of thermal conductivity, $(\text{Wm}^{-1}K^{-1})$.
A is the area of conducting material, (m^2).
For a concrete face area of 20 m^2 the thickness is 0.20 m and $k = 1.2$. At $x = 0$ m $T = 25°\text{K}$ and at $x = 0.20$ m $T = 5°\text{K}$. Find Q.

4 For an elastic spring the rate of working, $\dfrac{dW}{dx}$, is proportional to the extension, x, from the natural length, l. Write this as a differential equation. Now solve it using the following information. At the natural length there has been no work done. When $W = 2.5$ J the spring has been stretched by 0.1 m.

5 Storm tanks at a sewerage treatment plant store excess effluent following a heavy storm. V gallons is the volume of effluent stored in them. The rate of increase of volume, $\dfrac{dV}{dt}$, is related to time, t hours, after the start of the storm. We know that $\dfrac{dV}{dt} = k(t - 8)$. When $t = 0$ they still hold 200 000 gallons from the previous night's torrential downpour. Also the tanks are just empty 8 hours after the start of this storm. Show that $V = 3125t(t - 16) + 200000$.

Forming differential equations of the type $\dfrac{dy}{dx} = f(y)$

We know how to solve this type of differential equation. Using Example 15.13 and the Assignment we form the appropriate differential equation.

███████ **Example 15.13** ████████████████████████████████

We are interested in part of an electrical circuit. The rate of change of current, $\dfrac{di}{dt}$, is proportional to that current, i.

We are going to find the constant of proportionality and a particular solution. One boundary condition is that the initial current is 2.8 A. Also the current decays to only 1.0 A after 2.5 seconds.

We start with '$\dfrac{di}{dt}$ is proportional to i'

which we can write as

$$\frac{di}{dt} \propto i$$

> $\propto$ represents 'is proportional to'.

i.e. $\qquad \dfrac{di}{dt} = ki.$

> k is a constant of proportionality.

We divide through by i to link it with di,

i.e. $\qquad \dfrac{1}{i}\dfrac{di}{dt} = k.$

We integrate with respect to t,

i.e. $\qquad \displaystyle\int \dfrac{1}{i}\dfrac{di}{dt}\, dt = \int k\, dt$

> dts appear to cancel.

and include the boundary conditions to give

i.e. $\qquad \displaystyle\int_{2.8}^{1.0} \dfrac{1}{i}\, di = \int_{0}^{2.5} k\, dt$

so $\qquad \Big[\ln i\Big]_{2.8}^{1.0} = \Big[kt\Big]_{0}^{2.5}$

$$\ln 1.0 - \ln 2.8 = k2.5 + 0$$

$$-1.0296\ldots = 2.5k$$

i.e. $\qquad -0.41 = k$

To find the particular solution we amend the upper limits and replace k,
i.e.

$$\Big[\ln i\Big]_{2.8}^{i} = \Big[-0.41t\Big]_{0}^{t}$$

$$\ln i - \ln 2.8 = -0.41t + 0$$

i.e. $\qquad \ln\left(\dfrac{i}{2.8}\right) = -0.41t$

> Law of logarithms.
>
> Definition of logarithms.

$\therefore \qquad \dfrac{i}{2.8} = e^{-0.41t}$

i.e. $\qquad i = 2.8e^{-0.41t}$ is the particular solution.

▮▮▮ ASSIGNMENT ▮▮▮

For this look at our Assignment we return to the tank containing 500 litres of chemical solution. Our aim is to reduce the concentration of the chemical to only 10% of its original value. We are going to do this by flushing with water. This means the chemical solution will decrease from 100% (500 l) to 10% (50 l). We are interested in the time this will take. The inflow and outflow are both 40 l/min.

Let us consider the ratio of chemical solution to water at some general time, t. If there are x l of chemical solution then there must be $(500 - x)$ l of water, i.e.

chemical solution : water

is $\qquad x \qquad : \quad 500 - x$

i.e. $\qquad \dfrac{x}{500} \qquad : \quad \dfrac{500 - x}{500}$.

We divide through by 500 to give the ratio per litre. During the flushing within the tank let us look at 40 l of mixture. It contains $40 \times \dfrac{x}{500}$ l of chemical solution and $40 \times \dfrac{(500 - x)}{500}$ l of water. Every minute during the flushing 40 l of water flows **into (positive change)** the tank. Also $\dfrac{40x}{500}$ l of chemical solution flows **out (negative change)**. $\dfrac{dx}{dt}$ is the rate of change of chemical solution with respect to time,

i.e. $\qquad \dfrac{dx}{dt} = 0 - \dfrac{40x}{500}$

i.e. $\qquad \dfrac{dx}{dt} = \dfrac{-40x}{500}$.

We have formulated a differential equation. We can write that we have mathematically modelled the flow relating to this tank. Obviously we have already seen and solved this differential equation. There is no need to do so again.

■■■ EXERCISE 15.7 ■■■

1 The gradient of a curve is 3 times the dependent variable Write this as a differential equation. The curve passes through the point $(2, 6)$. Find the particular equation of this curve.

2 An object of mass 5.0 kg is subject to a force, F N. The force is proportional to the velocity, v ms^{-1}. Apply Newton's law, $F = ma$, to create a differential equation. Solve this differential equation for the velocity, v, in terms of time, t. You are given that the initial velocity of 10 ms^{-1} is halved in 20 seconds.

3 In this question we deal with Newton's law of cooling. The rate of cooling is proportional to the temperature difference of a body and its surrounding environment. In our case this is represented by the differential equation $\dfrac{dT}{dt} = -k(T - 15)$.

T is the temperature of the body, (°C).

t is the time, (minutes); and

k is the constant of proportionality to be found.

The temperature of the surrounding environment is 15°C. A body cools from 40°C to 25°C in 30 minutes. Find the particular solution of the differential equation.

4   In the last 50 years the population of a local town has doubled. During this period the rate of increase has been proportional to the actual population. Find the constant of proportionality, k. What would be the value of k if the population had trebled rather than doubled?

5   The radioactivity of thorium-228 is N. This is proportional to the rate of change of radioactivity, $\dfrac{dN}{dt}$. Write this as a differential equation. The half-life of thorium-228 is 1.9 years, i.e. after 1.9 years the initial value of N, N_0, is $\frac{1}{2}N_0$. What is the particular solution of this differential equation?

Forming differential equations of the type $\dfrac{dy}{dx} = f_1(x)\, f_2(y)$

Using Example 15.14 and the Assignment we are going to form appropriate differential equations.

Example 15.14

We have a light elastic spring of modulus $2g$ N and length 1.2 m. The upper end is fixed and the lower end carries a scale pan of mass 0.25 kg. We move the scale pan down a further 0.12 m and release it from rest. We are going to find an expression for the velocity of the scale pan after its release. $x = 0.12$ m and $v = 0\,\text{ms}^{-1}$ give us the boundary conditions.

T is the tension in the spring, (N).
λ is the modulus of elasticity, (N).
x is the extension from the natural length l, (m).

The light elastic spring obeys Hooke's law, $T = \dfrac{\lambda x}{l} = \dfrac{2gx}{1.2}$.

In Figs. 15.4 we draw three diagrams.

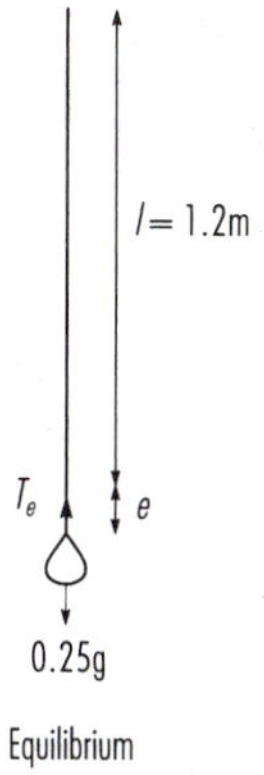

Fig. 15.4(a) shows the scale pan in equilibrium acted upon by two forces. They are the tension in the spring and the weight. The spring is slightly extended.

Using $\qquad\qquad F = ma$

we have $\quad T_e - 0.25g = 0.$

> T_e is the tension in equilibrium.

We substitute for T_e using Hooke's law

i.e. $\qquad \dfrac{2ge}{1.2} - 0.25g = 0$

> e is the extension in equilibrium.

$$\dfrac{2ge}{1.2} = 0.25g$$

Fig. 15.4(a)

so that $\qquad\qquad e = 0.25g \times \dfrac{1.2}{2g}$

i.e. $\qquad\qquad e = 0.15\,\text{m}$ is the extension from the natural length in the equilibrium position.

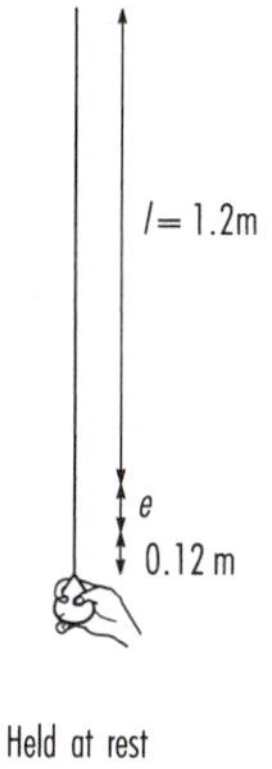

Fig. 15.4(b) shows the scale pan held at rest just before its release.

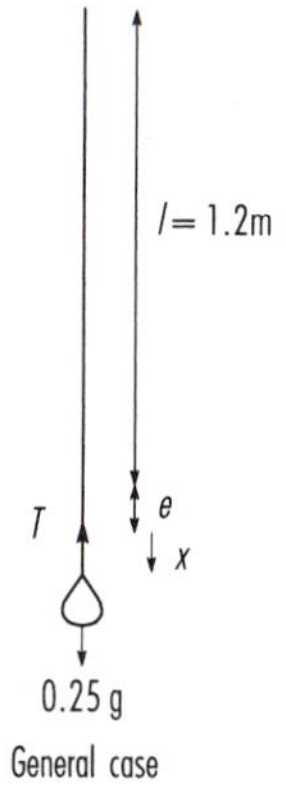

l = 1.2m

e

0.12 m

Held at rest

Fig. 15.4(b)

l = 1.2m

T

e

x

0.25 g

General case

Fig. 15.4(c)

Fig. 15.4(c) is the most important diagram. It shows the scale pan moving. Our diagram shows the instant when the spring is extended by $(x + e)$ m. Notice how we measure down from the natural length. We take this direction as positive.

Using $\qquad\qquad F = ma$

we have $\quad 0.25g - T = 0.25 \times$ acceleration.

We substitute for T using Hooke's law generally

i.e. $0.25g - \dfrac{2g(x + e)}{1.2} = 0.25 \times$ acceleration. —— ①

T_e ↑   ↓ *e*

0.25g

Equilibrium

l = 1.2m

Let us look at the acceleration. We know from Chapter 8 that we can use either $\dfrac{d^2x}{dt^2}$ or $\dfrac{dv}{dt}$. Because we are dealing with first order differential equations we choose $\dfrac{dv}{dt}$. However this gives us 3 variables, x, v and t. We only know how to deal with 2 variables. Using the function of a function rule,

$$\frac{dv}{dt} = \frac{dv}{dx} \times \frac{dx}{dt},$$

and $v = \dfrac{dx}{dt}$ we have

$$\frac{dv}{dt} = \frac{dv}{dx} \times v = v\frac{dv}{dx}.$$

In equation ① we substitute for $\dfrac{dv}{dt}$, i.e.

$$0.25g - \frac{2gx}{1.2} - \frac{2ge}{1.2} = 0.25v\frac{dv}{dx}$$

$$0.25g - \frac{2gx}{1.2} - 0.25g = 0.25v\frac{dv}{dx}$$

> Substituting $\dfrac{2ge}{1.2} = 0.25g$ from above.

i.e.

$$\frac{-2gx}{1.2} = 0.25v\frac{dv}{dx}.$$

> $g = 9.8 \text{ ms}^{-2}$.

We simplify the arithmetic to get

$$-65.\bar{3}x = v\frac{dv}{dx}.$$

We integrate with respect to x,

i.e.

$$\int -65.\bar{3}x \, dx = \int v\frac{dv}{dx} \, dx$$

and include the boundary conditions to give

> dxs appear to cancel.

i.e.

$$\int_{0.12}^{x} -65.\bar{3}x \, dx = \int_{0}^{v} v \, dv$$

so

$$\left[-65.\bar{3}\frac{x^2}{2}\right]_{0.12}^{x} = \left[\frac{v^2}{2}\right]_{0}^{v}$$

$$-65.\bar{3}\frac{x^2}{2} - -65.\bar{3}\frac{(0.12)^2}{2} = \frac{v^2}{2} - 0$$

$$65.\bar{3}(0.0144 - x^2) = v^2.$$

> Multiplying throughout by 2.

We take the square root of both sides to give

$$v = \pm\sqrt{65.\bar{3}\,(0.0144 - x^2)}.$$

Notice we include the $\pm$ sign. v is velocity and so direction is important. The positve square root shows the scale pan moving downwards (the direction we chose). The negative square root relates to the scale pan moving upwards.

■■■■ ASSIGNMENT ■■■■■■■■■■■■

For this final look at our Assignment we start once again with the tank containing 500 litres of chemical solution. As before, our aim is to reduce the chemical concentration by flushing with water. This time we are interested in the concentration after 20 minutes. We have an inflow of 40 l/min. After some general time, t, we will have introduced 40t l of water. With an outflow of 60 l/min we will have lost 60t l of mixture.

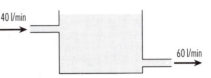

Using   chemical + water = mixture
         solution

we have y + x = $500 + 40t - 60t$

i.e. y + x = $500 - 20t$.

If there are y l of chemical solution then there must be $(500 - 20t - y)$ l of water, i.e.

         chemical solution :   water

is y : $500 - 20t - y$

i.e. $\dfrac{y}{500 - 20t}$: $\dfrac{500 - 20t - y}{500 - 20t}$.

We divide through by $(500 - 20t)$ to give the ratio per litre. During the flushing within the tank let us look at the changes. Every minute during the flushing 40 l of water flows **into (positive change)** the tank. Also 60 l of mixture flows **out (negative change)**. It contains $60 \times \dfrac{y}{500 - 20t}$ l of chemical solution. $\dfrac{dy}{dt}$ is the rate of change of chemical solution with respect to time, i.e.

$$\frac{dy}{dt} = 0 - \frac{60y}{500 - 20t}$$

i.e. $\dfrac{dy}{dt} = -\dfrac{60y}{500 - 20t}$.

We have formulated an alternative differential equation to model the flow through the tank. We have seen and already solved this equation earlier in the chapter. There is no need to do so again.

■■■■ EXERCISE 15.8 ■■■■■■■■■■■■

1 The gradient of a curve is twice the ratio of the independent variable to the dependent variable. Write this as a differential equation. For the curve passing through the point (2, 3) find the particular solution.

2 During fermentation the solids in solution have a mass of m kg. They decompose over time, t hours, according to

$$\frac{dm}{dt} + \frac{km}{(1+t)^3} = 0.$$

Make $\frac{dm}{dt}$ the subject of this differential equation.

Initially the mass is 4.5 kg. Separate the variables and integrate to find a particular solution including k. Also find the value of k if the mass is 4.0 kg after 18 hours.

3 For a gas the change in pressure, p, with respect to volume, V, is proportional to $\frac{p}{V}$. Write this as a differential equation. For hydrogen the constant of proportionality is -1.395. Find a general solution. When the volume doubles what is the ratio of the final pressure to initial pressure?

4 A strain of bacteria has a limited life expectancy. Their numbers, n, are related to time, t hours, by the differential equation

$$\frac{dn}{dt} = \left(\alpha - \frac{1}{40 - t}\right)n.$$

α is a constant. Separate the variables and integrate given that n is approximately 10^6 initially. Find the value of α if n trebles in the first 5 hours.

5 Force, F, acceleration, $v\frac{dv}{dx}$, and mass, m, are related by

$$9.8m - kx = mv\frac{dv}{dx}$$

where k is a constant. x is the positive displacement from the origin. Find a particular solution including k if $v = 12$ ms^{-1} at $x = 0$ and $v = 6$ ms^{-1} at some general displacement x. By now you should have a quadratic equation in x. k is a multiple of the mass m. Investigate the solution of this quadratic equation in x for different multiples. (**HINT**: substitute and solve for x when $k = m$, $k = 2m$, $k = 3m$, . . .)

16 Areas and Centroids

The objectives of this chapter are to:

1　Sketch a given regular area including a typical elemental (incremental) area.
2　Sum those elemental areas.
3　Find the area by definite integration.
4　Find the area of a composite shape using standard results.
5　Link together centroid, centre of gravity and centre of mass.
6　Sketch a given area including a typical elemental area whose centroid is known.
7　Find the moment of that elemental area about an axis in the plane of the area.
8　Find the first moment of area by definite integration.
9　Find the centroid by integration.
10　Find the centroid of a composite shape using standard results.

Introduction

In this chapter we start with the area ideas we introduced in Chapter 10. We apply them to different areas. Also we relate them to the y-axis just as easily as the x-axis. We gather together standard results to find the areas of composite shapes. Then we apply similar integration techniques to find the centroid of these areas.

■■■■ ASSIGNMENT ■■■■

The Assignment for this chapter is derived from circular work. In Fig. 16.1 we have part of a circle. The original piece was a circular rigid sheet of alloy. Notice we do *not* need the type and mass or thickness. We are looking at the top surface area. We assume it to be relatively thin and of constant density. The original circle has been cut into three equal pieces each of angle 120°. Then a small sector has been removed from each large sector.

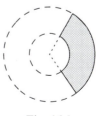

Fig. 16.1

Areas by integration

We recall some basic work from Chapter 10. Remember integration has many applications. Finding the area under a curve is one simple case.

Fig. 16.2 shows a general curve. We wish to find the area under the curve bounded by the x-axis and the lines $x = a$ and $x = b$. To do this we split the area into a series of thin parallel strips each of width δx.

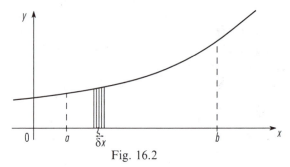

Fig. 16.2

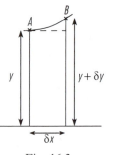

Fig. 16.3

Let us magnify one of those strips. Suppose A is the point (x, y). Remember a small change in x, δx, causes a small change in y, δy. Because B is a small distance away we know it is the point $(x + \delta x, y + \delta y)$. We approximate this strip to a rectangle. As the strip width gets narrower and narrower (i.e. as $\delta x \to 0$) the accuracy of the approximation improves. This means the extra area above the rectangle becomes very small (Fig. 16.3). Then

Area of thin strip $\approx$ Area of rectangle
$\approx y \, \delta x.$

> $\approx$ is 'equals approximately'.

The area between $x = a$ and $x = b$ is the sum of all such strips, i.e.

$$\text{Area} = \lim_{\delta x \to 0} \sum_{x=a}^{x=b} y \, \delta x.$$

In practice we would find it difficult to calculate this sum for the area. We can reach the same result more easily using the integration formula

$$\text{Area} = \int_a^b y \, dx.$$

Integration is the limit of the summation process.

We are going to repeat this technique. We use the same curve but this time find a different area. Notice how we use thin rectangular strips again, but this time horizontally rather than vertically.

Fig. 16.4 shows our general curve. We wish to find the area between the curve, the y-axis and the lines $y = c$ and $y = d$. To do this we split the area into a series of thin horizontal parallel strips each of width δy.

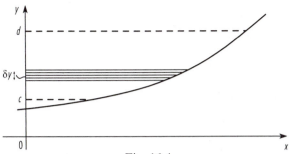

Fig. 16.4

We magnify one of those strips. Suppose A is the point (x, y). Remember a small change in x, δx, causes a small change in y, δy. Because B is a small distance away we know it is the point $(x + \delta x, y + \delta y)$. We approximate this strip to a rectangle. As the strip width gets narrower and narrower (i.e. as $\delta y \to 0$) the accuracy of the approximation improves. This means the extra area above the rectangle becomes very small. Then

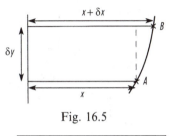

Fig. 16.5

Area of thin strip $\approx$ Area of rectangle

$\approx x\delta y$.

$\approx$ is 'equals approximately'.

The area between $y = c$ and $y = d$ is the sum of all such strips, i.e.

$$\text{Area} = \lim_{\delta y \to 0} \sum_{y=c}^{y=d} x\, \delta y$$

i.e.

$$\text{Area} = \int_c^d x\, dy.$$

Notice the similarities between these two techniques. The first one, from Chapter 10, gives us the standard result based on the horizontal axis. In the second one we simply reverse the roles of x and y. Then we base our area on the vertical axis.

Example 16.1

Find the area between the parabola $y^2 = 5x$, the vertical axis and the line $y = 4$.

We have included a sketch for this curve, Fig. 16.6.

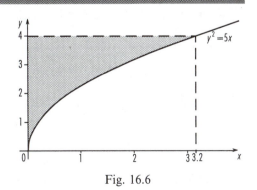

Fig. 16.6

Using our formula

$$\text{Area} = \int_c^d x \, dy$$

we substitute for $c = 0$, $d = 4$ and x to get

$$\text{Area} = \int_0^4 \frac{1}{5} y^2 \, dy.$$

Most of our integration rules have used the variable x. These rules apply to any variable. This means that

$$\text{Area} = \left[\frac{1}{15} y^3 \right]_0^4$$

$$= \frac{4^3}{15}$$

$$= 4.26 \text{ unit}^2.$$

Alternatively we could have applied a technique we used in Chapter 10, i.e
Shaded area = Area of rectangle − Unshaded area.
The rectangle has the 4 vertices (corners) (0, 0), (0, 3.2), (3.2, 4) and (0, 4).
The unshaded area is the area under the curve bounded by the x-axis and the line $x = 3.2$. Notice how we use the positive square root in the integral. This is the portion of the parabola above the x-axis.

$$\text{Shaded area} = (4 \times 3.2) - \int_0^{3.2} y \, dx$$

$$= 12.8 - \int_0^{3.2} 5^{1/2} x^{1/2} \, dx$$

$$\boxed{\begin{array}{l} y^2 = 5x \\ \therefore \quad y = \sqrt{5x}. \end{array}}$$

$$= 12.8 - \left[5^{1/2} \frac{x^{3/2}}{3/2} \right]_0^{3.2}$$

$$= 12.8 - \left[5^{1/2} \frac{(3.2)^{3/2}}{1.5} - 0 \right]$$

$$= 12.8 - 8.53$$

$$= 4.2\bar{6} \text{ unit}^2, \quad \text{as before.}$$

■■■■■ EXERCISE 16.1 ■■■■■

This Exercise is for you to use the formula $\text{Area} = \int_c^d x \, dy$ wherever possible. You may wish to check your answers as we demonstrated in Example 16.1.

1   Find the area bounded by the curve $y = x^3$ and the y-axis from $y = 0$ to $y = 8$.

2   $y = x^2 + 2$. Make x the subject of this relationship. Find the area between this curve to the right of the y-axis from $y = 3$ to $y = 11$. What problem arises if the limits of integration are $y = 0$ and $y = 1$?

3 You are given $y = \dfrac{4}{x}$. Find the values of y when $x = 0.5$ and $x = 1.5$.
From $x = 0.5$ to $x = 1.5$ find the area between the y-axis and the curve
$y = \dfrac{4}{x}$.

4 If $y = e^x$ make x the subject of this relationship. Find the area
bounded by this curve, the y-axis and the lines $y = 1$ and $y = 5$.

5 Sketch the circle $y^2 + x^2 = 25$. Remember it has centre $(0, 0)$ and is of
radius 5. Concentrate on the first quadrant. Briefly explain why both

of our area formulae, $\displaystyle\int_a^b x \, dy$ and $\displaystyle\int_c^d x \, dy$ give the same result.

Some standard techniques

Here we derive some standard results using integration. Each time we split
our regular shape into a series of elemental (incremental) areas. We find
the area of each element by using an earlier result. Then we add them
together. Remember that integration is the limit of the summation
process.

▬▬▬ Example 16.2 ▬▬▬▬▬▬▬▬▬▬▬▬▬▬▬▬▬▬▬

Find the formula for the area of a triangle.

In Fig. 16.7 we sketch a triangle
with a horizontal base. We choose
this because we use a horizontal
axis. If we wished to be awkward
we could have our axes skewed to
one side.

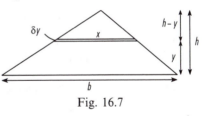

Fig. 16.7

Using the usual notation, our triangle has a base of length b and a vertical
height (altitude) h. We split our triangle into a series of thin strips parallel
to the base. Each of these strips approximates to a rectangle. The slight
variations at each end tend to zero in the limiting case. We look at a
typical element (rectangle).

Length of element = x
Thickness   = δy
Area   = $x \, \delta y$

The area of the triangle is the sum of all these thin strips. We add them
from the base to the vertex, i.e.

$$\text{Area} = \sum_{y=0}^{y=h} x \, \delta y.$$

In the limiting case as $\delta y \to 0$

$$\text{Area} = \int_0^h x \, dy.$$

Now x and y are linked by similar triangles. We compare bases to heights, i.e.

$$\frac{x}{b} = \frac{h-y}{h}$$

i.e.

$$x = \frac{b}{h}(h-y).$$

We substitute into our area formula to get

$$\text{Area} = \int_0^h \frac{b}{h}(h-y) \, dy$$

$$= \frac{b}{h} \int_0^h (h-y) \, dy$$

> We can remove $\frac{b}{h}$ from the integral. It is not variable.

$$= \frac{b}{h} \left[hy - \frac{y^2}{2} \right]_0^h$$

$$= \frac{b}{h} \left[h^2 - \frac{h^2}{2} - 0 \right]$$

$$= \frac{b}{h} \left[\frac{h^2}{2} \right]$$

$$= \frac{1}{2} bh.$$

This is a standard formula to find the area of a triangle. Alternatively, we can write '$\frac{1}{2} \times$ base $\times$ perpendicular height'. Remember there are other formulae. We met them in Volume 1.

▓▓▓ Example 16.3 ▓▓▓

Find the formula for the area of a trapezium.

In Fig. 16.8 we sketch a trapezium with a horizontal base. Again we choose this because we use a horizontal axis. Once again, if we wished to be awkward we could have our axes skewed to one side.

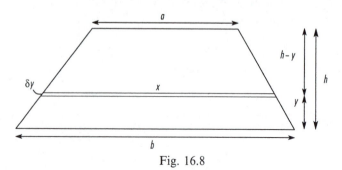

Fig. 16.8

Using the usual notation, our trapezium has parallel sides a and b with a vertical height (altitude) h. We split it into a series of thin strips parallel to the base. Each of these strips approximates to a rectangle. The slight variations at each end tend to zero in the limiting case. Let us look at a typical element (rectangle).

Length of element $= x$

Thickness $\qquad = \delta y$

Area $\qquad\qquad = x\,\delta y$

The area of the trapezium is the sum of all these thin strips. We add them from one parallel side to the other, i.e.

$$\text{Area} = \sum_{y=0}^{y=h} x\,\delta y.$$

In the limiting case as $\delta y \to 0$

$$\text{Area} = \int_0^h x\,dy.$$

Now x and y are linked by similar shapes. These are the two triangles on either side of the central portion (rectangle), in Fig. 16.9.

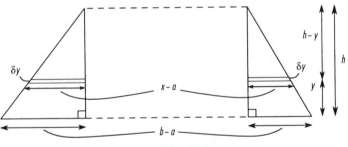

Fig. 16.9

We compare their combined bases to heights,

i.e. $\qquad \dfrac{x-a}{b-a} = \dfrac{h-y}{h}$

i.e. $\qquad x-a = \dfrac{(b-a)}{h}(h-y).$

You should check for yourself that this simplifies to

$$x = b + \frac{(a-b)}{h}y.$$

We substitute into our area formula to get

$$\text{Area} = \int_0^h \left(b + \frac{(a-b)}{h}y\right)dy$$

$$= \left[by + \frac{(a-b)}{h}\frac{y^2}{2}\right]_0^h$$

$$= \left[bh + \frac{(a-b)\,h^2}{h} \frac{1}{2} - 0 \right]$$

$$= bh + \frac{ah}{2} - \frac{bh}{2}$$

$$= \frac{1}{2}(a+b)h.$$

This is a standard formula to find the area of a trapezium. Alternatively, we can write '$\frac{1}{2}$ the sum of the parallel sides × perpendicular distance between them'.

Example 16.4

Find the formula for the area of a trapezium using previous results.

In Fig. 16.10 we sketch our trapezium in sections. We have a rectangular central section and two right-angled triangles. We can bring those triangles together. Then they have a base of $(b - a)$.

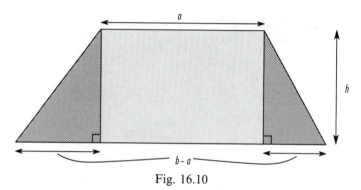

Fig. 16.10

Rather than using integration we apply our standard formulae.

Area of trapezium = Area of rectangle + Area of triangle(s)

$$= \qquad ah \qquad + \qquad \frac{1}{2}(b-a)h$$

$$= \frac{1}{2}(2a+b-a)h$$

$$= \frac{1}{2}(a+b)h, \text{ as before.}$$

Example 16.5

Find the formula for the area of a sector of a circle of radius r.

In Fig. 16.11 we sketch a sector. Notice the horizontal axis is a line of symmetry. The area below the axis is a mirror image of the area above. For our sector of angle α radians we have $\frac{1}{2}\alpha$ both above and below that axis.

We split our sector into a series of thin isosceles triangles. There is a slight variation with the curved base. This tends to zero in the limiting case. Let us look at a typical element (triangle) inclined at an angle θ to the axis. $\delta\theta$ is the angle between the sides of length r.

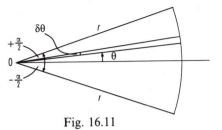

Fig. 16.11

$$\text{Area} = \frac{1}{2}r^2 \sin \delta\theta.$$

This is one of the three formulae for the area of a triangle. We have met it before in Volume 1. The area of the sector is the sum of all these thin isosceles triangles. We add them to get

$$\text{Area} = \sum_{\theta=-\alpha/2}^{\theta=\alpha/2} \frac{1}{2}r^2 \sin \delta\theta.$$

For small angles $\sin \delta\theta \to \delta\theta$ so that

$$\text{Area} = \sum_{\theta=-\alpha/2}^{\theta=\alpha/2} \frac{1}{2}r^2 \,\delta\theta.$$

In the limiting case as $\delta\theta \to 0$

$$\begin{aligned}
\text{Area} &= \int_{-\alpha/2}^{\alpha/2} \frac{1}{2}r^2 \,d\theta \\
&= \left[\frac{1}{2}r^2\theta\right]_{-\alpha/2}^{\alpha/2} \\
&= \left[\frac{1}{2}r^2\frac{\alpha}{2}\right] - \left[\frac{1}{2}r^2\frac{(-\alpha)}{2}\right] \\
&= \left[\frac{1}{2}r^2\frac{\alpha}{2} + \frac{1}{2}r^2\frac{\alpha}{2}\right] \\
&= \frac{1}{2}r^2\alpha,
\end{aligned}$$

which is a standard formula for the area of a sector.

▬▬▬ ASSIGNMENT ▬▬▬▬▬▬▬▬▬▬▬▬▬▬

We are going to find the area (i.e. top surface area) of our piece of alloy. The angle subtended at the centre of the circle is $120°$, i.e. $\frac{2}{3}\pi$ radians. The large radius is 0.80 m and the small radius is 0.30 m. In Fig. 16.12 we draw it with the horizontal axis as a line of symmetry. For $\alpha = \frac{2}{3}\pi$ we have $\frac{\alpha}{2} = \frac{\pi}{3}$ both above and below the axis.

We apply the integral from Example 16.5,

$$\text{Area} = \int_{-\alpha/2}^{\alpha/2} \frac{1}{2} r^2 \, d\theta.$$

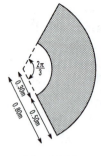

Area of large sector $= \displaystyle\int_{-\pi/3}^{\pi/3} \frac{1}{2}(0.80)^2 \, d\theta$

$r = 0.80.$

$$= \int_{-\pi/3}^{\pi/3} 0.32 \, d\theta.$$

Area of small sector $= \displaystyle\int_{-\pi/3}^{\pi/3} \frac{1}{2}(0.30)^2 d\theta$

Fig. 16.12

$r = 0.30.$

$$= \int_{-\pi/3}^{\pi/3} 0.045 \, d\theta.$$

We subtract these integrals, i.e.

Area of alloy = Area of large sector − Area of small sector

$$= \int_{-\pi/3}^{\pi/3} 0.32 \, d\theta - \int_{-\pi/3}^{\pi/3} 0.045 \, d\theta$$

$$= \int_{-\pi/3}^{\pi/3} (0.32 - 0.045) \, d\theta$$

$$= \int_{-\pi/3}^{\pi/3} 0.275 \, d\theta$$

$$= \left[0.275\theta \right]_{-\pi/3}^{\pi/3}$$

$$= 0.275\frac{\pi}{3} - 0.275\left(\frac{-\pi}{3}\right)$$

$$= 0.576 \text{ m}^2.$$

Examples 16.6

Apply Example 16.5 to find the areas of a i) quadrant, ii) semi-circle and iii) circle.

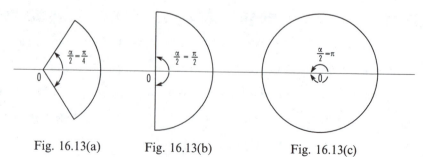

Fig. 16.13(a)   Fig. 16.13(b)   Fig. 16.13(c)

i) In Fig. 16.13(a) we have a quadrant of a circle. You can see it is a sector of a circle with an angle of $\alpha = \dfrac{\pi}{2}$ radians.

$$\text{Area} = \frac{1}{2}r^2\alpha = \frac{1}{2}r^2\frac{\pi}{2} = \frac{1}{4}\pi r^2.$$

Alternatively, we could work out this formula by integration. We need to use the area of a sector formula with different limits. Instead of $\dfrac{-\alpha}{2}$ and $\dfrac{\alpha}{2}$ we need $\dfrac{-\pi}{4}$ and $\dfrac{\pi}{4}$.

ii) In Fig. 16.13(b) we have a semi-circle. You can see it is a sector of a circle with an angle of $\alpha = \pi$ radians.

$$\text{Area} = \frac{1}{2}r^2\alpha = \frac{1}{2}r^2\pi = \frac{1}{2}\pi r^2.$$

Alternatively, we could work out this formula by integration. We need to use the area of a sector formula with different limits. Instead of $\dfrac{-\alpha}{2}$ and $\dfrac{\alpha}{2}$ we need $\dfrac{-\pi}{2}$ and $\dfrac{\pi}{2}$.

iii) In Fig. 16.13(c) we have a complete circle. This is the ultimate sector with an angle of $\alpha = 2\pi$ radians.

$$\text{Area} = \frac{1}{2}r^2\alpha = \frac{1}{2}r^2 2\pi = \pi r^2, \text{ as expected.}$$

Alternatively we could work out this formula by integration. We need to use the area of a sector formula with different limits. Instead of $\dfrac{-\alpha}{2}$ and $\dfrac{\alpha}{2}$ we need $-\pi$ and π.

ASSIGNMENT

We quickly return to our Assignment, this time applying the formula from Example 16.5, i.e.

$$\text{Area} = \frac{1}{2}r^2\alpha. \qquad \boxed{\alpha = \frac{2\pi}{3}.}$$

$$\text{Area of large sector} = \frac{1}{2}(0.80)^2\frac{2\pi}{3} = (0.80)^2\frac{\pi}{3}. \qquad \boxed{r = 0.80.}$$

$$\text{Area of small sector} = \frac{1}{2}(0.30)^2\frac{2\pi}{3} = (0.30)^2\frac{\pi}{3}. \qquad \boxed{r = 0.30.}$$

We subtract these integrals, i.e.

$$\text{Area of alloy} = \text{Area of large sector} - \text{Area of small sector}$$

$$= (0.80)^2\frac{\pi}{3} - (0.30)^2\frac{\pi}{3}$$

$$= 0.576 \text{ m}^2, \text{ as before.}$$

■ EXERCISE 16.2 ■

This Exercise is for you to try some integration applied to finding areas. You should check your answers using the standard formulae.

1 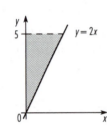 Find the area of the triangle bounded by the lines $y = 2x$, $y = 5$ and the y-axis.

2 Sketch the circle $x^2 + y^2 = 9$. Shade the semi-circle to the right of the vertical axis. Find the area of this semi-circle.

3 Find the area of the trapezium.

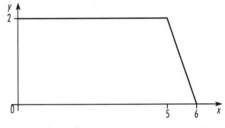

4 Show that the area of the circle $x^2 + y^2 = 16$ is 16π. For the general circle $x^2 + y^2 = r^2$ find the general formula πr^2.

5 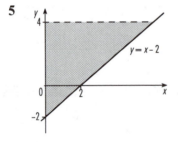 Find the area of the trapezium bounded by the lines $y = 4$, $y = x - 2$ and the coordinate axes.

Centroids

Our theory so far has been based on two-dimensional shapes. We assume these to have uniform thickness (or to be uniformly thin). This means the third dimension is negligible compared with the other two dimensions. Easy examples are a blank sheet of A4 paper and a sheet of 4 mm float glass. They are each an example of a **lamina**. **A lamina is a plane figure with its third dimension negligible compared with the other dimensions.**

Unless we state otherwise, each example uses one type of material only. This is to ensure uniform thickness and density. Then the mass (and weight) is proportional to the area. For simplicity we link together mass, weight and area. The centre of gravity is the point through which the weight appears to act. A body will balance in equilibrium about its centre of gravity. Alternatively, we may suspend a body in equilibrium by one string. Then the string and the weight lie in the same vertical line. The centre of gravity is the same position as the centre of mass. Because we are using mass and weight proportional to area these positions are the same as the **centroid** (centre of area).

This idea of balancing in equilibrium is important. We can link it to symmetry. For symmetry, one side of a shape is the mirror image in a line of symmetry of the other side. The sides 'balance' each other. This means the centroid always lies on a line of symmetry. Look for lines of symmetry before attempting any calculations.

In Fig. 16.14 we have a rectangle which has 2 lines of symmetry, *AB* and *CD*. The centroid must lie on them both, i.e. it must lie where they cross at *G*.

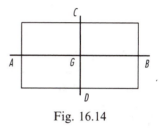

Fig. 16.14

The same idea applies to a square. Of course, all its sides are equal in length. Also it does have more than 2 lines of symmetry.

In Fig. 16.15 we have a circle which has many lines of symmetry. The centroid must lie on them all, i.e. it must lie where they cross at *G*.

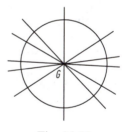

Fig. 16.15

Suppose we wish to find the moment of force (e.g. weight) about a given axis. **That moment is defined to be the product of that force and its perpendicular distance from the axis.** For the **first moment of area** we replace the force with an area. **About a given axis it is defined to be the product of that area and the perpendicular distance of its centroid from that axis.**

Let the area under a curve be *A* and the centroid have the coordinates $(\bar{x}, \bar{y})$. We show this in Fig. 16.16.

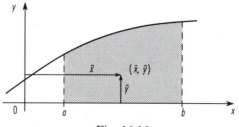

Fig. 16.16

We apply our definition for first moment of area. The first moment of area about the vertical axis is $\bar{x}A$. Similarly, the first moment of area about the horizontal axis is $\bar{y}A$. Now we use the definition to find the centroid.

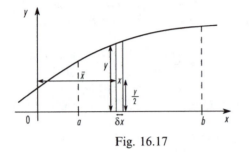

Fig. 16.17

In Fig. 16.17 we sketch a general curve and look at an area under it. We split the area into a series of thin vertical strips. Each strip approximates to a rectangle. The slight variation at the top tends to zero in the limiting case. We know the position of the centroid of a rectangle. It is halfway along each dimension. In this case it is at $(x + \frac{1}{2}\delta x, \frac{1}{2}y)$ which approximates to $(x, \frac{1}{2}y)$. Let us look at a typical element.

Length of element $= y$

Thickness $\quad\quad\; = \delta x$

Area $\quad\quad\quad\;\; = y\,\delta x$

Distance of centroid from the vertical axis $\;= x$.

Moment about the vertical axis $\quad\quad\quad = (y\,\delta x)(x) = xy\,\delta x$.

To find the total first moment of area about the vertical axis we sum all these thin strips, i.e.

$$\text{First moment of area} = \sum_{x=a}^{x=b} xy\,\delta x.$$

In the limiting case as $\delta x \to 0$

$$\text{First moment of area} = \int_a^b xy\,dx.$$

We link this integral with $\bar{x}A$ to give

$$\bar{x}A = \int_a^b xy\,dx$$

$$\bar{x} = \frac{\displaystyle\int_a^b xy\,dx}{\displaystyle\int_a^b y\,dx}.$$

$\boxed{A = \int_a^b y\,dx.}$

We look again at our general curve working to find $\bar{y}$.

Length of element $= y$

Thickness $\quad\quad\; = \delta x$

Area $\quad\quad\quad\;\; = y\,\delta x$

Distance of centroid from the horizontal axis $\;= \frac{1}{2}y$.

Moment about the horizontal axis $\quad\quad\quad = (y\,\delta x)(\frac{1}{2}y) = \frac{1}{2}y^2\,\delta x$.

To find the total first moment of area about the horizontal axis we sum all these thin strips, i.e.

First moment of area $= \sum\limits_{x=a}^{x=b} \frac{1}{2}y^2\,\delta x$.

In the limiting case as $\delta x \rightarrow 0$

First moment of area $= \int_a^b \frac{1}{2}y^2\,dx$.

We link this integral with $\bar{y}A$ to give

$$\bar{y}A = \int_a^b \frac{1}{2}y^2\,dx$$

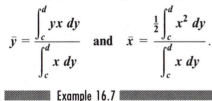

$$\bar{y} = \frac{\frac{1}{2}\int_a^b y^2\,dx}{\int_a^b y\,dx}.$$

We have derived 2 formulae, each involving 2 integrals. In our solutions we will first work out the integrals separately. Then we will divide those results as necessary.

These formulae are based on an area under the curve related to the horizontal axis. If the area is related to the vertical axis (Fig. 16.18) we have similar formulae.

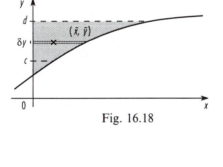

Fig. 16.18

We simply interchange x and y, i.e.

$$\bar{y} = \frac{\int_c^d yx\,dy}{\int_c^d x\,dy} \quad \textbf{and} \quad \bar{x} = \frac{\frac{1}{2}\int_c^d x^2\,dy}{\int_c^d x\,dy}.$$

Example 16.7

Find the centroid of the area under the curve $y = 4x - x^2$ above the horizontal axis.

We sketch the curve in Fig. 16.19. Notice it crosses the horizontal axis at $x = 0$ and $x = 4$. These are our limits of integration. Also the line $x = 2$ is a line of symmetry. This means the centroid lies on the line $x = 2$ (i.e. $\bar{x} = 2$). Thus we need only find $\bar{y}$. We work out the integrals separately.

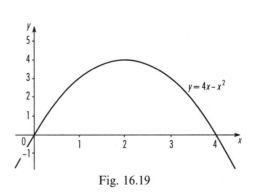

Fig. 16.19

$$\frac{1}{2}\int_a^b y^2 \, dx = \frac{1}{2}\int_0^4 (16x^2 - 8x^3 + x^4) \, dx$$

$$\begin{array}{l} b = 4 \\ y^2 = (4x - x^2)^2 \\ \quad = 16x^2 - 8x^3 + x^4. \\ a = 0. \end{array}$$

$$= \frac{1}{2}\left[\frac{16x^3}{3} - \frac{8x^4}{4} + \frac{x^5}{5}\right]_0^4$$

$$= \frac{1}{2}\left[\frac{16(4)^3}{3} - 2(4)^4 + \frac{(4)^5}{5}\right] - \frac{1}{2}\left[0\right]$$

$$= \frac{1}{2}[341.\bar{3} - 512 + 204.8]$$

$$= 17.06 \text{ unit}^3.$$

Also we have to find the area using the formula

$$\text{Area} = \int_a^b y \, dx$$

$$= \int_0^4 (4x - x^2) \, dx$$

$$\begin{array}{l} b = 4. \\ y = 4x - x^2. \\ a = 0. \end{array}$$

$$= \left[\frac{4x^2}{2} - \frac{x^3}{3}\right]_0^4$$

$$= \left[2(4)^2 - \frac{4^3}{3}\right] - \left[0\right]$$

$$= 10.\bar{6} \text{ unit}^2.$$

We divide these results to get

$$\bar{y} = \frac{17.0\bar{6}}{10.\bar{6}} = 1.6.$$

This means the centroid is at the point (2, 1.6).

   Let us glance again at Fig. 16.19. Notice how x varies from 0 to 4 with $x = 2$ as a line of symmetry (i.e. $\bar{x} = 2$). Also y varies from 0 to 4 yet the centroid is $\bar{y} = 1.6$, just below halfway between these limits. This is emphasised by the distribution of the area. There is more in the lower half than the upper half.

Example 16.8

Find the centroid of the area bounded by the curve $y^2 = 5x$ and the vertical axis between $y = 0$ and $y = 4$.

   We have seen this curve and mentioned area (4.2$\bar{6}$ unit^2) in Example 16.1, but re-sketch the curve in Fig. 16.20. Notice the area is related to the vertical axis so we use the amended formulae. Also, there is no line of symmetry to reduce the work. We need to calculate both $\bar{x}$ and $\bar{y}$.

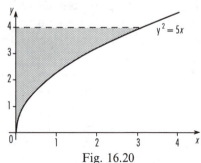

Fig. 16.20

Again we work out the integrals separately.

$$\frac{1}{2}\int_c^d x^2\, dy = \frac{1}{2}\int_0^4 \frac{y^4}{25}\, dy$$

$$= \frac{1}{2}\left[\frac{y^5}{25\times 5}\right]_0^4$$

$$= \frac{1}{2}\left(\left[\frac{4^5}{25\times 5}\right] - \left[0\right]\right)$$

$$= 4.096 \text{ unit}^3.$$

> $d = 4.$
> $5x = y^2$
> $x^2 = \left(\frac{y^2}{5}\right)^2 = \frac{y^4}{25}.$
> $c = 0.$

We divide this value by the area to get $\bar{x}$, i.e.

$$\bar{x} = \frac{4.096}{4.2\overline{6}} = 0.96.$$

$$\int_c^d yx\, dy = \int_0^4 \frac{y^3}{5}\, dy$$

$$= \left[\frac{y^4}{5\times 4}\right]_0^4$$

$$= \frac{4^4}{5\times 4} - 0$$

$$= 12.8 \text{ unit}^3.$$

> $d = 4.$
> $5x = y^2$
> $yx = y \times \frac{y^2}{5} = \frac{y^3}{5}.$
> $c = 0.$

We divide this value by the area to get $\bar{y}$, i.e.

$$\bar{y} = \frac{12.8}{4.2\overline{6}} = 3.$$

Together these results tell us the centroid is (0.96, 3).

Example 16.9

Find the centroid of the area bounded by the curve $y = x^2$ and the line $y = x + 2$.

In Fig. 16.21 we sketch these graphs, noting where they intersect.

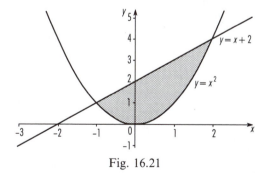

Fig. 16.21

Essentially we have a pair of simultaneous equations to solve for x only. These x values are our limits of integration.

$$y = 2 + x$$

and $y = x^2$

i.e $0 = 2 + x - x^2$ | Subtracting.

$0 = (1 + x)(2 - x)$ | Factorising.

Either $1 + x = 0$ or $2 - x = 0$

i.e. $x = -1, 2$.

To find $(\bar{x}, \bar{y})$ we need the values of 3 integrals. We could find the area under the line and under the curve. The area between them is the difference between these two areas. We concentrate on the difference. This means a typical strip width is also based on the difference, i.e. the strip width is $y = 2 + x - x^2$.

$$\int_a^b y \, dx = \int_{-1}^{2} (2 + x - x^2) \, dx$$

| $b = 2$.
 $y = 2 + x - x^2$.
 $a = -1$.

$$= \left[2x + \frac{x^2}{2} - \frac{x^3}{3} \right]_{-1}^{2}$$

$$= \left[2(2) + \frac{2^2}{2} - \frac{2^3}{3} \right] - \left[2(-1) + \frac{(-1)^2}{2} - \frac{(-1)^3}{3} \right]$$

$$= 3.\bar{3} - [-1.1\bar{6}]$$

$$= 3.\bar{3} + 1.1\bar{6}$$

$$= 4.5 \text{ unit}^2.$$

$$\int_a^b xy \, dx = \int_{-1}^{2} (2x + x^2 - x^3) \, dx$$

| $b = 2$.
 $xy = x(2 + x - x^2)$
 $= 2x + x^2 - x^3$.
 $a = -1$.

$$= \left[x^2 + \frac{x^3}{3} - \frac{x^4}{4} \right]_{-1}^{2}$$

$$= \left[2^2 + \frac{2^3}{3} - \frac{2^4}{4} \right] - \left[(-1)^2 + \frac{(-1)^3}{3} - \frac{(-1)^4}{4} \right]$$

$$= 2.\bar{6} - 0.41\bar{6}$$

$$= 2.25 \text{ unit}^3.$$

We divide this value by the area to get $\bar{x}$, i.e.

$$\bar{x} = \frac{2.25}{4.5} = 0.5.$$

$$\frac{1}{2} \int_a^b y^2 \, dx = \frac{1}{2} \int_{-1}^{2} (4 + 4x - 3x^2 - 2x^3 + x^4) \, dx$$

| $y = 2 + x - x^2$
 $y^2 = (2 + x - x^2)^2$.

$$= \frac{1}{2} \left[4x + 2x^2 - x^3 - \frac{2x^4}{4} + \frac{x^5}{5} \right]_{-1}^{2}$$

$$= \frac{1}{2} \left(\left[4(2) + 2(2)^2 - 2^3 - \frac{2^4}{2} + \frac{2^5}{5} \right] \right.$$

$$\left. - \left[4(-1) + 2(-1)^2 - (-1)^3 - \frac{(-1)^4}{2} + \frac{(-1)^5}{5} \right] \right)$$

$$= \frac{1}{2} (6.4 - -1.7)$$

$$= 4.05 \text{ unit}^3.$$

We divide this value by the area to get $\bar{y}$, i.e.

$$\bar{y} = \frac{4.05}{4.5} = 0.9.$$

Together these results tell us the centroid is (0.5, 0.9).

▰▰▰ EXERCISE 16.3 ▰▰▰

Where necessary you should apply the integral formulae to find the centroid $(\bar{x}, \bar{y})$ of the area in each question. If you see a line of symmetry use it to reduce your working. In some questions we suggest you make a sketch. You may find it useful to do this for each question.

1 By integration find the area under the curve $y = x^3$ from $x = 0$ to $x = 2$. Also find the centroid of this area.

2 For the curve $y = 3x^2$ from $x = 1$ to $x = 4$ find the area between the curve and the x-axis. Find the centroid of this area.

3 The curve $y = \sin \theta$ cuts the horizontal axis at $\theta = 0$ and $\theta = \pi$ radians. Find the centroid of this area bounded by the curve and horizontal axis.

4 By integration find the area under the curve $y = e^t$ from $t = 0$ to $t = 1$. Also find the centroid of this area.

5 Sketch the curve $y = (x - 2)^2$. Confirm it makes contact with the axes at (2, 0) and (0, 4). Find the centroid of the area bounded by the curve and both axes.

6 Sketch the circle $x^2 + y^2 = r^2$. Shade the area to the right of the vertical axis. By integration show the centroid of this shaded semi-circle is $\left(\frac{4r}{3\pi}, 0 \right)$. What is it for a similarly shaded semi-circle derived from the circle $x^2 + y^2 = 9$?

7 By integration find the area between the curve $y = x^2$ and the vertical axis to the right of that axis. Use $y = 0$ and $y = 4$ as your limits of integration. Also find the centroid of this area.

8

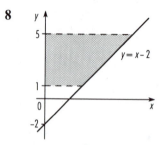

Find the area of the accompanying trapezium and its centroid by integration. It is bounded by the vertical axis, $y = x - 2$, $y = 1$ and $y = 5$.

9   On one set of axes sketch the line $y = 3$ and the curve $y = x(4 - x)$. Check they intersect at $(1, 3)$ and $(3, 3)$. Find the centroid of the area between them using these coordinates as your limits of integration.

10   The diagram shows the curves $y = x^2$ and $y = 8 - x^2$. Check they intersect at $(-2, 4)$ and $(2, 4)$. Find the centroid of the shaded area to the right of the vertical axis.

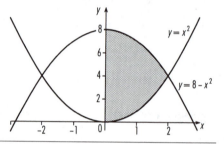

Some standard centroid techniques

We have already stated that a centroid lies on a line of symmetry. We identified this usefully for a rectangle, square and circle. Now we look at two other basic shapes. They are a triangle and a sector. Here we derive those standard results using integration. Each time we split our regular shape into a series of elemental (incremental) areas. We find the first moment of area of each element using an earlier result. Then we add them together.

━━━━━━ **Example 16.10** ━━━━━━

Find the centroid of a triangle.

We sketch a general triangle in Fig. 16.22. We can split it into a series of thin strips parallel to any side. Each strip approximates to a rectangle. The centroid of each rectangle is halfway along its length. Hence the centroid of the triangle must be along a line through all these halfway points. This line is called a **median. A median of a triangle is defined as the line joining a vertex to the mid-point of the opposite side.** Hence a triangle has 3 medians.

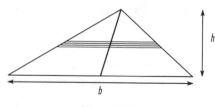

Fig. 16.22

We can repeat the method. We split the triangle into thin strips parallel to each side in turn (Figs. 16.23). Now we know the centroid lies on each median. A property of a triangle is that all the medians intersect one third up from a side. This means the centroid lies at the point of intersection.

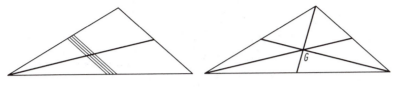

Figs. 16.23

In Example 16.11 we reach this result through integration.

Example 16.11

By integration find the centroid of a triangle.

In Fig. 16.24 again we sketch a general triangle. Using the usual notation, our triangle has a base of length b and a vertical height (altitude) h. We split our triangle into a series of thin strips parallel to the base. Each of these strips approxi-

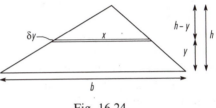

Fig. 16.24

mates to a rectangle. The slight variations at both ends tend to zero in the limiting case. We look at a typical element (rectangle).

Length of element $= x$
Thickness $\qquad = \delta y$
Area $\qquad\qquad = x\delta y$
Distance of centroid from the base $= y$
Moment about the base $\qquad = (x\delta y)(y) = xy\,\delta x.$

The first moment of area of the triangle is the sum of all these elemental moments. We add them to get

$$\text{First moment of area} = \sum_{y=0}^{y=h} xy\,\delta y.$$

In the limiting case as $\delta y \to 0$

$$\text{First moment of area} = \int_0^h xy\,dy.$$

Now x and y are linked by similar triangles. We compare bases to heights,

i.e. $\dfrac{x}{b} = \dfrac{h-y}{h}$

i.e. $\qquad\qquad x = \dfrac{b}{h}(h-y).$

We substitute into our first moment of area formula to get

First moment of area $= \displaystyle\int_0^h \frac{b}{h}(h-y)y\, dy$

We link this integral with $\bar{y}A$ where A is the area of the triangle, i.e.

$$\bar{y}A = \frac{b}{h}\int_0^h (hy - y^2)\, dy$$

> We can remove $\dfrac{b}{h}$ from the integral. It is *not* variable.

$$= \frac{b}{h}\left[\frac{hy^2}{2} - \frac{y^3}{3}\right]_0^h$$

$$= \frac{b}{h}\left[\frac{h^3}{2} - \frac{h^3}{3} - 0\right]$$

i.e. $\bar{y}\left(\dfrac{1}{2}bh\right) = \dfrac{b}{h}\left[\dfrac{h^3}{6}\right]$

> Area of triangle, $A = \dfrac{1}{2}bh$.

$\therefore$ $\bar{y} = \dfrac{bh^2}{6} \times \dfrac{2}{bh}$

i.e. $\bar{y} = \dfrac{h}{3}$,

i.e. the centroid is one third from the base. This is the standard formula for the first moment of area of a triangle. We can repeat the method using each side in turn as a base. This will confirm our result of Example 16.10, i.e. the centroid is one-third along a median from the side. Obviously we can think of this as two-thirds along a median from a vertex.

Example 16.12

Find the centroid of a sector of a circle of radius r.

In Fig. 16.25 we sketch a sector. Notice the horizontal axis is a line of symmetry. The area below the axis is a mirror image of the area above. For our sector of angle α radians we have $\dfrac{\alpha}{2}$ both above and below that axis.

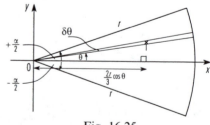

Fig. 16.25

We split our sector into a series of thin isosceles triangles. There is a slight variation with the curved base. This tends to zero in the limiting case. Let us look at a typical element (triangle) inclined at an angle θ to the axis. The centroid of our elemental triangle is $\dfrac{2}{3}r$ from the vertex at O. $\delta\theta$ is the angle between the sides of length r.

Area $= \dfrac{1}{2}r^2 \sin\delta\theta = \dfrac{1}{2}r^2\delta\theta.$

> As $\delta\theta \to 0$.

Perpendicular distance of centroid from the vertical axis

$$= \frac{2}{3} r \cos \theta.$$

Moment about the vertical axis $= \left(\frac{1}{2} r^2 \delta\theta \right) \left(\frac{2}{3} r \cos \theta \right)$

$$= \frac{1}{3} r^3 \cos \theta \, \delta\theta.$$

The first moment of area of the sector is the sum of all these elemental moments. We add them to get

First moment of area $= \displaystyle\sum_{\theta=-\alpha/2}^{\theta=\alpha/2} \frac{1}{3} r^3 \cos \theta \, \delta\theta.$

In the limiting case as $\delta\theta \rightarrow 0$

First moment of area $= \displaystyle\int_{-\alpha/2}^{\alpha/2} \frac{1}{3} r^3 \cos \theta \, d\theta.$

We link this integral with $\bar{x}A$ where A is the area of the sector, i.e.

$$\bar{x}A = \left[\frac{1}{3} r^3 \sin \theta \right]_{-\alpha/2}^{\alpha/2}$$

$$= \left[\frac{1}{3} r^3 \sin \frac{\alpha}{2} \right] - \left[\frac{1}{3} r^3 \sin \left(\frac{-\alpha}{2} \right) \right]$$

$$= \frac{1}{3} r^3 \sin \frac{\alpha}{2} + \frac{1}{3} r^3 \sin \frac{\alpha}{2}$$

i.e. $\bar{x} \left(\frac{1}{2} r^2 \alpha \right) = \frac{2}{3} r^3 \sin \frac{\alpha}{2}$ $\boxed{\text{Area of sector, } A = \frac{1}{2} r^2 \alpha.}$

$$\therefore \quad \bar{x} = \frac{2}{3} r^3 \sin \frac{\alpha}{2} \times \frac{2}{r^2 \alpha}$$

i.e $\bar{x} = \dfrac{4r}{3\alpha} \sin \dfrac{\alpha}{2},$

which is a standard formula for the first moment of area of a sector.

We can find another, more important, result by applying this formula for the sector. A semi-circle is a sector with $\alpha = \pi$, i.e. $\frac{\alpha}{2} = \frac{\pi}{2}$ and $\sin \frac{\alpha}{2} = \sin \frac{\pi}{2} = 1$. This means the centroid of a semi-circle is $\dfrac{4r}{3\pi}$ from its straight edge along the line of symmetry (Fig. 16.26)

We now have some useful standard results for a rectangle, triangle and semi-circle. We apply these in the next section.

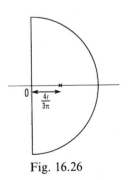

Fig. 16.26

Composite shapes

Many shapes are made up of several standard shapes. For example you can think of a trapezium as a rectangle between two triangles. Alternatively, you can remove two different triangles from a larger rectangle to create that same trapezium (Figs. 16.27).

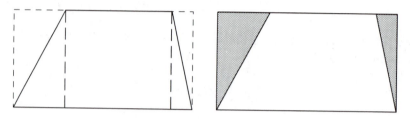

Figs. 16.27

Just as you can add and subtract areas, so you can add and subtract first moments of area. We have two simple rules (formulae):

Moment of the Sum = Sum of the Moments.

and

Moment of the Remainder = Moment of the Whole − Moment of the Section Removed.

We demonstrate these rules in the next Examples and the Assignment.

Example 16.13

Fig. 16.28 shows a composite shape. It is made up from an isosceles triangle bonded to a semi-circle of equal base. We are going to find the centroid of the shape.

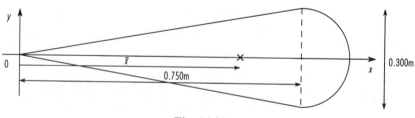

Fig. 16.28

The centroid lies on the line Ox, a line of symmetry. We know how to find the centroid of a triangle and that of a semi-circle. Also we move the length of the triangle before adding the standard distance for the semi-circle's centroid, $h + \dfrac{4r}{3\pi}$. Our method is in stages; the areas, the centroid distances and the first moments of area. Once you have seen the process we shall set out the information in a table. This is shorter and just as orderly. We note the first few decimal places keeping all of them in the calculator's memory.

Area of triangle $= \dfrac{1}{2} \times 0.75 \times 0.30 = 0.1125$ m^2.

$\boxed{A = \dfrac{1}{2}bh.}$

Distance of centroid from $Oy = \dfrac{2}{3} \times 0.75 = 0.50$ m.

First moment of area $= 0.1125 \times 0.50 = 0.056\ldots$ m^3.

Area of semi-circle $= \dfrac{1}{2}\pi \times 0.15^2 = 0.035\ldots$ m^2.

$\boxed{A = \dfrac{1}{2}bh.}$

Distance of centroid from $Oy = 0.75 + \left(\dfrac{4 \times 0.15}{3\pi}\right) = 0.813\ldots$ m.

First moment of area $= 0.035\ldots \times 0.813\ldots = 0.028\ldots$ m^3.

Total area $= 0.1125 + 0.035\ldots = 0.147\ldots$ m^2.

Distance of centroid from $Oy = \bar{x}$ m.

First moment of area $= 0.147\ldots\bar{x}$ m^3.

We use our formula

  Moment of the Sum = Sum of the Moments,

i.e. $\qquad 0.147\ldots\bar{x} = 0.056\ldots + 0.028\ldots$

$$\bar{x} = \frac{0.085\ldots}{0.147\ldots}$$

$$= 0.575 \text{ m is the distance of our composite shape}$$

from the Oy axis.

Let us gather together our information in a table. We use 4 columns. You can see we multiply the entries of columns ② and ③ to get the final column.

Shape	Area	Distance of centroid from the Oy axis	First moment of area
	0.1125	0.50	0.056...
(+)	0.035...	0.813...	0.028...
Total	0.147...	$\bar{x}$	0.147...$\bar{x}$

As before we use our formula

Moment of the Sum = Sum of the Moments,

i.e. $\qquad 0.147\ldots\bar{x} = 0.056\ldots + 0.028\ldots$

to give $\qquad \bar{x} = 0.57$ m.

Example 16.14

Fig. 16.29 shows the same type of composite shape we have just considered. It is made from an isosceles triangle bonded to a semi-circle of equal base. This time we are going to look at it algebraically. We wish to hang it by 1 string from A, a point on the join. Also we want the line of symmetry to be horizontal. This means we need the centroid to be vertically below A, i.e. $\bar{x} = h$.

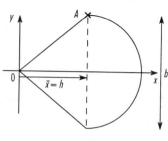

Fig. 16.29

Shape	Area	Distance of centroid from the Oy axis	First moment of area
(triangle)	$\dfrac{1}{2}bh$	$\dfrac{2}{3}h$	$\dfrac{1}{2}bh \times \dfrac{2}{3}h$
(+) (semi-circle)	$\dfrac{1}{2}\pi\left(\dfrac{b}{2}\right)^2$	$h + \dfrac{4(b/2)}{3\pi}$	$\dfrac{1}{8}\pi b^2\left(h + \dfrac{2b}{3\pi}\right)$
Total	$\dfrac{1}{2}bh + \dfrac{1}{8}\pi b^2$	$\bar{x} = h$	$\left(\dfrac{1}{2}bh + \dfrac{1}{8}\pi b^2\right)h$

We use our formula

  Moment of the Sum = Sum of the Moments,

i.e. $\left(\dfrac{1}{2}bh + \dfrac{1}{8}\pi b^2\right)h = \left(\dfrac{1}{2}bh \times \dfrac{2h}{3}\right) + \dfrac{1}{8}\pi b^2\left(h + \dfrac{2b}{3\pi}\right).$

We cancel this in stages to get

$$\left(h + \dfrac{\pi b}{4}\right)h = \dfrac{2}{3}h^2 + \dfrac{1}{4}\pi b\left(h + \dfrac{2b}{3\pi}\right) \qquad \boxed{\begin{array}{l}\text{Expanding the}\\ \text{brackets.}\end{array}}$$

i.e. $h^2 + \dfrac{\pi bh}{4} = \dfrac{2}{3}h^2 + \dfrac{\pi bh}{4} + \dfrac{\pi b}{4} \times \dfrac{2b}{3\pi}$ $\boxed{\dfrac{\pi b}{4} \times \dfrac{2b}{3\pi} = \dfrac{b^2}{6}.}$

$$h^2 - \dfrac{2h^2}{3} = \dfrac{b^2}{6}$$

$$\dfrac{1}{3}h^2 = \dfrac{b^2}{6}$$

i.e. $h^2 = \dfrac{b^2}{2}$

so that $h = \dfrac{b}{\sqrt{2}},$ $\boxed{\begin{array}{l}\text{Square root of}\\ \text{both sides.}\end{array}}$

i.e. this is the length of the triangle in terms of its base for a horizontal line of symmetry.

ASSIGNMENT

In Fig. 16.30 we redraw our shape formed by removing a small sector from a large sector. The radii are 0.30 m and 0.80 m respectively. As before, we use a horizontal line of symmetry.

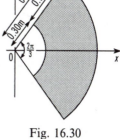

We recall some earlier formulae.

Area of sector $= \dfrac{1}{2}r^2\alpha,$

Distance of centroid from the vertex $= \dfrac{4r}{3\alpha}\sin\dfrac{\alpha}{2}.$

Fig. 16.30

We apply these formulae to the large sector (radius of 0.80 m) and the small sector (radius of 0.30 m). Our values are gathered in the table below. We quote the early decimal places and retain more in the calculator's memory.

Shape	Area	Distance of centroid from the Oy axis	First moment of area
◁	0.670...	0.441...	0.295...
(−) ◁	0.094...	0.165...	0.015...
Remainder	0.575...	$\bar{x}$	0.575... × $\bar{x}$

We use our formula

Moment of the Remainder = Moment of the Whole
  −Moment of the Section Removed,

i.e. $0.575...\bar{x} = 0.295... - 0.015...$

$$\bar{x} = \frac{0.280...}{0.575...}$$

i.e. $\bar{x} = 0.49$ m is the distance of the centroid from the vertex along the line of symmetry. Alternatively, we could write 0.49 m from the Oy axis along the Ox axis.

Example 16.15

Fig. 16.31 shows a trapezium with a circular hole drilled in it. We are going to find the centroid of the shape twice. For the first method we remove two triangles and a circle from a large rectangle. The second time we add together two triangles and a small rectangle, and subtract the circle.

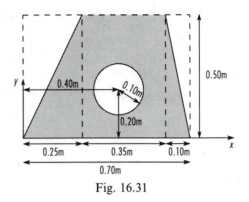

Fig. 16.31

There is no line of symmetry. We need to find the centroid $(\bar{x}, \bar{y})$ by calculation. Both methods use the area and centroid formulae from earlier in the chapter. Let us gather together our information in a table using the usual 4 columns.

Method 1 to find $\bar{x}$.

Shape	Area	Distance of centroid from the Oy axis	First moment of area
▭	0.70×0.50	0.35	0.35×0.35
$(-)$ ◸	$\dfrac{1}{2} \times 0.25 \times 0.50$	$\dfrac{1}{3} \times 0.25$	$0.0625 \times 0.08\bar{3}$
$(-)$ ◺	$\dfrac{1}{2} \times 0.10 \times 0.50$	$0.60 + \left(\dfrac{2}{3} \times 0.10\right)$	$0.0250 \times 0.\bar{6}$
$(-)$ ◯	$\pi(0.10)^2$	0.40	$0.031 \ldots \times 0.40$
Remainder	$0.35 - 0.0625 -$ $0.0250 - 0.031 \ldots$ $= 0.231 \ldots$	$\bar{x}$	$0.231 \ldots \bar{x}$

As before we use our formula

Moment of the Remainder = Moment of the Whole
                                          −Moment of the Section Removed,

i.e. $0.231 \ldots \bar{x} = (0.35 \times 0.35) - (0.0625 \times 0.08\bar{3})$
                                $- (0.0250 \times 0.\bar{6}) - (0.031 \ldots \times 0.40)$

          $0.231 \ldots \bar{x} = 0.1225 - 0.0052 \ldots - 0.01\bar{6} - 0.0125$

∴ $\bar{x} = \dfrac{0.088 \ldots}{0.231 \ldots}$

i.e. $\bar{x} = 0.38$ m is the distance of the centroid from the vertical axis.

Method 1 to find $\bar{y}$.

Shape	Area	Distance of centroid from the Ox axis	First moment of area
	0.70×0.50	0.25	0.35×0.25
$(-)$	$\frac{1}{2} \times 0.25 \times 0.50$	$\frac{2}{3} \times 0.50$	$0.0625 \times 0.\bar{3}$
$(-)$	$\frac{1}{2} \times 0.10 \times 0.50$	$\frac{2}{3} \times 0.50$	$0.0250 \times 0.\bar{3}$
$(-)$	$\pi(0.10)^2$	0.20	$0.031\ldots \times 0.20$
Remainder	$0.231\ldots$ as before	$\bar{y}$	$0.231\ldots\bar{y}$

Again we use our formula

Moment of the Remainder = Moment of the Whole
−Moment of the Section Removed,

i.e.
$$0.231\ldots\bar{y} = (0.35 \times 0.25) - (0.0625 \times 0.\bar{3})$$
$$-(0.0250 \times 0.\bar{3}) - (0.031\ldots \times 0.20)$$
$$0.231\ldots\bar{y} = 0.0875 - 0.0208\ldots$$
$$-0.0083\ldots - 0.0062\ldots$$

∴
$$\bar{y} = \frac{0.052\ldots}{0.231\ldots}$$

i.e.
$$\bar{y} = 0.23 \text{ m is the distance of the centroid from the horizontal axis.}$$

Method 2 to find $\bar{x}$.

Shape	Area	Distance of centroid from the Oy axis	First moment of area
	$\frac{1}{2} \times 0.25 \times 0.50$	$\frac{2}{3} \times 0.25$	$0.0625 \times 0.1\bar{6}$
$(+)$	0.35×0.50	$0.25 + 0.175$	0.175×0.425
$(+)$	$\frac{1}{2} \times 0.10 \times 0.50$	$0.60 + \left(\frac{1}{3} \times 0.10\right)$	$0.0250 \times 0.6\bar{3}$
$(-)$	$\pi(0.10)^2$	0.40	$0.031\ldots \times 0.40$
Remainder	$0.231\ldots$ as before	$\bar{x}$	$0.231\ldots\bar{x}$

This time our remainder is a combination of additions and a subtraction.

i.e. $0.231\ldots\bar{x} = (0.0625 \times 0.1\bar{6}) + (0.175 \times 0.425)$

$$+ (0.0250 \times 0.6\bar{3}) - (0.031\ldots \times 0.40)$$

$0.231\ldots\bar{x} = 0.0104\ldots + 0.0743\ldots + 0.0158\ldots - 0.0125\ldots$

$$\therefore \qquad \bar{x} = \frac{0.88\ldots}{0.231\ldots}$$

i.e. $\bar{x} = 0.38$ m is the distance of the centroid from the vertical axis, as before.

Method 2 to find $\bar{y}$.

Shape	Area	Distance of centroid from the Ox axis	First moment of area
△	$\frac{1}{2} \times 0.25 \times 0.50$	$\frac{1}{3} \times 0.50$	$0.0625 \times 0.1\bar{6}$
(+) ▢	0.35×0.50	0.25	0.175×0.25
(+) ◁	$\frac{1}{2} \times 0.10 \times 0.50$	$\frac{1}{3} \times 0.50$	$0.0250 \times 0.1\bar{6}$
(−) ◯	$\pi(0.10)^2$	0.20	$0.031\ldots \times 0.20$
Remainder	$0.231\ldots$ as before	$\bar{y}$	$0.231\ldots\bar{y}$

Again, our remainder is a combination of additions and a subtraction.

i.e. $0.231\ldots\bar{y} = (0.0625 \times 0.1\bar{6}) + (0.175 \times 0.25)$

$$+ (0.0250 \times 0.1\bar{6}) - (0.031\ldots \times 0.20)$$

$0.231\ldots\bar{y} = 0.0104\ldots + 0.0437\ldots + 0.0041 - 0.0062\ldots$

$$\therefore \qquad \bar{y} = \frac{0.052\ldots}{0.231\ldots}$$

i.e. $\bar{y} = 0.23$ m is the distance of the centroid from the horizontal axis, as before.

▬▬▬ EXERCISE 16.4 ▬▬▬▬▬▬▬▬

In this Exercise we draw the composite shapes with their dimensions. In each case you need to find the centroid. Use any lines of symmetry if appropriate.

1 This is a semi-circular annulus. The larger radius is 1.00 m and the smaller one is 0.50 m. Find the centroid.

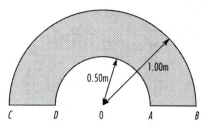

2

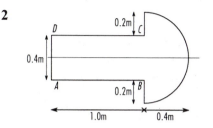

The diagram shows a plastic rectangle bonded to a semi-circle of the same material. Find the centroid of the composite shape.

3 From a isosceles triangular lamina of base 1.00 m and height 0.90 m has been removed a smaller isosceles triangle. The smaller triangle has the same base and half the height. Find the centroid of the remainder.

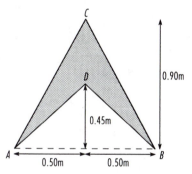

4

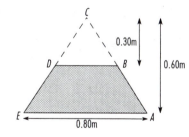

An isosceles triangle has been sliced into two sections. The height of the top section is half the original height. Find the centroid of the remaining trapezium.

5 A steel cross-section is a rectangle with a semi-circle removed. The dimensions of the rectangle are 1.2 m × 0.6 m. The diagram shows the semi-circle to have a diameter of 0.6 m. Where is the centroid?

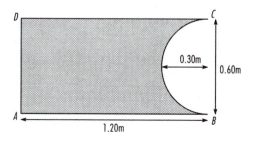

6   Our diagram shows a square with adjacent isosceles triangles pressed from a large sheet of brass. Where is the line of symmetry? Find the centroid of the brass pressing.

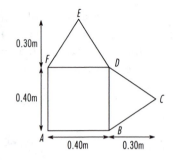

7

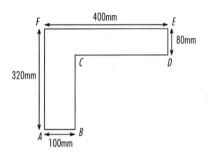

A semi-circle has been stamped out of the triangular sheet of muntz metal as shown. Find the position of the centroid of the remainder.

8   The diagram shows an L-shaped piece of aluminium alloy. It is to be used for strengthening a corner joint. Find its centroid.

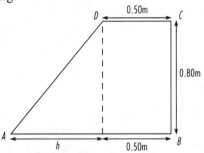

9   The rectangle and right-angled triangle are chromium steel, from part of machinery in paper-making.

   a)   Find the centroid of the composite shape if $h = 0.6\,\text{m}$.

   b)   What must be the value of h if the centroid lies on the join between the two chromium steel sections?

10   In each case, where is the centroid? The radius of the larger circular plate is always 1.0 m. The centres of any drilled holes are on a circle, with the same centre, of radius 0.7 m. Each hole is of radius 0.2 m.

a) The first diagram shows two pairs of diametrically opposed holes. The centres of the holes form a square.

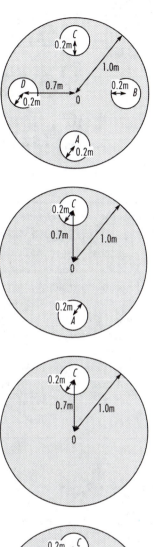

b) The second diagram shows one pair of diametrically opposed holes.

c) The third diagram shows one hole only.

d) The final diagram shows two adjacent holes.

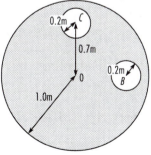

17 Volumes of Revolution and Centroids

Introduction

In this chapter we extend the area and first moment of area ideas of two dimensions. Here we look at volume and first moment of volume. We generate our volumes by rotating an area through 360° about an axis. Again we relate them to the y-axis just as easily as the x-axis. We gather together standard results to find the volumes of composite shapes. Then we apply integration techniques to find the centroid of these volumes.

We also link some two- and three-dimensional results using Pappus' theorem. This is the second of the two theorems we first saw in Volume 1, Chapter 16.

■■■■■■ ASSIGNMENT ■■■■■■■■■■■■■■■■■■■■■■

The Assignment for this chapter is based on cylinders. In Fig. 17.1 effectively we have two cylinders of different lengths and radii bonded together. Because we are in three dimensions we are interested in their volumes rather than their masses.

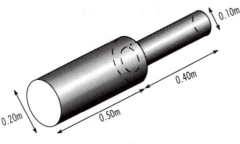

Fig. 17.1

Volumes of revolution by integration

We extend the ideas we used when finding the area under a curve by integration. Fig. 17.2 shows a general curve with an elemental strip of width δx and length y. It is bounded by the x-axis and the lines $x=a$ and $x=b$.

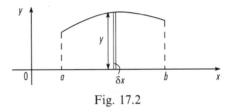

Fig. 17.2

We generate a volume by rotating the curve through 360° (2π radians) about the horizontal axis (Fig. 17.3). Our elemental strip traces out a thin 'disc' of radius y. As usual, as the strip width gets narrower and narrower (i.e. as $\delta x \to 0$) the accuracy of the approximation improves. This means our elemental shape gets closer and closer to a true disc. Our elemental disc (thin cylinder) has the following features.

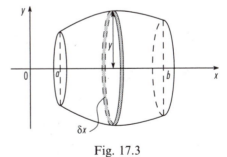

Fig. 17.3

Area of disc $= \pi y^2$
Thickness $= \delta x$
Volume $= \pi y^2 \delta x$.

The volume generated between $x=a$ and $x=b$ is the sum of all such elemental discs, i.e.

$$\text{Volume} = \underset{\delta x \to 0}{\text{Lim}} \sum_{x=a}^{x=b} \pi y^2 \delta x.$$

In practice we would find it difficult to calculate this sum for the volume. We can reach the same result more easily using the integration formula

Volume $= \pi \displaystyle\int_a^b y^2 \, dx.$

> π is a constant multiplier. We can remove it from the integration.

Remember integration is the limit of the summation process.

████████ **Examples 17.1** ████████

Find the volume generated by rotating the parabola $y^2 = 5x$ once about the horizontal axis from i) $x = 0$ to $x = 3.5$, and ii) $x = 1$ to $x = 3$.

We have included a sketch for this curve and generated volume in Fig. 17.4.

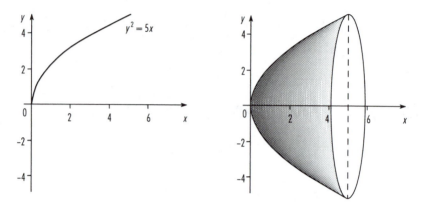

Figs. 17.4

i) Using our formula

$$\text{Volume} = \pi \int_a^b y^2 \, dx$$

we substitute for $a = 0$, $b = 3.5$ and y to get

$$\text{Volume} = \pi \int_0^{3.5} 5x \, dx$$

$$= \pi \left[\frac{5x^2}{2} \right]_0^{3.5}$$

$$= \pi \left[\frac{5(3.5)^2}{2} - 0 \right]$$

$$= 30.625\pi \quad \text{or} \quad 96.21 \text{ unit}^3.$$

ii) We use the same integration formulae, this time with $a = 1$ and $b = 3$.

$$\text{Volume} = \pi \int_1^3 5x \, dx$$

$$= \pi \left[\frac{5x^2}{2} \right]_1^3$$

$$= \pi \left[\frac{5(3)^2}{2} - \frac{5(1)^2}{2} \right]$$

$$= 20\pi \quad \text{or} \quad 62.83 \text{ unit}^3.$$

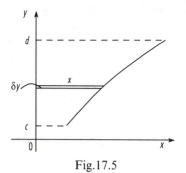

Fig.17.5

We can repeat this technique. Again we use the same curve but this time find a different volume. Notice how we again use thin strips, horizontally rather than vertically. Fig. 17.5 shows a general curve with an elemental strip of width δy and length x. It is bounded by the y-axis and the lines $y = c$ and $y = d$.

We generate a volume by rotating the curve through 360° (2π radians) about the vertical axis (Fig. 17.6). Our elemental strip traces out a thin

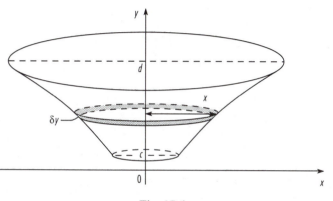

Fig. 17.6

'disc' of radius x. As usual, as the strip width gets narrower and narrower (i.e. as $\delta y \to 0$) the accuracy of the approximation improves. This means our elemental shape gets closer and closer to a true disc. Our elemental disc (thin cylinder) has the following features.

Area of disc $= \pi x^2$
Thickness $= \delta y$
Volume $= \pi x^2 \delta y$.

The volume generated between $y = c$ and $y = d$ is the sum of all such elemental discs, i.e.

$$\text{Volume} = \lim_{\delta y \to 0} \sum_{y=c}^{y=d} \pi x^2 \delta y.$$

In practice we would find it difficult to calculate this sum for the volume. We can reach the same result more easily using the integration formula

$$\text{Volume} = \pi \int_c^d x^2 \, dy.$$

> π is a constant multiplier. We can remove it from the integration.

Example 17.2

Find the volume generated by rotating the parabola $y^2 = 5x$ once about the vertical axis from $y = 0$ to $y = 4$.

We have included a sketch for this curve and generated volume, Fig. 17.7.

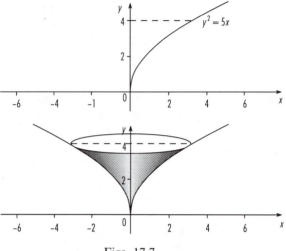

Figs. 17.7

Using our amended formula

$$\text{Volume} = \pi \int_c^d x^2 \, dy$$

we substitute for $c = 0$, $d = 4$ and x to get

$$\text{Volume} = \pi \int_0^4 \frac{y^4}{25} \, dy$$

$$= \pi \left[\frac{y^5}{25 \times 5} \right]_0^4$$

$$= \pi \left[\frac{4^5}{125} - 0 \right]$$

$$= 8.192\pi \quad \text{or} \quad 25.74 \text{ unit}^3.$$

$$x = \frac{y^2}{5}$$

$$\therefore \quad x^2 = \left(\frac{y^2}{5} \right)^2 = \frac{y^4}{25}.$$

EXERCISE 17.1

This Exercise is for you to use the integration formulae for volumes of revolution. Generally we do suggest you sketch the curve though we mention it only in Question 3.

1   Find the volume of revolution generated by rotating the curve $y = x^2$ about the x-axis. Use $x = 0$ and $x = 2$ as your limits of integration.

2   $y = \sqrt{x + 1}$. Rotate this curve once about the horizontal axis. Find the volume generated between $x = 4$ and $x = 9$.

3 You are given $y = \sin\theta$. Sketch this curve between $\theta = 0$ and $\theta = \pi$ radians. Rotate the area under the curve about the horizontal axis and find the volume generated.

4 If $xy = 2$ make y the subject of this relationship. Between $x = 1$ and $x = 2$ find the volume generated by rotating this portion of the curve about the x-axis.

5 Find the volume of revolution generated by rotating the curve $y = x^2$ about the y-axis. Use $y = 0$ and $y = 1$ as your limits of integration.

6 $y^2 = x$. Rotate this curve once about the vertical axis. Find the volume generated between $y = 1$ and $y = 4$.

7 If $xy = 5$ make x the subject of this relationship. Between $y = 2.5$ and $y = 5$ find the volume generated by rotating this portion of the curve about the y-axis.

8 $y = \dfrac{x+1}{x}$. Simplify this equation. Rotate the curve once about the x-axis. Find the volume generated between $x = 1.5$ and $x = 2.5$.

9 Find the volume of revolution generated by rotating the curve $y = e^t$ about the horizontal axis. Use $t = 0$ and $t = 1.5$ as your limits of integration.

10 $y = \ln x$. Rotate this curve once about the vertical axis. Find the volume generated between $y = 1.2$ and $y = 1.4$.

Application of Pappus' theorem

In Volume 1, Chapter 16, we looked at Pappus' theorems. We return to the first of those. By rotating a plane area (i.e. a cross-sectional area) about an axis we generate a volume. We combine the area and the position of the centroid in the formula

Volume = Area × Distance travelled by the centroid.

There is an important feature for the volume we generate. The axis about which we rotate must *not* cut through the cross-sectional area.

Example 17.3

We apply Pappus' theorem to find the volume of a cylinder.

In Fig. 17.8 we rotate a rectangle about a horizontal axis. This generates a cylinder of the same length as the rectangle. The radius of the cylinder is the length of the other side of the rectangle.

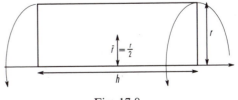

Fig. 17.8

The area of the rectangle $= hr$.

The distance of the rectangle's centroid from the axis $= \dfrac{r}{2}$.

The centroid traces out the circumference of a circle $= 2\pi\left(\dfrac{r}{2}\right)$.

Using Pappus' theorem,

Volume = Area × Distance travelled by the centroid

we get   Volume $= hr \times 2\pi\left(\dfrac{r}{2}\right)$

i.e.   Volume $= \pi r^2 h$ is the standard formula for the cylinder.

Example 17.4

We apply Pappus' theorem to find the centroid of a semi-circle.

In Fig. 17.9 we rotate a semi-circle about a horizontal axis. This generates a sphere of the same radius as the semi-circle.

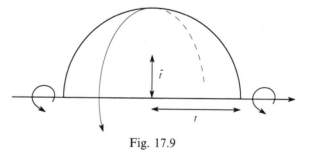

Fig. 17.9

The area of the semi-circle $= \dfrac{1}{2}\pi r^2$.

The distance of the semi-circle's centroid from the axis $= \bar{r}$.

The centroid traces out the circumference of a circle $= 2\pi\bar{r}$.

Using Pappus' theorem,

Volume = Area × Distance travelled by the centroid

we get $\dfrac{4}{3}\pi r^3 = \dfrac{1}{2}\pi r^2 \times 2\pi\bar{r}$

$\therefore$ $2\pi\bar{r} = \dfrac{4}{3}\pi r^3 \times \dfrac{2}{\pi r^2}$ | Dividing by $\dfrac{1}{2}\pi r^2$, invert and multiply.

$2\pi\bar{r} = \dfrac{8r}{3}$

i.e. $\bar{r} = \dfrac{4r}{3\pi}$ | Dividing by 2π.

is the standard formula for the centroid of a semi-circle.

████ **Example 17.5** ████

We apply Pappus' theorem to find the volume of a cone.

In Fig. 17.10 we rotate a triangle about a horizontal axis to generate a cone.

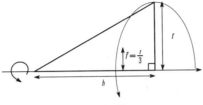

Fig. 17.10

The area of the triangle $= \dfrac{1}{2}hr$.

The distance of the triangle's centroid from the axis $= \dfrac{r}{3}$.

The centroid traces out the circumference of a circle $= 2\pi\left(\dfrac{r}{3}\right)$.

Using Pappus' theorem,

$$\text{Volume} = \text{Area} \times \text{Distance travelled by the centroid}$$

we get $\text{Volume} = \dfrac{1}{2}hr \times 2\pi\left(\dfrac{r}{3}\right)$

i.e. $\text{Volume} = \dfrac{1}{3}\pi r^2 h$ is the standard formula for the cone.

████ **ASSIGNMENT** ████

Let us look at our composite body of two cylinders. We find the volume using the formula $\pi r^2 h$. Notice we show the diameters rather than the radii in Fig. 17.11.

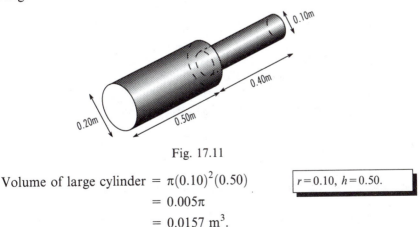

Fig. 17.11

Volume of large cylinder $= \pi(0.10)^2(0.50)$ $r = 0.10,\ h = 0.50.$

$= 0.005\pi$

$= 0.0157 \text{ m}^3.$

Volume of small cylinder $= \pi(0.05)^2(0.40)$ | $r = 0.05$, $h = 0.40$.

$$= 0.001\pi$$
$$= 0.0031 \text{ m}^3.$$

Total volume $=$ Sum of these volumes

$$= 0.0157 + 0.0031$$
$$= 0.0188 \text{ m}^3.$$

Some standard techniques

Here we derive some standard results using integration. Each time we split our regular shape into a series of elemental (incremental) volumes. We find the volume of each element using an earlier result. Then we add them together. Remember that integration is the limit of the summation process.

Example 17.6

Find the formula for the volume of a cone.

We have used the result in Volume 1 and have confirmed it using Pappus' theorem. Here we find it using integration.

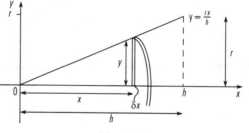

Fig. 17.12

We generate a volume by rotating the triangle through 360° (2π radians) about the horizontal axis (Fig. 17.12). Our elemental strip traces out a thin 'disc' of radius y. As usual, as the strip width gets narrower and narrower (i.e. as $\delta x \to 0$) the accuracy of the approximation improves. This means our elemental shape gets closer and closer to a true disc. Our elemental disc (thin cylinder) has the following features.

Area of disc $= \pi y^2$
Thickness $= \delta x$
Volume $= \pi y^2 \delta x$.

The volume generated between $x = 0$ and $x = h$ is the sum of all such elemental discs, i.e.

$$\text{Volume} = \lim_{\delta x \to 0} \sum_{x=0}^{x=h} \pi y^2 \delta x.$$

In practice we would find it difficult to calculate this sum for the volume. We can reach the same result more easily using the integration formula

$$\text{Volume} = \pi \int_0^h y^2 \, dx$$

$$= \pi \int_0^h \frac{r^2}{h^2} x^2 \, dx$$

$$= \pi \frac{r^2}{h^2} \int_0^h x^2 \, dx$$

$$= \pi \frac{r^2}{h^2} \left[\frac{x^3}{3} \right]_0^h$$

$$= \pi \frac{r^2}{h^2} \left[\frac{h^3}{3} - 0 \right]$$

$$= \frac{1}{3} \pi r^2 h \text{ which is the standard formula for the cone.}$$

> Substituting, $y = \frac{r}{h} x$
>
> $\therefore \quad y^2 = \left(\frac{r}{h} x \right)^2 = \frac{r^2}{h^2} x^2.$
>
> $\pi \dfrac{r^2}{h^2}$ is a constant multiplier. We can remove it from the integration.

Example 17.7

Find the formula for the volume of a sphere.

We have used the result in Volume 1 and have applied it in Pappus' theorem. Here we find it using integration.

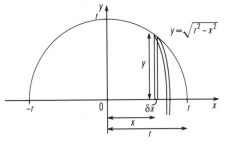

Fig. 17.13

We generate a volume by rotating the semi-circle through $360°$ (2π radians) about the horizontal axis (Fig. 17.13). Our elemental strip traces out a thin 'disc' of radius y. As usual, as the strip width gets narrower and narrower (i.e. as $\delta x \to 0$) the accuracy of the approximation improves. This means our elemental shape gets closer and closer to a true disc. Our elemental disc (thin cylinder) has the following features.

Area of disc $= \pi y^2$
Thickness $= \delta x$
Volume $= \pi y^2 \delta x.$

The volume generated between $x = -r$ and $x = r$ is the sum of all such elemental discs, i.e.

$$\text{Volume} = \underset{\delta x \to 0}{\text{Lim}} \sum_{x=-r}^{x=r} \pi y^2 \delta x.$$

In practice we would find it difficult to calculate this sum for the volume. We can reach the same result more easily using the integration formula

$$\text{Volume} = \pi \int_{-r}^{r} y^2 \, dx$$

$$= \pi \int_{-r}^{r} (r^2 - x^2) \, dx \qquad \boxed{\text{Substituting } y^2 = r^2 - x^2.}$$

$$= \pi \left[r^2 x - \frac{x^3}{3} \right]_{-r}^{r}$$

$$= \pi \left(\left[r^3 - \frac{r^3}{3} \right] - \left[r^2(-r) - \frac{(-r)^3}{3} \right] \right)$$

$$= \pi \left(r^3 - \frac{r^3}{3} + r^3 - \frac{r^3}{3} \right)$$

$$= \frac{4}{3}\pi r^3 \text{ is the standard formula for a sphere.}$$

■ EXERCISE 17.2 ■

This Exercise is for you to practice applying integration to find volumes.

1   What is the equation of the straight line inclined to the horizontal axis in the diagram? Rotate the shaded triangle about the horizontal axis. What shape do you generate? Using integration find the volume. Check your answer using a standard formula.

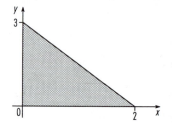

2   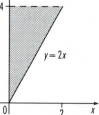 The diagram shows a triangle formed by the line $y = 2x$ and the vertical axis from $y = 0$ to $y = 4$. Rotate the triangle about the vertical axis. By integration find the volume generated.

3   Sketch the circle $x^2 + y^2 = 9$. Shade the semi-circle to the right of the vertical axis. Rotate it about the vertical axis. By integration find the volume of the sphere you generate. Suppose you shade the semi-circle above the horizontal axis. If you rotate it about the horizontal axis why is your volume the same as before?

4   What is the equation of the straight line in the accompanying diagram? Rotate the shaded triangle about the vertical axis. What shape do you generate? Using integration find the volume. Check

your answer using a standard formula. Why is this volume different from your answer in Question 1? How might you alter the equation to make those generated volumes the same?

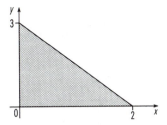

5   The diagram shows the line $y = 4$ and another line, OA, inclined to the x-axis. What is the equation of that line? The line $y = 4$ between $x = -4$ and $x = 6$ is rotated about the x-axis to generate a cylinder. What is its volume? OA is also rotated about the x-axis. What shape is generated and what is its volume? By subtraction find the volume generated when the shaded area is rotated about the x-axis.

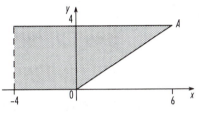

Composite volumes

We can add and subtract volumes just as we can add and subtract areas. We need a little more care with our equations.

Remember, we integrate y^2, i.e. $\pi \int_a^b y^2 \, dx$, for volumes but only y, i.e. $\int_a^b y \, dx$, for areas.

Example 17.8

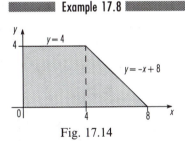

Fig. 17.14

We rotate the shaded area (Fig. 17.14) through 360° about the horizontal axis to generate a volume. This composite volume is a cylinder joined to a cone. Obviously we can find the total volume using the standard formulae. You should check our answer by applying these for yourself.

The equations of the lines are $y = 4$ and $y = -x + 8$. Notice they are drawn (and hence defined) between different values of x. This means our limits of integration will be different. We have to treat them separately.

$$\text{Volume } \textcircled{1} \; = \; \pi \int_0^4 y^2 \, dx \; = \; \pi \int_0^4 16 \, dx$$

$$\begin{array}{|l|} \hline y = 4, \\ y^2 = 4^2 = 16. \\ \hline \end{array}$$

$$= \; \pi \left[16x \right]_0^4 \; = \; \pi[16(4) - 0]$$

$$= \; 64\pi \text{ unit}^3.$$

$$\text{Volume } ② = \pi \int_4^8 y^2 \, dx$$

$$= \pi \int_4^8 (x^2 - 16x + 64) \, dx$$

$$\boxed{\begin{array}{l} y = -x + 8, \\ \therefore\ y^2 = (-x + 8)^2. \end{array}}$$

$$= \pi \left[\frac{x^3}{3} - \frac{16x^2}{2} + 64x \right]_4^8$$

$$= \pi \left(\left[\frac{8^3}{3} - 8(8)^2 + 64(8) \right] - \left[\frac{4^3}{3} - 8(4)^2 + 64(4) \right] \right)$$

$$= \pi([170.\bar{6} - 512 + 512] - [21.\bar{3} - 128 + 256])$$

$$= 21.\bar{3}\pi \text{ unit}^3.$$

$$\therefore\ \text{Total volume} = 64\pi + 21.\bar{3}\pi$$

$$= 85.\bar{3}\pi \quad \text{or} \quad 268.1 \text{ unit}^3.$$

Example 17.9

We rotate the shaded area (Fig. 17.15) through 360° about the horizontal axis to generate a volume. There is no standard formula for this volume. We find the volume by integration. In this case the area to be rotated is between the same values of x, i.e. $x = -1$ to $x = 1$. This means we have the same limits of integration.

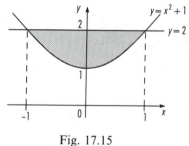

Fig. 17.15

$$\text{Volume} = \pi \int_a^b y^2 \, dx.$$

When we apply this formula we substitute for y^2. We do this for the line and the curve separately. Next we subtract before finally integrating.

We rotate the line $y = 2$ about the horizontal axis to get volume ①, i.e.

$$\text{Volume } ① = \pi \int_{-1}^1 y^2 \, dx$$

$$= \pi \int_{-1}^1 4 \, dx.$$

$$\boxed{\begin{array}{l} y = 2, \\ \therefore y^2 = 2^2 = 4. \end{array}}$$

We rotate the curve $y = x^2 + 1$ about the horizontal axis to get volume ②, i.e.

$$\text{Volume } ② = \pi \int_{-1}^1 y^2 \, dx$$

$$= \pi \int_{-1}^1 (x^4 + 2x^2 + 1) \, dx.$$

$$\boxed{\begin{array}{l} \therefore\ y = x^2 + 1, \\ \therefore\ y^2 = (x^2 + 1)^2. \end{array}}$$

Now we are able to subtract these volume integrals, i.e.

Total volume $= \pi \displaystyle\int_{-1}^{1} (4 - x^4 - 2x^2 - 1)\, dx$

$$= \pi \left[4x - \frac{x^5}{5} - \frac{2x^3}{3} - x \right]_{-1}^{1}$$

$$= \pi \left(\left[4(1) - \frac{1^5}{5} - \frac{2(1)^3}{3} - 1 \right] \right.$$

$$\left. - \left[4(-1) - \frac{(-1)^5}{5} - \frac{2(-1)^3}{3} - (-1) \right] \right)$$

$$= \pi([4 - 0.2 - 0.\bar{6} - 1] - [-4 + 0.2 + 0.\bar{6} + 1])$$

$$= 4.2\bar{6}\pi \quad \text{or} \quad 13.40 \text{ unit}^3.$$

EXERCISE 17.3

1 The shaded area in the diagram lies between the lines $y = 2$, $y = 2 - 2x$ and $x = 1$. This area is rotated about the x-axis. By integration find the volume generated.

2 Check the line $y = 2x$ and the curve $y = x^2$ intersect at the points $(0, 0)$ and $(2, 4)$. Sketch these graphs and shade the area between them. Rotate this area about the x-axis. Using integration find the volume generated.

3 The diagram shows the shaded area between parts of the curves $y = x^2$ and $y^2 = x$. Find the coordinates of their points of intersection. This area is rotated about the horizontal axis. Find the volume generated using the volume integration formula.

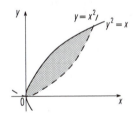

4 Check the line $y = 3$ and the curve $y = x(4 - x)$ intersect at the points $(1, 3)$ and $(3, 3)$. Sketch these graphs and shade the area between them. Rotate this area about the x-axis. Using integration find the volume generated.

5 The diagram shows the shaded area between parts of the curves $y = x^3$ and $y^2 = 32x$. What are the coordinates of their points of intersection? This area is rotated about the horizontal axis. Find the volume generated using the volume integration formula.

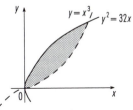

Centroids

In the previous chapter we introduced centroids. We linked them with centre of mass and centre of gravity. All those ideas continue to apply as does the idea of balancing in equilibrium. Again we can link it to symmetry. Here we often generate a volume by rotation. The axis of rotation is an axis of symmetry. Thus we may have either a line of symmetry or an axis of symmetry. Also we may have a plane of symmetry through the body. The part of the body on one side of the plane (of symmetry) is a mirror image of the other part. Remember that the centroid always lies on a line, or axis or plane, of symmetry. Look for a line, or axis or plane, of symmetry before attempting any calculations.

In Fig. 17.16 we have a rectangular block. We label three planes of symmetry, AB, CD and EF. The centroid must lie on them all, i.e. it must lie where they cross at G.

The same idea applies to a cube. Of course, all its sides are equal.

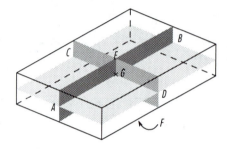

Fig. 17.16

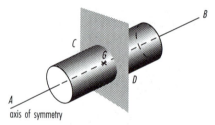

axis of symmetry

Fig. 17.17

In Fig. 17.17 we have a cylinder with an axis of symmetry, AB. Also we have a plane of symmetry, CD. This is the circular plane at right-angles to the axis of symmetry. The centroid must lie on them both, i.e. it must lie where they cross at G.

In Fig. 17.18 we have a sphere which has many axes and planes of symmetry. The centroid must lie on them all, i.e. it must lie where they cross at G.

Fig. 17.18

We can find similar formulae to the ones we have for centroids of areas. Here we simply derive one for $\bar{x}$, the distance from the y-axis. Remember, we are dealing with volumes. We are interested in the **first moment of volume. About a given axis it is defined to be the product of that volume and the perpendicular distance of its centroid from that axis.**

Let the volume be V and the distance of the centroid from the y-axis be $\bar{x}$. We show this in Fig. 17.19.

We apply our definition for first moment of volume. The first moment of volume about the vertical axis is $\bar{x}V$. We use the definition to find the centroid.

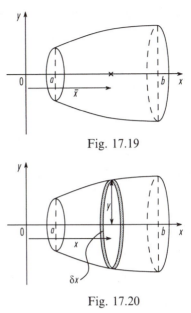

We generate a volume by rotating the area through 360° (2π radians) about the horizontal axis (Fig. 17.20). Our elemental strip traces out a thin 'disc' of radius y. As usual, as the strip width gets narrower and narrower (i.e. as $\delta x \rightarrow 0$) the accuracy of the approximation improves. This means our elemental shape gets closer and closer to a true disc. Our elemental disc (thin cylinder) has the following features.

Fig. 17.19

Fig. 17.20

Area of disc $= \pi y^2$
Thickness $= \delta x$
Volume $= \pi y^2 \delta x$.
Distance of centroid from the vertical axis $= x$.
Moment about the vertical axis $= (\pi y^2 \delta x)(x)$.

To find the total first moment of volume about the vertical axis we sum all these thin discs, i.e.

$$\text{First moment of volume} = \sum_{x=a}^{x=b} \pi x y^2 \delta x.$$

In practice we would find it difficult to calculate this sum for the first moment of volume. We can reach the same result more easily using the integration formula. In the limiting case as $\delta x \rightarrow 0$

$$\text{First moment of volume} = \pi \int_a^b x y^2 \, dx.$$

We link this integral with $\bar{x}V$ to give

$$\bar{x}V = \pi \int_a^b x y^2 \, dx$$

$$\bar{x} = \frac{\pi \int_a^b x y^2 \, dx}{\pi \int_a^b y^2 \, dx}$$

Dividing by V,
$V = \pi \int_a^b y^2 \, dx$.

i.e.

$$\bar{x} = \frac{\int_a^b x y^2 \, dx}{\int_a^b y^2 \, dx}.$$

πs cancel.

Example 17.10

Find the centroid of the volume generated by rotating the shaded area in Fig. 17.21 about the *x*-axis.

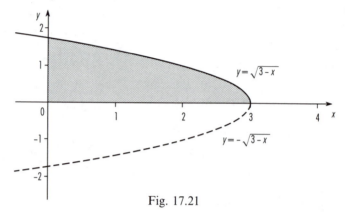

Fig. 17.21

The equation of the curve is $y^2 = 3 - x$. We shade the portion above the *x*-axis, i.e. $y = \sqrt{3 - x}$, between $x = 0$ and $x = 3$. The centroid lies on the *x*-axis because it is an axis of symmetry. We work out the integrals separately.

$$\int_a^b xy^2 \, dx = \int_0^3 (3x - x^2) \, dx$$

$$= \left[\frac{3x^2}{2} - \frac{x^3}{3} \right]_0^3$$

$$= \left[\frac{3(3)^2}{2} - \frac{3^3}{3} - 0 \right]$$

$$= 4.5.$$

$$b = 3.$$
$$y = \sqrt{3 - x},$$
$$\therefore \quad y^2 = 3 - x$$
$$\text{and} \quad xy^2 = x(3 - x).$$
$$a = 0.$$

We also find the value of the other integral.

$$\int_a^b y^2 \, dx = \int_0^3 (3 - x) \, dx$$

$$= \left[3x - \frac{x^2}{2} \right]_0^3$$

$$= \left[3(3) - \frac{3^2}{2} - 0 \right]$$

$$= 4.5.$$

We divide these results to get

$$\bar{x} = \frac{4.5}{4.5} = 1.$$

This means the centroid of the volume generated is at the point (1, 0).

Our formula finds the centroid of a volume. It is generated when an area is rotated about the horizontal axis. Alternatively, we know we can generate a volume by rotating an area about the vertical axis shown in Fig. 17.22. All we do is amend our formula. We interchange x and y to get

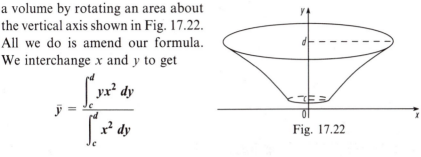

$$\bar{y} = \frac{\int_c^d yx^2\, dy}{\int_c^d x^2\, dy}$$

Fig. 17.22

▓▓▓▓▓ **Example 17.11** ▓▓▓▓▓▓▓▓▓▓▓▓▓▓▓▓▓▓▓▓▓▓▓▓▓▓▓▓

Find the centroid of the volume generated by rotating the shaded area in Fig. 17.23 about the y-axis.

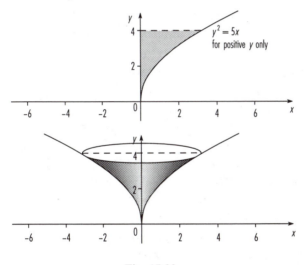

Fig. 17.23

The equation of the curve is $y^2 = 5x$. We have seen it before in Example 17.2. We shade the portion to the right of the y-axis, i.e. $y = \sqrt{5x}$, between $y = 0$ and $y = 4$. The centroid lies on the y-axis because it is an axis of symmetry. Now

$$\int_c^d yx^2\, dy = \int_0^4 \frac{y^5}{25}\, dy$$

$$= \left[\frac{y^6}{25 \times 6}\right]_0^4$$

$$= \left[\frac{4^6}{25 \times 6} - 0\right]$$

$$= 27.30\overline{6}.$$

$$b = 4$$
$$5x = y^2$$
$$\therefore \quad x^2 = \left(\frac{y^2}{5}\right)^2$$
$$\text{and} \quad yx^2 = y\left(\frac{y^4}{25}\right)$$
$$a = 0.$$

This time we recall the value of the other integral from Example 17.2. There we found the volume and included π in our answer. Here we only need the value 8.192. Remember, in our derivation of the centroid formula the πs cancelled.

We divide these results to get

$$\bar{y} = \frac{27.30\bar{6}}{8.192} = 3.\bar{3}.$$

This means the centroid of the volume generated is at the point $(0, 3.\bar{3})$.

■■■■ EXERCISE 17.4 ■■■■

In this Exercise use the appropriate formula to find the centroid. Each question involves a volume of revolution. This means finding either $\bar{x}$ or $\bar{y}$ along the axis of symmetry.

1   The area under the curve $y = x^3$ from $x = 0$ to $x = 2$ is rotated about the x-axis. Find the centroid of this generated volume.

2   The area under the curve $y = 2x^2$ from $x = 1$ to $x = 4$ is rotated about the x-axis. Find the centroid of this volume of revolution.

3   The curve $y = e^t$ cuts the vertical axis at $t = 0$. Sketch this curve. Shade the area bounded by the curve, both axes and the line $t = 1$. This area is rotated about the horizontal axis. Where is the centroid of the volume this generates?

4   Sketch the curve $y = (x - 3)^2$. Confirm it makes contact with the axes at $(3, 0)$ and $(0, 9)$. The area bounded by the curve and both axes is rotated about the x-axis. Find the coordinates of the centroid of the volume of revolution.

5   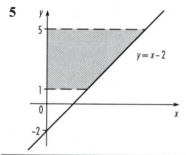   The area of the shaded trapezium is rotated about the vertical axis. It is bounded by the vertical axis, $y = x - 2$, $y = 1$ and $y = 5$. This generates a solid frustum of a cone. Find the centroid of that frustum.

Some standard centroid techniques

We know that a centroid lies on a line (or axis or plane) of symmetry. We identified this usefully for a rectangular block, cube, cylinder and sphere. Now we look at 2 other basic centroids of volumes. They relate to a cone and a hemisphere. Here we derive those standard results using integration.

Each time we split our regular shape into a series of elemental (incremental) volumes. We find the first moment of volume of each element using an earlier result. Then we add them together. All our results depend on generating a volume by rotating an area. This means all our bodies are solid. Whether a body is solid or hollow does affect the position of its centroid.

Example 17.12

By integration find the centroid of a cone.

We generate a volume by rotating the triangle through 360° (2π radians) about the horizontal axis (Fig. 17.24). Hence the centroid must lie on the horizontal axis which is an axis of symmetry. Our elemental strip traces out a thin 'disc' of radius *y*. As usual, as the strip width gets narrower and narrower (i.e. as δ*x* → 0) the accuracy of the approximation improves. This means our elemental shape gets closer and closer to a true disc. Our elemental disc has the following features.

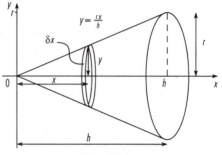

Area of disc $= \pi y^2$
Thickness $= \delta x$
Volume $= \pi y^2\,\delta x$.

Fig. 17.24

Distance of the centroid from the vertical axis = *x*.
Moment about the vertical axis = $(\pi y^2 \delta x)(x)$.

To find the total first moment of volume about the vertical axis we sum all these thin discs, i.e.

$$\text{First moment of volume} = \sum_{x=0}^{x=h} \pi x y^2\,\delta x.$$

In practice we would find it difficult to calculate this sum for the first moment of volume. We can reach the same result more easily using the integration formula. In the limiting case as δ*x* → 0

$$\text{First moment of volume} = \pi \int_0^h x y^2\,dx$$

$$= \pi \int_0^h x \frac{r^2 x^2}{h^2}\,dx.$$

Substituting, $y = \dfrac{r}{h} x$

$\therefore\ y^2 = \left(\dfrac{r}{h} x\right)^2 = \dfrac{r^2}{h^2} x^2.$

We link this integral with $\bar{x} V$ to give

$$\bar{x} V = \pi \frac{r^2}{h^2} \int_0^h x^3\,dx$$

$\pi \dfrac{r^2}{h^2}$ is a constant multiplier. We can remove it from the integration.

$$\bar{x}V = \pi \frac{r^2}{h^2} \left[\frac{x^4}{4} \right]_0^h$$

$$= \pi \frac{r^2}{h^2} \left[\frac{h^4}{4} - 0 \right]$$

i.e. $\qquad \bar{x}\left(\frac{1}{3}\pi r^2 h \right) = \frac{1}{4}\pi r^2 h^2$

> Volume of a cone,
> $V = \frac{1}{3}\pi r^2 h.$

$\therefore \qquad\qquad \bar{x} = \frac{1}{4}\pi r^2 h^2 \times \frac{3}{\pi r^2 h}$

i.e. $\qquad \bar{x} = \frac{3}{4}h$ is the distance of the centroid of a solid cone from its vertex along the axis of symmetry. Alternatively, it is $\frac{1}{4}h$ from the base.

Example 17.13

By integration find the centroid of a hemisphere.

We generate a volume by rotating a quadrant of a circle through 360° (2π radians) about the horizontal axis (Fig. 17.25). Hence the centroid must lie on the horizontal axis, which is an axis of symmetry. Our elemental strip traces out a thin 'disc' of radius y. As usual, as the strip width gets narrower and narrower (i.e. as $\delta x \to 0$) the accuracy of the approximation improves. This means our elemental shape gets closer and closer to a true disc. Our elemental disc has the following features.

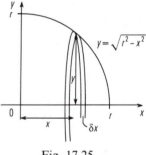

Fig. 17.25

Area of disc $= \pi y^2$
Thickness $\quad = \delta x$
Volume $\qquad = \pi y^2 \delta x.$

Distance of the centroid from the vertical axis $= x$.

Moment about the vertical axis $= (\pi y^2 \delta x)(x)$.

To find the total first moment of volume about the vertical axis we sum all these thin discs, i.e.

First moment of volume $= \sum_{x=0}^{x=r} \pi x y^2 \delta x.$

In practice we would find it difficult to calculate this sum for the first moment of volume. We can reach the same result more easily using the integration formula. In the limiting case as $\delta x \to 0$

First moment of volume $= \pi \int_0^r x y^2 \, dx.$

We link this integral with $\bar{x}V$ to give

$$\bar{x}V = \pi \int_0^r (xr^2 - x^3)\,dx$$

$$y = \sqrt{r^2 - x^2},$$
$$y^2 = r^2 - x^2,$$
$$\therefore \quad xy^2 = x(r^2 - x^2).$$

$$= \pi\left[\frac{r^2x^2}{2} - \frac{x^4}{4}\right]_0^r$$

$$= \pi\left[\frac{r^4}{2} - \frac{r^4}{4} - 0\right]$$

i.e.

$$\bar{x}\left(\frac{2}{3}\pi r^3\right) = \frac{1}{4}\pi r^4$$

Volume of hemisphere,
$$V = \frac{1}{2} \times \frac{4}{3}\pi r^3.$$

$$\therefore \qquad \bar{x} = \frac{1}{4}\pi r^4 \times \frac{3}{2\pi r^3}$$

i.e. $\qquad \bar{x} = \dfrac{3r}{8}$ is the distance of the centroid of a solid hemisphere from its plane face along the axis of symmetry.

◼◼◼ ASSIGNMENT ◼◼◼

Remember our Assignment is a composite body of two cylinders. We know the position of the centroid of each cylinder **separately**. It is half way along the axis of symmetry. Here we set up the initial steps for you to complete as a simple exercise. It is based on the larger cylinder.

We generate a volume by rotating the rectangle through 360° (2π radians) about the horizontal axis (Fig. 17.26). The equation for y is $y = 0.10$. Our limits of integration are $x = 0$ and $x = 0.50$. From here you should continue with the standard method to get $\bar{x} = 0.25$.

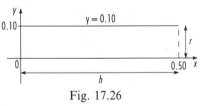

Fig. 17.26

Composite bodies

Many bodies are made up of several standard parts. The two-dimensional ideas for centroids of areas can be applied in three dimensions to volumes. Just as you can add and subtract volumes so you can add and subtract first moments of volume. We repeat two simple rules (formulae).

Moment of the Sum = Sum of the Moments.

and

Moment of the Remainder = Moment of the Whole − Moment of the Section Removed.

We demonstrate these rules for **solid** bodies in the next Example and the Assignment.

◼◼◼◼ **ASSIGNMENT** ◼◼◼◼

This is the first of two looks at the Assignment in this section. We are going to find the centroid from the larger plane circular end. Remember, the volume of a cylinder is given by $V = \pi r^2 h$. As usual, the radius is r and the height (or length) is h. The centroid of a cylinder is halfway along its axis of symmetry, i.e. $\frac{1}{2}h$. As we did in the previous chapter we display our method in a table. Also we retain more decimal places in the calculator's memory.

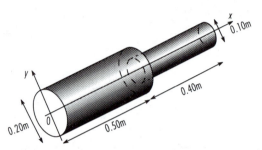

Fig. 17.27

Shape	Volume	Distance of centroid from the Oy axis	First moment of volume
⬭	$\pi(0.10)^2 0.50$	0.25	$0.005\pi \times 0.25$
(+) ⬭	$\pi(0.05)^2 0.40$	$0.50 + 0.20$	$0.001\pi \times 0.70$
Total	$0.005\pi + 0.001\pi$ $= 0.006\pi.$	$\bar{x}$	$0.006\pi\bar{x}$

We use our formula

Moment of the Sum = Sum of the Moments

i.e. $\qquad 0.006\pi\bar{x} = (0.005\pi \times 0.25) + (0.001\pi \times 0.70)$

$\qquad 0.006\bar{x} = 0.00125 + 0.0007$ $\boxed{\pi\text{s cancel.}}$

i.e. $\qquad \bar{x} = \dfrac{0.00195}{0.006}$

to give $\qquad \bar{x} = 0.325$ m,

i.e. the centroid of our composite body is 0.325 m from the larger plane circular end along the axis of symmetry.

Example 17.14

Fig. 17.28 shows a composite body. It is a hemisphere with a cone machined from it. The radius of each base is 0.48 m. The height of the conical space is 0.36 m. We are going to find the centroid of this body.

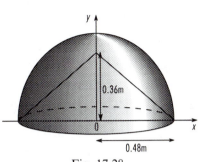

Fig. 17.28

The centroid lies on the axis Oy, an axis of symmetry. We know the following details.

Volume of hemisphere $= \dfrac{2}{3}\pi r^3$.

Distance of centroid along Oy from the base $= \dfrac{3}{8}r$.

Volume of cone $= \dfrac{1}{3}\pi r^2 h$.

Distance of centroid along Oy from the base $= \dfrac{1}{4}h$.

As before our method is in stages: the volumes; the centroid distances; and the first moments of volume. Again we use a table of values. We note the first few decimal places, keeping all of them in the calculator's memory.

Shape	Volume	Distance of centroid from the Ox axis	First moment of volume
(hemisphere)	$\dfrac{2}{3}\pi(0.48)^3$	$\dfrac{3}{8} \times 0.48$	$0.073\ldots\pi \times 0.18$
$(-)$ (cone)	$\dfrac{1}{3}\pi(0.48)^2(0.36)$	$\dfrac{1}{4} \times 0.36$	$0.027\ldots\pi \times 0.09$
Remainder	$0.073\ldots\pi -$ $0.027\ldots\pi$ $= 0.046\ldots\pi.$	$\bar{y}$	$0.046\ldots\pi\bar{y}$

We use our formula

Moment of the Remainder = Moment of the Whole
        − Moment of the Section Removed,

i.e. $0.046\ldots\pi\bar{y} = (0.073\ldots\pi \times 0.18) - (0.027\ldots\pi \times 0.09)$

     $0.046\ldots\bar{y} = 0.013\ldots - 0.002\ldots$ $\boxed{\pi\text{s cancel.}}$

i.e. $$\bar{y} = \frac{0.0107\ldots}{0.0460\ldots}$$

to give $$\bar{y} = 0.23 \text{ m},$$

i.e. the centroid of our composite body is 0.23 m from the opening along the axis of symmetry.

ASSIGNMENT

This time we have machined a cylindrical hole along the complete length. It is of diameter 0.06 m. As before the volume of a cylinder is given by $V = \pi r^2 h$. As usual the radius is r and the height (or length) is h. The centroid of a cylinder is halfway along its axis of symmetry, i.e. $\frac{1}{2}h$. As we did in the previous chapter we display our method in a table. Also we retain more decimal places in the calculator's memory.

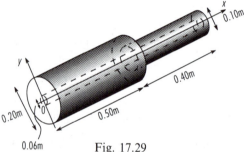

0.06m   Fig. 17.29

Shape	Volume	Distance of centroid from the Oy axis	First moment of volume
	$\pi(0.10)^2 0.50$	0.25	$0.005\pi \times 0.25$
(+)	$\pi(0.05)^2 0.40$	$0.50 + 0.20$	$0.001\pi \times 0.70$
(−)	$\pi(0.03)^2 0.90$	0.45	$0.00081\pi \times 0.45$
Remainder	$0.005\pi + 0.001\pi$ $- 0.00081\pi$ $= 0.00519\pi.$	$\bar{x}$	$0.00519\pi\bar{x}$

We use our formula

Moment of the Remainder = Moment of the Whole
                          − Moment of the Section Removed,

i.e. $0.00519\pi\bar{x} = (0.005\pi \times 0.25) + (0.001\pi \times 0.70)$
                               $- (0.00081\pi \times 0.45)$

$$0.00519\bar{x} = 0.00125 + 0.0007 - 0.0003\dots$$

i.e. $\qquad \bar{x} = \dfrac{0.00158\dots}{0.00519}$ | πs cancel. |

to give $\qquad \bar{x} = 0.31$ m,

i.e. the centroid of our composite body is 0.31 m from the larger plane circular end along the axis of symmetry.

▄▄▄▄ EXERCISE 17.5 ▄▄▄▄▄▄▄▄▄▄

In this Exercise we draw the composite bodies with their dimensions. In each case you need to find the centroid along the axis of symmetry.

1   Our diagram shows a solid cone of height 0.720 m and base radius 0.240 m joined to a solid hemisphere. The flat surfaces coincide exactly. What is the distance of the centroid from the vertex of the cone?

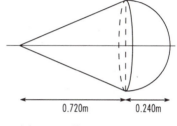

0.720m   0.240m

2

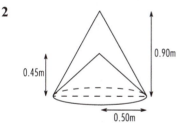

0.90m

0.45m

0.50m

We start with a solid cone of base radius 0.50 m and height 0.90 m. From the cone we machine out a conical space with the same base radius but only half the height. Where, along the axis of symmetry, from the original base, is the centroid?

3   A cylinder has a radius of 0.30 m and a height of 0.90 m. Fitting exactly over one flat end is a hemisphere and over the opposite end is a cone. The overall length of this solid composite body is

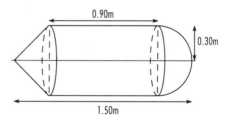

0.90m

0.30m

1.50m

1.50 m. Find the distance of the centroid from the vertex of the cone.

4

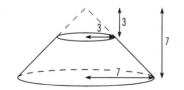

3   3

7

7

The diagram shows a frustum of a cone. The original cone had a radius and a height each of 7 units. The smaller cone that has been removed has a radius and a height each of 3 units. Find the distance of the centroid from the larger plane end of the frustum.

5 The accompanying figure shows a cylinder of length 1.20 m and radius 0.30 m. A cylindrical hole of length 0.45 m and radius 0.10 m has been drilled from one end along the axis of symmetry. Find the distance of the centroid from the opposite end.

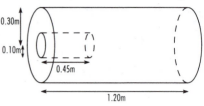

6 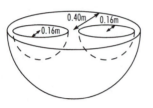 The radius of the hemisphere is 0.40 m. The diagram includes two smaller hemispherical hollows of radii 0.16 m. Where is the centroid of the remaining body?

7 A cone of height 1.20 m and base radius 0.50 m is partially bored out. The cylindrical space is 0.40 m long and of radius 0.20 m. Where is the centroid from the common base?

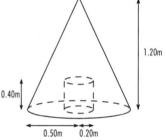

8 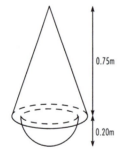 A solid cone of height 0.75 m has a base radius of 0.25 m. It is bonded onto a solid hemisphere of radius 0.20 m, creating a constant overlap of 0.05 m. Find the position of the centroid.

9 The diagram shows a plastic cylinder centrally bonded to a hemisphere of the same material. Find the position of the centroid of this solid body from the circular plane face.

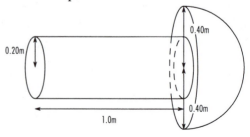

10 Our body started as a solid cylinder of height 1.250 m and radius 0.350 m. A hemisphere of radius 0.400 m was bonded to one end of the cylinder. Now there is a constant overlap of 0.05 m between the cylinder and hemiphere. Finally a cylindrical plug of radius 0.150 m and length 1.30 m was removed around the axis of symmetry. Find the distance of the centroid from the opening.

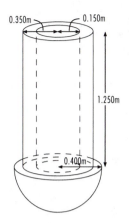

18 Second Moments of Area

The objectives of this chapter are to:

1 Sketch a given area including a typical elemental (incremental) area parallel to a specified axis in the plane of the area.
2 Define second moment of area.
3 Find the second moment of the elemental area about the specified axis.
4 Find the second moment of area of standard shapes about a specified axis.
5 Find the radius of gyration.
6 Find the second moment of area of composite shapes about a common axis.
7 State and apply the parallel axes theorem.
8 State and apply the perpendicular axes theorem.

Introduction

In Chapter 16 we looked at first moments of area. It involved looking at elemental areas and adding together their moments. In the limiting case we simplified the process using integration. Again we look at two-dimensional shapes. We assume these to have uniform thickness (or be uniformly thin). This means the third dimension is negligible compared with the other two dimensions. Easy examples are a blank sheet of A4 paper and a sheet of 4 mm float glass. They are each an example of a **lamina. A lamina is a plane figure with its third dimension negligible compared with the other dimensions**. Unless we describe otherwise each example uses one type of material only. This is to ensure uniform thickness and density. Then the mass (and weight) is proportional to the area. For simplicity we link together mass, weight and area. The centre of gravity is the same position as the centre of mass. Because we are using mass and weight proportional to area these positions are the same as the **centroid** (centre of area).

We extend the ideas and techniques of Chapter 16. Those formulae used the product of area and distance. For second moments of area we use the product of area and distance squared. The second moment of area has various applications. It can be applied in the theory of bending beams, torsion of shafts, centres of pressure and moments of inertia. For example, in moments of inertia we replace area with mass in the general theory.

402

ASSIGNMENT

The Assignment for this chapter applies second moment of area to moment of inertia. Remember that energy is conserved where no external forces act upon the system. We look at changes in mechanical energy. These involve a loss of potential energy and a consequent gain of kinetic energy. We look at the moment of inertia about different axes. This is for a rod and a rod together with an extra mass at one end, i.e. a compound pendulum.

Second moment of area

For any **second moment of area** we need to state the axis to which we are referring. **About any given axis it is defined as the product of that area and the square of the perpendicular distance from that axis**. We can apply this definition to various shapes. The definition tells us the essential features. The position of the axis of rotation (i.e. the position of the pivot) is important. Also, we need to consider the distribution of the area (mass for moment of inertia), i.e. the shape.

In Fig. 18.1 we split the area into a series of strips parallel to the axis shown, YY. The area of a typical strip is A and its typical distance from YY is x. Then the second moment of area about YY is Ax^2. We can do this for each strip in the general area. The second moment of area of the shape about the axis YY (I_{YY}) is the sum of all these elemental second moments, i.e.

$$I_{YY} = \sum Ax^2.$$

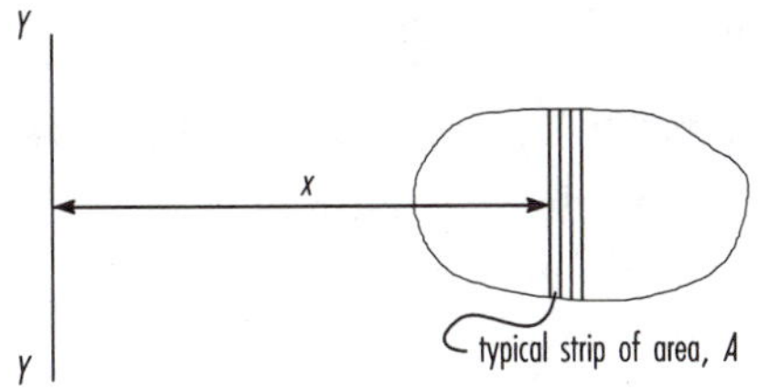

Fig. 18.1

Example 18.1

Find the second moment of area of a rectangle about an edge.

In Fig. 18.2 we sketch a rectangle of length l and breadth b. Notice we state the edge as the axis, YY. We split the area into a series of thin vertical (elemental) strips.

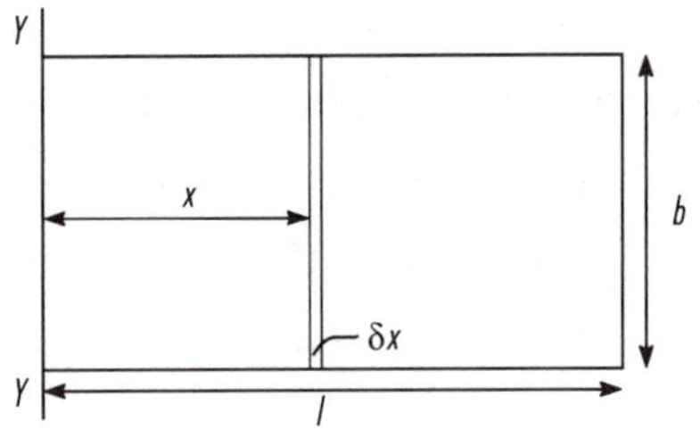

Fig. 18.2

Length of element $= b$
Thickness $= \delta x$
Area $= b\delta x$
Distance of elemental strip from the axis $= x$.
Second moment of area about that axis $= (b\delta x)(x^2)$.

To find the total second moment of area about the axis we sum all these elemental second moments, i.e.

$$\text{Second moment of area} = \sum_{x=0}^{x=l} bx^2 \, \delta x.$$

In the limiting case as $\delta x \to 0$

$$\text{Second moment of area} = b \int_0^l x^2 \, dx$$

$$= b \left[\frac{x^3}{3} \right]_0^l$$

$$= b \left[\frac{l^3}{3} - 0 \right]$$

$$= \frac{1}{3} b l^3$$

i.e.
$$I_{YY} = \frac{1}{3} A l^2.$$

> Area of rectangle, $A = bl$.

There are other versions of this formula. If the length of the rectangle is $2d$ rather than l we have

$$\text{Second moment of area} = \frac{4}{3} A d^2.$$

> $l = 2d,$
> $\therefore \quad l^2 = (2d)^2 = 4d^2.$

We have found the second moment of area about one edge of a rectangle. We can do so about any edge. In Examples 18.2 and 18.3 we do this for different rectangles.

▬▬ Examples 18.2 ▬▬▬▬▬▬▬▬▬▬▬▬▬▬▬▬

We find the second moment of area about the edges of two separate rectangles. The dimensions in mm are given in Figs. 18.3. However, you may prefer to use cm. This means the size of the seond moment of area value (cm^4) is easier to handle. $1 \, \text{cm} = 10 \, \text{mm}$, so that $1 \, \text{cm}^4 = 10^4 \, \text{mm}^4 = 1000 \, \text{mm}^4$.

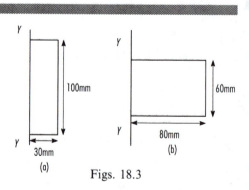

Figs. 18.3

We are going to apply the formula we have just derived, $I = \frac{1}{3} A l^2$. For each rectangle we have drawn the axis YY along the left edge.

(a)   Area $= 30 \times 100 = 3000 \, \text{mm}^2$.

$$I_{YY} = \frac{1}{3} \times 3000 \times 30^2 = 900\,000 \, \text{mm}^4.$$

> 90 cm^4.

(b)   Area $= 80 \times 60 = 4800 \, \text{mm}^2$.

$$I_{YY} = \frac{1}{3} \times 4800 \times 80^2 = 10\,240\,000 \, \text{mm}^4.$$

> 1024 cm^4.

■■■■■■■■ **Examples 18.3** ■■

We find the second moment of area about the edges of two separate rectangles. The dimensions in mm are given in Figs. 18.4. Again we are going to apply the formula we have derived previously, $I = \frac{1}{3} Al^2$. For each rectangle we have drawn the axis XX along the bottom edge.

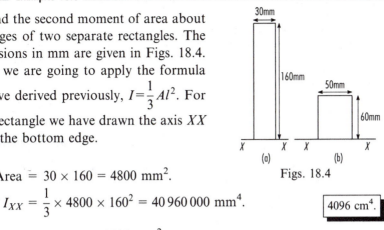

Figs. 18.4

(a) Area $= 30 \times 160 = 4800$ mm^2.

$$I_{XX} = \frac{1}{3} \times 4800 \times 160^2 = 40\,960\,000 \text{ mm}^4.$$

$$\boxed{4096 \text{ cm}^4.}$$

(b) Area $= 50 \times 60 = 3000$ mm^2.

$$I_{XX} = \frac{1}{3} \times 3000 \times 60^2 = 3\,600\,000 \text{ mm}^4.$$

$$\boxed{360 \text{ cm}^4.}$$

Now we turn our attention away from an edge. This time we derive a formula related to an axis through the centroid and parallel to an edge.

■■■■■■■■ **Example 18.4** ■■

Find the second moment of area of a rectangle about an axis through its centre, parallel to an edge.

In Fig. 18.5 we sketch a rectangle of length l and breadth b with the centrally placed axis. Again, we split the area into a series of thin vertical (elemental) strips. The early method is similar to that in Example 18.1 with a slight change of limits, i.e.

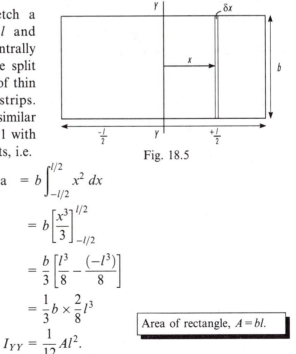

Fig. 18.5

Second moment of area $= b \displaystyle\int_{-l/2}^{l/2} x^2 \, dx$

$$= b \left[\frac{x^3}{3} \right]_{-l/2}^{l/2}$$

$$= \frac{b}{3} \left[\frac{l^3}{8} - \frac{(-l^3)}{8} \right]$$

$$= \frac{1}{3} b \times \frac{2}{8} l^3$$

$$\boxed{\text{Area of rectangle, } A = bl.}$$

i.e. $I_{YY} = \dfrac{1}{12} Al^2.$

Again, there are other versions of this formula. If the length of the rectangle is $2d$ rather than l we have

$$\text{Second moment of area} = \frac{4}{12} A d^2$$

$$= \frac{1}{3} A d^2.$$

$$\begin{aligned} & l = 2d, \\ \therefore \quad & l^2 = (2d)^2 = 4d^2. \end{aligned}$$

███ ASSIGNMENT ███

We have a brass rod of length 1.5 m. It has a square cross-section of side 25 mm. The brass has a density of $8180 \, \text{kg m}^{-3}$. The volume is a mixture of units. Let us standardise these on metres to be consistent with the density.

$$\text{Volume} = 1.5 \times 0.025^2 = 9.375 \times 10^{-4} \, \text{m}^3.$$

$$\therefore \quad \text{Mass} = 8180 \times 9.375 \times 10^{-4} = 7.66 \ldots \, \text{kg}.$$

Our theory for second moment of area is based on a rectangular lamina. The formulae apply even though the rod is not a lamina. We may think of the rod as a series of rectangular laminas that have been laminated. We are interested in moment of inertia. This means we replace the area with mass in the formulae.

In Fig. 18.6 our rod is horizontal, bisected by a vertical axis, i.e. an axis through the centroid. For this uniform rod we have the centroid (centre of mass or centre of gravity) at the mid-point. It is in equilibrium, i.e. it will remain at rest unless acted upon by an external force. This means we have no change of energy.

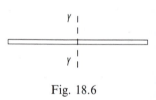

Fig. 18.6

In Fig. 18.7 we have the pivot at the left hand end. From an initial horizontal position it will fall, rotating about the axis through the end A. Because the height of A does *not* change we use it as the position of zero potential energy.

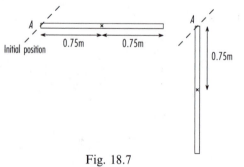

Fig. 18.7

Potential energy $= mgh$ J,

Kinetic energy $= \dfrac{1}{2} I \omega^2$ J.

As usual we use m for mass (kg),

g for the acceleration due to gravity, 9.81 ms^{-2},

h for height (m),

I for moment of inertia (kg m^2),

ω for angular speed (rads^{-1}).

$$I = \frac{1}{3} m l^2 = \frac{1}{3} \times 7.66 \ldots \times 1.5^2 = 5.75 \ldots \text{ kg m}^2.$$

Initially the rod is held horizontally at rest,

i.e.   P.E. $= 0$,

K.E. $= 0$.

For the instant the rod is vertical its centroid has fallen 0.75 m and the rod is moving,

i.e.   P.E. $= 7.66... \times 9.81 \times (-0.75) = -56.42 \ldots$ J,

K.E. $= \dfrac{1}{2} \times 5.75 \ldots \times \omega^2 \qquad = 2.87 \ldots \times \omega^2$ J.

The principle of conservation of energy allows us to write

Initial energy $=$ Final energy

i.e. $0 + 0 = -56.42 \ldots + 2.87 \ldots \times \omega^2$

to give $\omega^2 = \dfrac{56.42 \ldots}{2.87 \ldots}$

i.e. $\omega = \sqrt{19.62}$ | Square root of both sides. |

$= 4.43$ rads^{-1}

is the angular speed when the rod is vertical.

▬▬▬ **Example 18.5** ▬▬▬

Find the second moment of area of a triangle about an edge.

In Fig. 18.8 we sketch a triangle of vertical height (altitude) h and base b. Notice we state the edge as the axis. If we twist our triangle we twist our

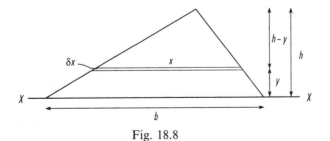

Fig. 18.8

axes. We split the area into a series of thin (elemental) strips parallel to the base. Each of these strips approximates to a rectangle. The slight variations at each end tend to zero in the limiting case. We look at a typical element (rectangle).

Length of element $= x$

Thickness $\qquad = \delta y$

Area $\qquad\qquad = x\delta y$

Distance of elemental strip from the axis $= y$.

Second moment about that axis $= (x\delta y)(y^2) = xy^2\delta y$.

The second moment of area of the triangle about the axis is the sum of all these elemental moments. We add them to get

$$\text{Second moment of area } = \sum_{y=0}^{y=h} xy^2\delta y.$$

In the limiting case as $\delta y \to 0$

$$\text{Second moment of area } = \int_0^h xy^2 \, dy.$$

Now x and y are linked by similar triangles. We compare bases to heights,

i.e. $\quad \dfrac{x}{b} = \dfrac{h-y}{h}$

i.e. $\quad x = \dfrac{b}{h}(h-y).$

We substitute into our second moment of area formula to get

$$\text{Second moment of area } = \int_0^h \frac{b}{h}(h-y)y^2 \, dy$$

We can remove $\dfrac{b}{h}$ from the integral. It is *not* variable.

$$= \frac{b}{h}\int_0^h (hy^2 - y^3) \, dy$$

$$= \frac{b}{h}\left[\frac{hy^3}{3} - \frac{y^4}{4}\right]_0^h$$

$$= \frac{b}{h}\left[\frac{h^4}{3} - \frac{h^4}{4} - 0\right]$$

$$= \frac{b}{h}\left[\frac{h^4}{12}\right]$$

$$= \frac{1}{12}bh^3$$

Area of triangle, $A = \dfrac{1}{2}bh.$

i.e. $\qquad\qquad I_{XX} = \dfrac{1}{6}Ah^2.$

▰▰▰▰▰ **Example 18.6** ▰▰▰▰▰

Find the second moment of area of a semi-circle about its straight edge.

In Fig. 18.9 we sketch a semi-circle of radius r.

We split the area into a series of thin (elemental) strips parallel to the straight edge. There are a slight variations from an elemental rectangle at the ends. These tend to zero in the limiting case. We look at a typical element (rectangle).

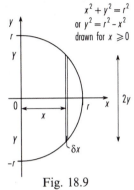

Fig. 18.9

Length of element $= 2y$
Thickness $\quad\quad = \delta x$
Area $\quad\quad\quad\quad = 2y\delta x$
Distance of elemental strip from the axis $= x$.
Second moment about that axis $= (2y\delta x)(x^2) = 2yx^2\delta x$.

The second moment of area of the semi-circle about the axis is the sum of all these elemental moments. We add them to get

$$\text{Second moment of area} \quad = \sum_{x=0}^{x=r} 2yx^2\delta x.$$

In the limiting case as $\delta x \to 0$

$$\text{Second moment of area} \quad = \int_0^r 2yx^2 \, dx. \quad\text{——— ①}$$

We are attempting to integrate with respect to x. However we have a mixture of variables, x and y, and the integral is rather complicated. We need to make a substitution for x, y and dx using trigonometry.

Let $\quad\quad\quad\quad x = r\cos\theta$

$\therefore \quad\quad\quad\quad \dfrac{dx}{d\theta} = -r\sin\theta$

i.e. $\quad\quad\quad\quad dx = -r\sin\theta \, d\theta$ 〔Multiplying through by $d\theta$.〕

Also we need to change our limits using

$\quad\quad\quad\quad\quad x = r\cos\theta.$

For $x=r$, $\quad\quad r = r\cos\theta_U$ 〔θ_U is the new upper limit.〕

i.e. $\quad\quad\quad\quad 1 = \cos\theta_U$

so that $\quad\quad \theta_U = 0.$

For $x=0$, $\quad\quad 0 = r\cos\theta_L$ 〔θ_L is the new lower limit.〕

so that $\quad\quad \theta_L = \dfrac{\pi}{2}.$

Also $\quad\quad\quad y^2 = r^2 - x^2$

$\quad\quad\quad\quad\quad = r^2 - r^2\cos^2\theta = r^2(1 - \cos^2\theta) = r^2\sin^2\theta$

$\therefore \quad\quad\quad\quad y = r\sin\theta.$ 〔Positive square root of both sides.〕

In ① we replace x, y, dx and the limits in our integral to get

$$\text{Second moment of area} = \int_{\pi/2}^{0} 2(r\sin\theta)(r\cos\theta)^2(-r\sin\theta\,d\theta)$$

$$= -\int_{\pi/2}^{0} 2r^4(\sin\theta\cos\theta)^2\,d\theta$$

$$\boxed{\sin 2\theta = 2\sin\theta\cos\theta.}$$

$$= -2r^4\int_{\pi/2}^{0} \left(\frac{1}{2}\sin 2\theta\right)^2 d\theta$$

$$= -\frac{2}{4}r^4\int_{\pi/2}^{0} \sin^2 2\theta\,d\theta$$

$$\boxed{\begin{array}{l}\cos 4\theta = 1 - 2\sin^2 2\theta \\ 2\sin^2 2\theta = 1 - \cos 4\theta.\end{array}}$$

$$= -\frac{1}{2}r^4\int_{\pi/2}^{0} \left(\frac{1 - \cos 4\theta}{2}\right) d\theta$$

$$= -\frac{1}{4}r^4\int_{\pi/2}^{0} (1 - \cos 4\theta)\,d\theta$$

$$= -\frac{1}{4}r^4\left[\theta - \frac{1}{4}\sin 4\theta\right]_{\pi/2}^{0}$$

$$= -\frac{1}{4}r^4\left(\left[0 - \frac{1}{4}\sin 0\right] - \left[\frac{\pi}{2} - \frac{1}{4}\sin\frac{4\pi}{2}\right]\right)$$

$$= -\frac{1}{4}r^4\left(-\frac{\pi}{2}\right)$$

i.e.
$$I_{YY} = \frac{1}{8}\pi r^4$$

$$\boxed{\begin{array}{l}\text{Area of semi-circle,} \\ A = \frac{1}{2}\pi r^2.\end{array}}$$

or
$$I_{YY} = \frac{1}{4}Ar^2.$$

There are other versions of this formula. We may use the diameter, d, rather than the radius, r, i.e.

$$\text{Second moment of area} = \frac{\pi}{128}d^4 \text{ or } \frac{1}{16}Ad^2.$$

$$\boxed{d = 2r.}$$

━━━━━━ **Example 18.7** ━━━━━━

We find the second moment of area about the straight edge of a semi-circle of radius 80 mm.

We have drawn the semi-circle with a horizontal diameter in Fig. 18.10. The diameter is labelled XX. We apply our formula

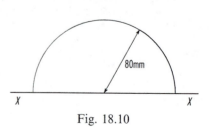

Fig. 18.10

$$I_{XX} = \frac{1}{8}\pi r^4$$

to get $\quad I_{XX} = \frac{1}{8}\pi(80^4) = 5\,120\,000\pi \text{ mm}^4.$

$\boxed{512\pi \text{ cm}^4.}$

Radius of gyration

We have derived various formulae for the second moment of area of different shapes. They involve area, A, some distance squared and a scalar multiplier. There are alternative versions of these formulae involving the **radius of gyration, k**. Generally we write

$$\text{Second moment of area} = Ak^2.$$

In Examples 18.8 we show the alternatives based on our formulae so far.

Examples 18.8

i) For a rectangle about an edge,

$$\text{Second moment of area} = \frac{1}{3}Al^2.$$

Comparing this with Ak^2 we have $k^2 = \frac{1}{3}l^2,$

i.e. the radius of gyration is $k = \dfrac{l}{\sqrt{3}}.$

ii) For a rectangle about an axis through its centroid parallel to an edge,

$$\text{Second moment of area} = \frac{1}{12}Al^2.$$

Comparing this with Ak^2 we have $k^2 = \frac{1}{12}l^2,$

i.e. the radius of gyration is $k = \dfrac{l}{2\sqrt{3}}.$

$\boxed{\begin{aligned}\sqrt{12} &= \sqrt{4 \times 3}\\ &= \sqrt{4} \times \sqrt{3}\\ &= 2\sqrt{3}.\end{aligned}}$

Composite shapes

We know how to add areas and how to add first moments of areas. The same ideas apply to second moments of areas provided they refer to the **same axis**.

Second Moment of the Total Area = Sum of the Second Moments of the Part Areas.

For two shapes, P and Q, we may write

$$I_{P+Q} = I_P + I_Q.$$

We can extend this to many shapes, i.e.

$$I = \sum_{i=1}^{n} I_i.$$

The same principles apply to subtraction. Remember, subtraction is simply negative addition.

We demonstrate these ideas in the next Examples, using work from earlier in the chapter.

Example 18.9

We find the second moments of area of the L-shape about XX and YY.

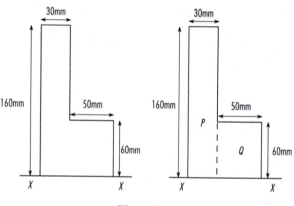

Figs. 18.11

In Figs. 18.11 we draw the original L-shape with the dimensions in mm. Also we split it into two rectangles, P and Q. Each has an edge adjacent to the XX axis. We apply the formula for the second moment of area about the edge of a rectangle, $I = \frac{1}{3} A l^2$. Now look back to Examples 18.3. We use those results to write

$$I_P = 4096 \text{ cm}^4$$

and $I_Q = 360 \text{ cm}^4.$

The L-shape is composed of these two rectangles so we may add the results to get

$$I_{XX} = 4096 + 360 = 4456 \text{ cm}^4,$$

i.e. the second moment of area of the L-shape about the axis XX is 4456 cm^4.

In Figs. 18.12 we split the L-shape into two different rectangles, R and S. Each has an edge adjacent to the YY axis. We apply the same formula using results from Examples 18.2,

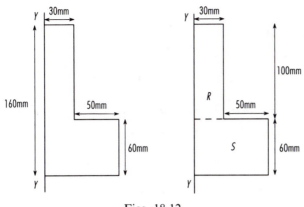

Figs. 18.12

i.e. $I_R = 90 \text{ cm}^4$

and $I_S = 1024 \text{ cm}^4$.

The L-shape is composed of these two rectangles so we may add the results to get

$$I_{YY} = 90 + 1024 = 1114 \text{ cm}^4,$$

i.e. the second moment of area of the L-shape about the axis YY is 1114 cm^4.

There are other ways of finding these second moments of area. We will look at this L-shape again later in the chapter.

Example 18.10

We find the second moment of area of the border shape about YY.

In Fig. 18.13 we draw the axis YY vertically bisecting our shape. We find the second moment of area separately of the large and small rectangles. The border is the difference of these 2 areas. Hence its second moment of area is the subtraction of these second moment values. We use our formula for a rectangle, $\frac{1}{12}Al^2$, from Example 18.4.

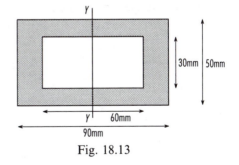

Fig. 18.13

$$I_{\text{large}} = \frac{1}{12}(50 \times 90)90^2 = 3\,037\,500 \text{ mm}^4.$$

$$I_{\text{small}} = \frac{1}{12}(60 \times 30)60^2 = 540\,000 \text{ mm}^4.$$

Hence $I_{YY} = I_{\text{large}} - I_{\text{small}}$

gives $I_{YY} = 3\,037\,500 - 540\,000$

$= 2\,497\,500 \text{ mm}^4 \text{ or } 249.75 \text{ cm}^4$

as the second moment of area of the border about YY.

Example 18.11

We find the second moment of area of a circle about a diameter. This circle has a radius of 80 mm.

We use the formula from Example 18.6 and the result from Example 18.7, i.e.

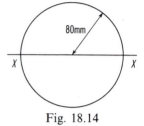

$$I_{XX} = \frac{1}{8}\pi r^4$$

Fig. 18.14

and $I_{XX} = 512\pi \text{ cm}^4$ for each semi-circle.

Thinking of the circle as the sum of 2 semi-circles we write

$I_{\text{circle}} = 2 \times I_{XX}$

$= 2 \times 512\pi$

$= 1024\pi \text{ cm}^4$ is the second moment of this circle about a diameter.

ASSIGNMENT

This time we have our brass rod with a mass of 2 kg attached to the free end. The mass is $7.66\ldots + 2 = 9.66\ldots$ kg. In Fig. 18.15 we have the pivot at the left-hand end. From an initial horizontal position it will fall, rotating about the axis through the end A. Because the height of A does *not* change we use it as the position of zero potential energy.

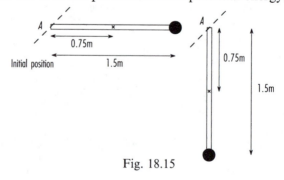

Fig. 18.15

As before, Potential energy $= mgh$ J,

$$\text{Kinetic energy} = \frac{1}{2}I\omega^2 \text{ J}$$

with the usual interpretations for the letters.

In this section we are able to use some of our previous Assignment calculations. Now we have a composite shape so that

$$I = I_{rod} + I_{mass}$$

gives $\quad I = 5.75\ldots + (2 \times 1.5^2) = 10.25\ldots \text{kg m}^2.$

Initially the rod is held horizontally at rest,

i.e. $\qquad$ P.E. $= 0,$

$\qquad$ K.E. $= 0.$

For the instant the rod is vertical its centroid has fallen 0.75 m and the extra mass has fallen 1.5 m. Also the rod is moving,

i.e. $\qquad$ P.E. $= -56.42\ldots + 2 \times 9.81 \times (-1.5) = -85.85\ldots$ J,

$\qquad$ K.E. $= \dfrac{1}{2} \times 10.25\ldots \times \omega^2 \qquad = 5.12\ldots \times \omega^2$ J.

The principle of conservation of energy allows us to write

$\qquad$ Initial energy $=$ Final energy

i.e. $\qquad 0 + 0 = -85.85\ldots + 5.12\ldots \times \omega^2$

to give $\qquad \omega^2 = \dfrac{85.85\ldots}{5.12\ldots}$

i.e. $\qquad \omega = \sqrt{16.74\ldots}$

$\qquad \omega = 4.09 \text{ rads}^{-1}$ $\qquad$ | Square root of both sides. |

is the angular speed when the rod is vertical.

■■■■ EXERCISE 18.1 ■■■■

We have drawn a series of composite shapes. They are labelled with a variety of axes *XX* and *YY*. In each case find the second moment of area about the axis/axes shown.

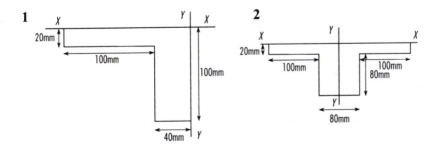

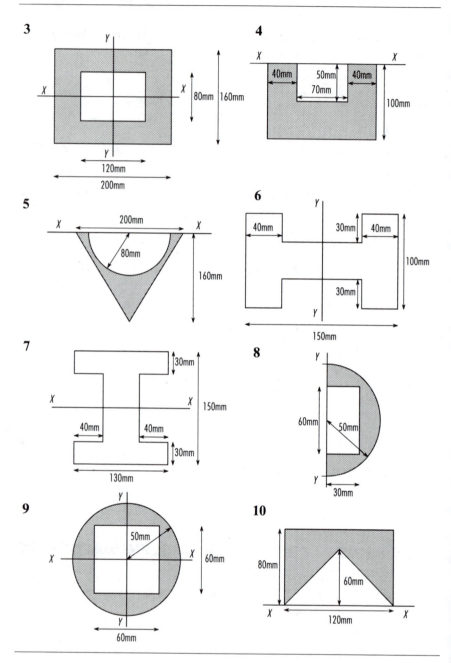

Parallel axes theorem

This theorem connects second moments of area. In Fig. 18.16 we see two parallel axes (*GG* and *ZZ*) associated with the general area *A*. Notice the axis *GG* passes through the centroid *C*. Also the general parallel axis *ZZ* is a distance *d* from *GG*.

The parallel axes theorem states

$$I_{ZZ} = I_{GG} + Ad^2.$$

This is an easy theorem to apply. Remember we need one of the axes, GG, to pass through the centroid C.

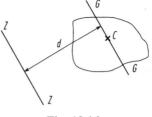

Fig. 18.16

<hr>

██████ **Example 18.12** ████████████████████████████████

We apply the parallel axes theorem to the rectangle in Fig. 18.17. We know from Example 18.4 that $I_{GG} = \frac{1}{12}Al^2$ where A is the area of the rectangle of length l. In this case the distance between the axes YY and GG is $\frac{l}{2}$, i.e. $d = \frac{l}{2}$.

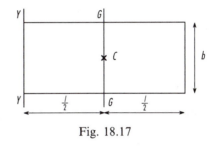

Fig. 18.17

We apply the parallel axes theorem,

$$I_{ZZ} = I_{GG} + Ad^2$$

to get

$$I_{YY} = \frac{1}{12}Al^2 + A\left(\frac{l}{2}\right)^2$$

$$= \frac{1}{12}Al^2 + \frac{1}{4}Al^2$$

$$= \frac{1}{3}Al^2$$

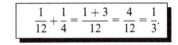

$$\frac{1}{12} + \frac{1}{4} = \frac{1+3}{12} = \frac{4}{12} = \frac{1}{3}.$$

which agrees with our result in Example 18.1.

<hr>

██████ **Example 18.13** ████████████████████████████████

We apply the parallel axes theorem to the triangle in Fig. 18.18.

We know the position of the centroid, C, and so draw an axis GG through C.

We apply the parallel axes theorem twice,

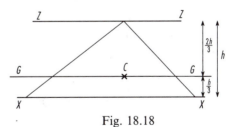

Fig. 18.18

$$I_{ZZ} = I_{GG} + Ad^2$$

to get

$$I_{ZZ} = I_{GG} + A\left(\frac{2}{3}h\right)^2$$

and

$$I_{XX} = I_{GG} + A\left(\frac{1}{3}h\right)^2.$$

Subtracting these equations, we eliminate I_{GG},

i.e. $\qquad I_{ZZ} - I_{XX} = \dfrac{4}{9}Ah^2 - \dfrac{1}{9}Ah^2$

and substitute from Example 18.5, $I_{XX} = \dfrac{1}{6}Ah^2$,

to get $\quad I_{ZZ} - \dfrac{1}{6}Ah^2 = \dfrac{1}{3}Ah^2$

$$\dfrac{4}{9} - \dfrac{1}{9} = \dfrac{3}{9} = \dfrac{1}{3}.$$

i.e. $\qquad\qquad I_{ZZ} = \dfrac{1}{6}Ah^2 + \dfrac{1}{3}Ah^2$

$$= \dfrac{1}{2}Ah^2$$

$$\dfrac{1}{6} + \dfrac{1}{3} = \dfrac{1+2}{6} = \dfrac{3}{6} = \dfrac{1}{2}.$$

is the second moment of area of the triangle about an axis through the vertex parallel to the edge.

Example 18.14

Find the second moment of area of the L-shape about XX.

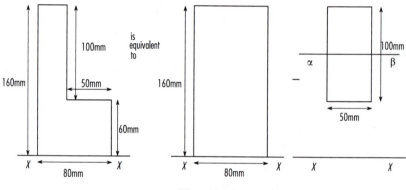

Figs. 18.19

In Figs. 18.19 we construct our L-shape by removing a small rectangle from a large rectangle. We look at the rectangles separately and then bring those results together.

For the large rectangle,

$$I_{LXX} = \dfrac{1}{3}(160 \times 80)160^2$$

$$= 1.092\ldots \times 10^8 \text{ mm}^4.$$

$I = \dfrac{1}{3}Al^2.$

$1.092\ldots \times 10^4 \text{ cm}^4.$

For the small rectangle,

$$I_{S\alpha\beta} = \dfrac{1}{12}(50 \times 100)100^2$$

$$= 4\,166\,666.\bar{6} \text{ mm}^4.$$

$I = \dfrac{1}{12}Al^2.$

$416.\bar{6} \text{ cm}^4.$

Also, applying the parallel axes theorem,

$$I_{SXX} = I_{S\alpha\beta} + Ad^2$$

we get $\quad I_{SXX} = 4\,166\,666.\bar{6} + (50 \times 100)110^2$

$$= 4\,166\,666.\bar{6} + 60\,500\,000$$

$$= 64\,666\,666.\bar{6} \text{ mm}^4.$$

$6466.\bar{6} \text{ cm}^4.$

Finally we bring together these results by subtraction,

i.e. $\quad I_{XX} = I_{LXX} - I_{SXX}$

to get $\quad I_{XX} = 10922.\bar{6} - 6466.\bar{6}$

$$= 4456 \text{ cm}^4.$$

This answer agrees with Example 18.9.

ASSIGNMENT

This time we have our brass rod with a circular disc of radius 100 mm and thickness 5 mm. The centre of the disc is attached to the free end of the rod. Because the disc is relatively thin compared to the radius it approximates to a lamina. We standardise our units, recalling the brass has a density of 8180 kg m^3.

Volume of disc $= \pi \times 0.1^2 \times 0.005 = 1.5707\ldots \times 10^{-4} \text{ m}^3.$

$\therefore \qquad$ Mass $= 8180 \times 1.5707\ldots \times 10^{-4} = 1.28\ldots \text{ kg.}$

In Fig. 18.20 we have the pivot at the left hand end. From an initial horizontal position it will fall, rotating about the axis through the end A.

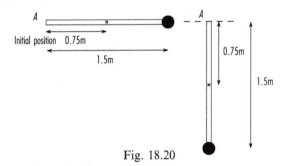

Fig. 18.20

The pivot is parallel to the horizontal diameter of the disc. Because the height of A does *not* change we use it as the position of zero potential energy.

As before, $\quad$ Potential energy $= mgh$ J,

$$\text{Kinetic energy} = \frac{1}{2}I\omega^2 \text{ J}$$

with the usual interpretations for the letters.

In this section we are able to use some of our previous Assignment calculations. Also we have looked at a semi-circle and a circle as the sum of 2 semi-circles. Using the ideas of Example 18.11 we write

$$I_{circle} = 2 \times \frac{1}{8}\pi r^4 = \frac{1}{4}Ar^2.$$

Area of a circle, $A = \pi r^2$.

We apply the parallel axes theorem for the disc because our pivot is at A, i.e.

$$I_{disc} = \frac{1}{4}Ar^2 + Ad^2$$

and replace area, A, with mass to get

$$I_{disc} = \left(\frac{1}{4} \times 0.1^2 + 1.5^2\right)1.28\ldots = 2.89\ldots \text{ kg m}^2.$$

Now we have a composite shape so that

$$I = I_{rod} + I_{disc}$$

gives $I = 5.75\ldots + 2.89\ldots = 8.64\ldots \text{ kg m}^2.$

Initially the rod is held horizontally at rest,

i.e.   P.E. = 0,

K.E. = 0.

For the instant the rod is vertical its centroid has fallen 0.75 m and the disc has fallen 1.5 m. Also the rod is moving,

i.e.   P.E. = $-56.42\ldots + 1.28\ldots \times 9.81 \times (-1.5) = -75.33\ldots$ J,

K.E. = $\frac{1}{2} \times 8.64\ldots \times \omega^2$ $= 4.32\ldots \times \omega^2$ J.

The principle of conservation of energy allows us to write

Initial energy = Final energy

i.e. $0 + 0 = -75.33\ldots + 4.32\ldots \times \omega^2$

to give $\omega^2 = \dfrac{75.33\ldots}{4.32\ldots}$

i.e. $\omega = \sqrt{17.42\ldots}$

Square root of both sides.

$\omega = 4.17 \text{ rads}^{-1}$

is the angular speed when the rod is vertical.

◼◼◼ EXERCISE 18.2 ◼◼◼

We have drawn a series of composite shapes. They are labelled with a variety of different axes XX and YY. In each case find the second moment of area about the axis shown.

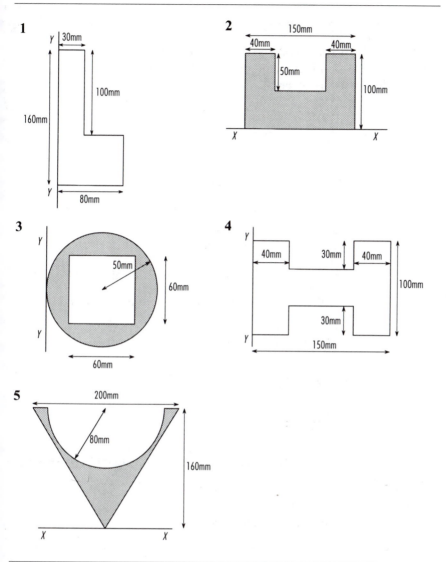

Perpendicular axes theorem

This theorem applies only to laminas. In Fig. 18.21 we have a general area A. O and the axes Ox and Oy lie in the plane of the lamina. The axis Oz is perpendicular to that plane.

The **perpendicular axes theorem** states

$$I_{Oz} = I_{Ox} + I_{Oy}.$$

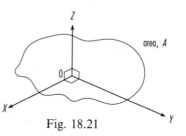

Fig. 18.21

Example 18.15

We apply the perpendicular axes theorem to the rectangle in Fig. 18.22. We find the second moment of area about an axis through a vertex perpendicular to the plane of the rectangle.

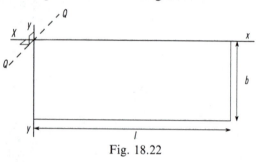

Fig. 18.22

We know from Example 18.1 that $I_{YY} = \frac{1}{3}Al^2$ and can apply this to deduce $I_{XX} = \frac{1}{3}Ab^2$. The axis QQ is perpendicular to both axes XX and YY. We apply our theorem to write

$$I_{QQ} = I_{XX} + I_{YY}$$

i.e. $$I_{QQ} = \frac{1}{3}Ab^2 + \frac{1}{3}Al^2$$

$$= \frac{1}{3}A(b^2 + l^2).$$

Example 18.16

We find the second moment of area of a circle about an axis through its centre perpendicular to the plane of the circle.

We know from Example 18.11 that $I_{YY} = \frac{1}{4}\pi r^4$ and by symmetry $I_{XX} = I_{YY}$. The axis QQ is perpendicular to both axes XX and YY. We apply our theorem to write

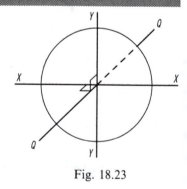

Fig. 18.23

$$I_{QQ} = I_{XX} + I_{YY}$$

i.e. $$I_{QQ} = 2 \times \frac{1}{4}\pi r^4$$

$$= \frac{1}{2}\pi r^4 \quad \text{or} \quad \frac{1}{2}Ar^2.$$

Area of a circle, $A = \pi r^2$.

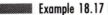

Example 18.17

We find the second moment of area of a circle about an axis through its centre perpendicular to the plane of the circle using integration.

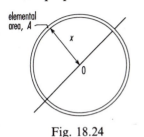

A circle of centre O is a series of concentric rings. Let us look at a typical ring of radius x. We may split it into elemental areas, A. The second moment of area of each elemental area about an axis through O is Ax^2.

Fig. 18.24

Now let us look at that series of rings making up the circular disc, Fig. 18.25.

Each elemental ring is approximately of length $2\pi x$, i.e. the circumference.

Length of element $= 2\pi x$
Thickness $= \delta x$
Area $= 2\pi x\delta x$

Fig. 18.25

Distance of element from the perpendicular axis through $O = x$.

Second moment of area about that axis $= (2\pi x\delta x)(x^2)$.

To find the total second moment of area about the axis we sum all these elemental second moments, i.e.

Second moment of area $= \displaystyle\sum_{x=0}^{x=r} 2\pi x^3\,\delta x.$

In the limiting case as $\delta x \to 0$

$$\text{Second moment of area} = 2\pi \int_0^r x^3\,dx$$

$$= 2\pi \left[\frac{x^4}{4}\right]_0^r$$

$$= 2\pi \left[\frac{r^4}{4} - 0\right]$$

$$= \frac{\pi}{2}r^4$$

> Area of a circle, $A = \pi r^2$.

i.e. $\qquad I_{QQ} = \dfrac{1}{2}Ar^2$

to agree with Example 18.16.

ASSIGNMENT

Again, we have our brass rod with the same circular disc of mass 1.28...kg. The centre of the disc is attached to the free end of the rod. In Fig. 18.26 we have the pivot at the left-hand end, A. This time the pivot is perpendicular to the plane of the disc. Because the height of A does *not* change we use it as the position of zero potential energy. Again, we are going to look at the change of energy. We change our positions from previous Assignment calculations.

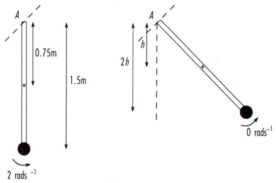

Fig. 18.26

As before,   Potential energy $= mgh$ J,

$$\text{Kinetic energy} = \frac{1}{2}I\omega^2 \text{ J}$$

with the usual interpretations for the letters.

Again we are able to use some of our previous calculations.

$$I_{\text{circle}} = \frac{1}{2}Ar^2.$$

> Area of a circle, $A = \pi r^2$.

We apply the parallel axes theorem for the disc because our pivot is at A, i.e.

$$I_{\text{disc}} = \frac{1}{2}Ar^2 + Ad^2$$

and replace area, A, with mass to get

$$I_{\text{disc}} = \left(\frac{1}{2} \times 0.1^2 + 1.5^2\right)1.28\ldots = 2.89\ldots \text{ kg m}^2.$$

Now we have a composite shape so that

$$I = I_{\text{rod}} + I_{\text{disc}}$$

gives $I = 5.75\ldots + 2.89\ldots = 8.64\ldots \text{ kg m}^2$.

We compare the rod in the vertical position to a point of instantaneous rest. Suppose its angular velocity is 2 rads^{-1} in that vertical position, i.e.   P.E. $= -56.42\ldots + 1.28\ldots \times 9.81 \times (-1.5) = -75.33\ldots$ J,

K.E. $= \dfrac{1}{2} \times 8.64\ldots \times 2^2$ J. $= 17.29\ldots$ J.

For the instant the rod is at rest,

i.e.   P.E. $= 7.66\ldots \times 9.81 \times (-h) + 1.28\ldots \times 9.81 \times (-2h)$

$= -100.44\ldots \times h$ J,

K.E. $= 0$.

The principle of conservation of energy allows us to write

$-75.33\ldots + 17.29\ldots = -100.44\ldots \times h + 0$

to give $\dfrac{-58.03\ldots}{-100.44\ldots} = h$

i.e. $h = 0.58$ m

and $2h = 1.16$ m.

This means the centre of the rod is 0.58 m below A and the centre of the disc is 1.16 m below A. We can find the angle of inclination for this position of instantaneous rest (Fig. 18.27).

$$\cos\alpha = \frac{1.16}{1.5} = 0.77\overline{3},$$

i.e. $\alpha = 39°$ is the approximate angle of inclination of the rod.

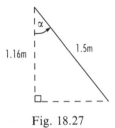

Fig. 18.27

■ EXERCISE 18.3 ■

We have drawn a series of composite shapes. They are labelled with a variety of axes XX and YY. In each case we draw an axis QQ through their point of intersection. QQ is perpendicular to the plane of the shape and the axes XX and YY. Find the second moment of area about the axis in each case.

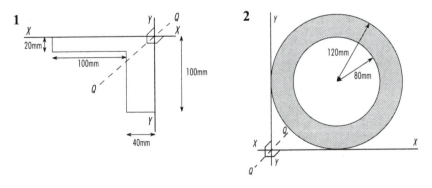

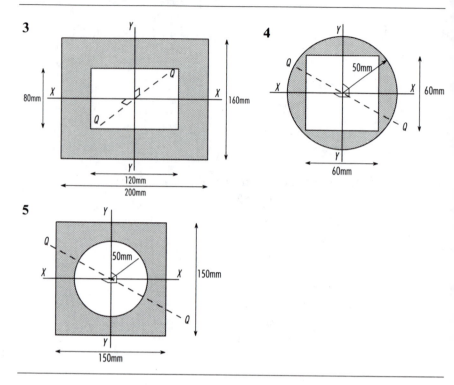

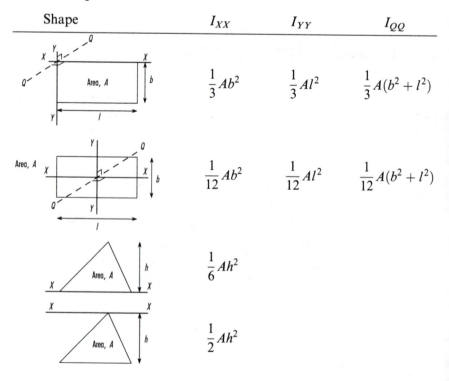

Summary of Second Moments

Shape	I_{XX}	I_{YY}	I_{QQ}
Area, A (rectangle, X at edge)	$\dfrac{1}{3}Ab^2$	$\dfrac{1}{3}Al^2$	$\dfrac{1}{3}A(b^2+l^2)$
Area, A (rectangle, X through centre)	$\dfrac{1}{12}Ab^2$	$\dfrac{1}{12}Al^2$	$\dfrac{1}{12}A(b^2+l^2)$
Area, A (triangle, X at base)	$\dfrac{1}{6}Ah^2$		
Area, A (triangle, X at apex)	$\dfrac{1}{2}Ah^2$		

Shape	I_{XX}	I_{YY}	I_{QQ}
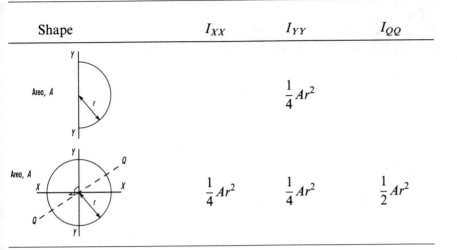		$\frac{1}{4}Ar^2$	
	$\frac{1}{4}Ar^2$	$\frac{1}{4}Ar^2$	$\frac{1}{2}Ar^2$

Answers

Chapter 1

Exercise 1.1

1	200	**2**	-233	**3**	0	**4**	-80
5	152	**6**	$-1.58\bar{3}$	**7**	152	**8**	108
9	248	**10**	-0.312				

Exercise 1.2

1	139	**2**	-6	**3**	-216	**4**	41
5	-72						

Exercise 1.3

1 $x=11, y=-9, z=8$

2 $x=-2, y=12, z=5$

3 $x=3, y=4, z=1$

4 $x=15, y=10, z=7$

5 $x=4, y=10, z=3$

6 $x=-7, y=-2, z=4$

7 $x=-1, y=2.5, z=-2$

8 $x=24, y=4, z=5$

9 $x=2, y=-1, z=3.5$

10 $x=3, y=-0.5, z=1.5$

Exercise 1.4

1 $\begin{pmatrix} 9 & 3 & 5 \\ 5 & 24 & -12 \\ 3 & -3 & 8 \end{pmatrix}$

2 $\begin{pmatrix} -5 & 9 & 19 \\ 5 & -6 & -2 \\ 3 & 5 & -24 \end{pmatrix}$

3 $\begin{pmatrix} 13 & -3 & 25 \\ 16 & 26 & -11.5 \\ 1 & 2 & 22 \end{pmatrix}$

4 $\begin{pmatrix} -15 & 22 & 36 \\ -2 & 2 & -3 \\ 28 & 35 & -14 \end{pmatrix}$

5 $\begin{pmatrix} 5 & -4 & 2 \\ 12 & -14 & -1 \\ -22 & -25 & -34 \end{pmatrix}$

6 $\begin{pmatrix} 5 & -9 & -19 \\ -5 & 6 & 2 \\ -3 & -5 & 24 \end{pmatrix}$

7 $\begin{pmatrix} 9 & 3 & 5 \\ 5 & 24 & -12 \\ 3 & -3 & 8 \end{pmatrix}$

8 $\begin{pmatrix} -32 & 44 & -26 \\ -47 & 10 & -3.5 \\ 56 & 45 & -22 \end{pmatrix}$

9 $\begin{pmatrix} 11 & 9 & 17 \\ 10 & 33 & -19 \\ 6 & -2 & 0 \end{pmatrix}$

10 $\begin{pmatrix} -8 & 10 & 27 \\ -1.5 & 9 & -0.75 \\ 24 & 32.5 & 17 \end{pmatrix}$

11 $\begin{pmatrix} 17 & -16 & -24 \\ 7 & 7 & -4 \\ -25 & -34 & 6 \end{pmatrix}$ **12** $\begin{pmatrix} 4 & 18 & -62 \\ -28 & 33 & -18.5 \\ 9 & -22 & -18 \end{pmatrix}$

13 $\begin{pmatrix} 3 & -9 & 14 \\ 8.5 & -2.5 & 4 \\ -3.5 & 4.5 & 18 \end{pmatrix}$ **14** $\begin{pmatrix} -55 & 80 & 18 \\ -68 & 49 & -11.5 \\ 131 & 131 & 24 \end{pmatrix}$

15 $\begin{pmatrix} 23 & -34 & 4 \\ 24 & 2 & 4 \\ -32 & -25 & 42 \end{pmatrix}$

Exercise 1.5

1 $\begin{pmatrix} -5 \\ 18 \\ 72 \end{pmatrix}$ **2** $\begin{pmatrix} -3 \\ 25 \\ -2 \end{pmatrix}$ **3** $\begin{pmatrix} 15 \\ 48 \\ 59 \end{pmatrix}$ **4** $\begin{pmatrix} -8.5 \\ 28.5 \\ 59.5 \end{pmatrix}$

5 $\begin{pmatrix} 34 & 17 & -29 \\ 74 & 9 & 88 \\ -98 & -11 & -12 \end{pmatrix}$ **6** $\begin{pmatrix} 7 & 9 & 8 \\ -11 & -7 & -12 \\ 107 & 175 & 90 \end{pmatrix}$

7 $\begin{pmatrix} -43 & 44 & 104 \\ 0 & 5 & -14 \\ -37 & -6 & 40 \end{pmatrix}$ **8** $\begin{pmatrix} -22 \\ -11 \\ 33 \end{pmatrix}$

9 $11 \begin{pmatrix} -23 \\ -5 \\ 6 \end{pmatrix}$ **10** $\begin{pmatrix} 119 \\ -5 \\ 47 \end{pmatrix}$

Exercise 1.6

1 $\begin{pmatrix} 2 & 5 & 6 \\ -1 & 4 & 0 \\ 2 & 2 & 8 \end{pmatrix}, 44$ **2** $\begin{pmatrix} 0 & 1 & 2 \\ 11 & -2 & 4 \\ 5 & -6 & 7 \end{pmatrix}, -169$

3 $\begin{pmatrix} 3 & 8 & 1 \\ 10 & -6 & 5 \\ 0 & 6 & 7 \end{pmatrix}, -716$ **4** $\begin{pmatrix} 0 & 1 & 7 \\ 3 & 2 & 31 \\ -17 & 3 & -1 \end{pmatrix}, -223$

5 $\begin{pmatrix} 5 & 11 & 1 \\ 2 & -6 & 6 \\ 6 & 5 & 4 \end{pmatrix}, 84$

Exercise 1.7

1 $-\dfrac{1}{34} \begin{pmatrix} -7 & -6 & 3 \\ -2 & 8 & -4 \\ 27 & 28 & -31 \end{pmatrix}$ **2** $\dfrac{1}{210} \begin{pmatrix} -18 & 6 & -2 \\ 0 & 0 & 70 \\ 51 & 18 & -6 \end{pmatrix}$

3 $\dfrac{1}{48} \begin{pmatrix} -3 & -10 & 5 \\ 12 & 8 & -4 \\ -15 & 14 & 17 \end{pmatrix}$ **4** $-\dfrac{1}{184} \begin{pmatrix} -60 & 32 & -22 \\ 22 & -24 & 5 \\ 20 & -72 & 38 \end{pmatrix}$

5 $\dfrac{1}{288}\begin{pmatrix} 48 & 16 & -32 \\ -24 & 40 & 10 \\ 24 & -40 & 26 \end{pmatrix}$

Exercise 1.8

1 $x=6,\ y=5,\ z=-4$

2 $x=5,\ y=1,\ z=5$

3 $x=5,\ y=2,\ z=1$

4 $x=0.5,\ y=1,\ z=1.5$

5 $x=-0.5,\ y=1,\ z=2$

6 $x=3,\ y=6.5,\ z=4.5$

7 $x=-2,\ y=1,\ z=2$

8 No solution

9 $x=-10,\ y=7,\ z=29$

10 No solution

Exercise 1.9

1 $x=14,\ y=4,\ z=3$

2 $I_1=3,\ I_2=2,\ I_3=6$

3 $x=1,\ y=3,\ z=2$; Equations ② and ③ are inconsistent

4 $x=15,\ y=25,\ z=25$

5 $I_1=4,\ I_2=2,\ I_3=1$

Chapter 2

Exercise 2.1

1 $-2\pm j1.73$

2 $0.75\pm j2.22$

3 $-0.50\pm j0.87$

4 $0.1\bar{6}\pm j1.28$

5 $0.29\pm j0.25$

Exercise 2.2

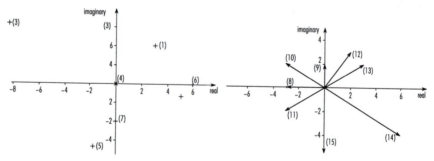

Exercise 2.3

1 $6.5-j2.5$

2 $8.5+j0.5$

3 $4-j3$

4 $-5.5-j0.5$

5 $5.5+j0.5$

6 $1.5-j4.5$

7 $9.5+j5.5$

8 $-4+j3$

9 $-8.5+j0.5$

10 $0.5-j10.5$

Exercise 2.4

1 $15+j12$

2 $-6-j2$

3 $6-j14$

4 $-3 + j6$ **5** $1.5 + j9$ **6** $6 + j36$

7 $6 + j36$ **8** $-8 + j10$ **9** $7 + j22$

10 $13 + j19$ **11** $15 - j5$ **12** $8 - j36$

13 25 **14** $j169$ **15** $39 + j57$

16 $-22 + j7$ **17** $1 + j83$ **18** $92 + j213$

19 $21 + j33$ **20** $5 - j2$

Exercise 2.5

1 $0.08 + j0.44$ **2** $0.24 - j1.06$ **3** $0.20 - j0.10$

4 $0.02 - j0.26$ **5** $-1.40 + j3.80$ **6** $0.80 + j0.10$

7 $7 - j55$ **8** $21.50 - j63.50$ **9** $2 - j3$

10 $0.21 + j0.07$

Exercise 2.6

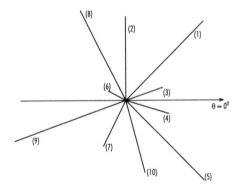

Exercise 2.7

1 $18\underline{/58°}$ **2** $75\underline{/45°}$ **3** $18\underline{/65°}$ **4** $9\underline{/-117°}$

5 $5.5\underline{/-161°}$ **6** $18\underline{/-160°}$ **7** $12\underline{/149°}$ **8** $2.5\underline{/-75°}$

9 $52.5\underline{/82°}$ **10** $38.4\underline{/-67°}$

Exercise 2.8

1 $2\underline{/32°}$ **2** $1.\overline{6}\underline{/146°}$ **3** $0.48\underline{/-146°}$ **4** $8\underline{/-128°}$

5 $3.2\underline{/133°}$ **6** $1.4\underline{/-65°}$ **7** $5.8\overline{3}\underline{/45°}$ **8** $0.4\overline{6}\underline{/44°}$

9 $0.\overline{740}\underline{/-170°}$ **10** $18.\overline{6}\underline{/92°}$

Exercise 2.9

1 $1.75 + j0.40$ **2** $-2.04 + j4.01$ **3** $0.60 - j2.43$

4 $-8.06 - j8.06$ **5** $0 - j2.7$ **6** $5.39\underline{/21.80°}$

7 $9.22\underline{/-49.40°}$ **8** $3.16\underline{/108.43°}$ **9** $5\underline{/-126.87°}$

10 $21.21\underline{/-135°}$

Exercise 2.10

1	$7\underline{/30°}$, $7\underline{/-150°}$	**2**	$11\underline{/67.5°}$, $11\underline{/-112.5°}$
3	$3.69\underline{/-12°}$, $3.69\underline{/168°}$	**4**	$2.49\underline{/-59°}$, $2.49\underline{/121°}$
5	$5\underline{/90°}$, $5\underline{/-90°}$	**6**	$2.24\underline{/26.57°}$, $2.24\underline{/-153.43°}$
7	$3.61\underline{/78.69°}$, $3.61\underline{/-101.31°}$	**8**	$5\underline{/-8.13°}$, $5\underline{/171.87°}$
9	$4.05\underline{/-71.22°}$, $4.05\underline{/108.78°}$	**10**	$3\underline{/45°}$, $3\underline{/-135°}$

Exercise 2.11

1 $0.022 - j0.012$, $0.025\underline{/-29.74°}$

2 $60 + j50$, $78.10\underline{/39.81°}$; $64.03\underline{/51.34°}$, $20\underline{/0°}$, $16.4\underline{/11.53°}$

3 $8 + j2$, $2.94 - j1.74$

4 $13 + j5.\bar{6}$ A, 14.18 A

5 $1.\bar{3}\underline{/65°}$; $1.15\underline{/32.5°}$, $1.15\underline{/-147.5°}$

6 $5.83\underline{/59.04°}$, $250.05\underline{/88.85°}$; $3.82 \times 10^{-3}\underline{/73.9°}$, $(1.06 + j3.67)10^{-3}$

7 $81.05\underline{/75.7°}$, $230\underline{/0°}$, $2.84\underline{/-75.7°}$

8 $1.79 + j1.71$

9 $0.54 \times 10^{3}\underline{/-14.8°}$, $0.54 \times 10^{3}\underline{/165.2°}$

10 $103.06\underline{/-90°}$, $-j103.06$; $101.41\underline{/-90°}$, $-j101.41$; $102.23\underline{/\pm90°}$, $\pm j102.23$

Chapter 3

Exercise 3.1

1	0.44 rad	**2**	4.84 rad	**3**	−1.05 rad	**4**	1.90 rad
5	6.63 rad	**6**	84.22°	**7**	315°	**8**	123.19°
9	75°	**10**	77.35°				

Exercise 3.2

1	i)	1,	ii)	3,	iii)	120°, 2.09 rad,	iv)	0.48 Hz
2	i)	5,	ii)	2,	iii)	180°, 3.14 rad,	iv)	0.32 Hz
3	i)	1,	ii)	3,	iii)	120°, 2.09 rad,	iv)	0.48 Hz
4	i)	2,	ii)	3,	iii)	120°, 2.09 rad,	iv)	0.48 Hz
5	i)	1,	ii)	1,	iii)	360°, 6.28 rad,	iv)	0.16 Hz

7 $y = 4\sin 15.7t$ **8** $y = 2\sin 6t$ **14** π rad

Exercise 3.4

1 $y = 4\cos\left(t + \dfrac{\pi}{3}\right)$ **2** $y = 3\sin\left(2t - \dfrac{\pi}{6}\right)$ **3** $y = \sin(x + 90°)$

4 i_1 leads i_2 by $\dfrac{\pi}{6}$ **5** v_1 leads v_2 by $\dfrac{7\pi}{12}$

Exercise 3.6

	Amplitude	Phase angle	
1	16.97	$45°$	$y = 16.97 \sin(x + 45°)$
2	7.21	$33.7°$	$y = 7.21 \sin(2t + 33.7°)$
3	7.83	$26.6°$	$y = 7.83 \sin(3x + 26.6°)$
4	1.93	$\dfrac{\pi}{12}$	$i = 1.93 \sin\left(2t + \dfrac{\pi}{12}\right)$
5	1.22	$\dfrac{\pi}{24}$	$v = 1.22 \sin\left(t + \dfrac{\pi}{24}\right)$

Exercise 3.7

3 -6.2 **4** 0, 1.7 rad

Chapter 4

Exercise 4.1

1	2.366	**2**	1	**3**	-2	**4**	0.577
5	1.155	**6**	-1.064	**7**	2.670	**8**	-12.64
9	-1.013	**10**	-1.082				

14 $\dfrac{1}{0}$ is not defined; 180° and 360° for cosecant and cotangent, 270° for secant

15 Yes

Exercise 4.2

1	$-\sin\theta$	**2**	$\sin\theta$	**3**	$-\cos\theta$	**4**	$\sin\theta$
5	$-\sin\theta$	**6**	$\sin\theta$	**7**	$\sin\theta$	**8**	$-\cot\theta$
9	$-\tan\theta$	**10**	$\cos\theta$				

Exercise 4.3

1 $\dfrac{\sqrt{3}-1}{2\sqrt{2}}$ **2** $\dfrac{\sqrt{3}-1}{2\sqrt{2}}$ **3** $\dfrac{\sqrt{3}-1}{2\sqrt{2}}$

4 $\dfrac{\sqrt{3}-1}{\sqrt{3}+1}$ **5** $\dfrac{1+\sqrt{3}}{2\sqrt{2}}$ **6** $\dfrac{\cos x + \sin x}{\sqrt{2}}$

7 $\dfrac{\cos x - \sqrt{3}\sin x}{2}$ **8** $\sqrt{2}(\cos x + \sin x)$ **9** $\sin x + \sqrt{3}\cos x$

10 $\sqrt{2}(\cos x - \sin x)$ **11** $\dfrac{44}{125}$ **12** $-\dfrac{220}{221}$

13 $-\dfrac{44}{125}$ **14** $-\dfrac{56}{33}$ **15** $-\dfrac{44}{117}$

Exercise 4.4

1 $\dfrac{240}{289}$ **2** $-\dfrac{24}{7}$ **3** $\dfrac{161}{289}$ **4** $-\dfrac{120}{119}$

5 $\dfrac{76}{1445}$ **6** $\dfrac{416}{87}$ **7** $-\dfrac{527}{336}$ **8** $-\dfrac{336}{625}$

9 $\dfrac{44}{125}$ **10** $-\dfrac{117}{125}$ **15** $\sqrt{3}$

Exercise 4.5

4 $\sqrt{3}$ **5** $\sqrt{1 + t^2}$ **6** $\dfrac{4t^3}{1 - t^4}$ **7** $1 + t^2$

8 $\dfrac{1}{t^2}$ **9** $2 \sin A \cos A, \dfrac{4t(1 - t^2)}{(1 + t^2)^2}$

10 $\dfrac{2 \tan A}{1 - \tan^2 A}, \dfrac{4t(1 - t^2)}{1 - 6t^2 + t^4}$

Exercise 4.6

3 i) $-\cot A$,   ii) $\cot A$ **5** $\dfrac{\sin 3\alpha}{\cos 5\alpha}$

Exercise 4.7

1 7.21, 33.69°, $7.21 \sin(2x - 33.69°)$
2 16.97, 45°
3 $7 \sin 3t - 3.5 \cos 3t$, 7.83, 0.46 rad
4 5.32, 0.72 rad, $5.32 \sin(2t + 0.72)$, 5.32
5 15.81, 1.25 rad, $15.81 \sin(10\pi t + 1.25)$, 15.81

Exercise 4.8

1 66.93°, 113.07°   **2** None   **3** 77.86°, 257.86°
4 2.42, 5.56 rad   **5** 0.26, 2.88 rad

Exercise 4.9

1 24.30°, 65.70°, 204.30°, 245.70°   **2** ±24.18°, ±95.82°, ±144.18°
3 26.82°, 153.18°, 206.82°, 333.18°   **4** 0.21, 1.37, 3.35, 4.51 rad
5 0.22, 1.04 rad

Exercise 4.10

1 $7.21 \sin(2x + 33.69°)$; 15.33°, 40.98°, 195.33°, 220.98°
2 7.07, 45°, −137.74°, −87.26°; 42.26°, 92.74°
3 $3.5 \sin 3t + 5 \cos 3t$, $6.10 \sin(3t + 0.96)$; 0.55, 1.95 rad
4 0.93 s
5 0.04 s

Exercise 4.11

1 $0°, 60°, 180°, 300°, 360°$ **2** $71.45°, 288.55°$

3 $\pm54.74°, \pm125.26°$ **4** $180°$

5 $71.57°, 116.57°, 251.57°, 296.57°$ **6** $0, \pi, 2\pi$

7 $67.98°, 292.02°$ **8** $0°, 138.59°, 221.41°, 360°$

9 $70.67°, 250.67°$ **10** $25.1°, 154.9°, 205.1°, 334.9°$

Chapter 5

Exercise 5.1

1 $a^8 + 8a^7x + 28a^6x^2 + 56a^5x^3 + 70a^4x^4 + 56a^3x^5 + 28a^2x^6 + 8ax^7 + x^8$

2 $a^5 - 5a^4x + 10a^3x^2 - 10a^2x^3 + 5ax^4 - x^5$

3 $a^6 - 6a^5x + 15a^4x^2 - 20a^3x^3 + 15a^2x^4 - 6ax^5 + x^6$

4 $16r^4 + 32r^3s + 24r^2s^2 + 8rs^3 + s^4$

5 $x^5 - 15x^4y + 90x^3y^2 - 270x^2y^3 + 405xy^4 - 243y^5$

Exercise 5.2

1 $a^5 + 5a^4b + 10a^3b^2 + 10a^2b^3 + 5ab^4 + b^5$

2 $1 + 5b + 10b^2 + 10b^3 + 5b^4 + b^5$

3 $1 - 5b + 10b^2 - 10b^3 + 5b^4 - b^5$

4 $x^4 + 16x^3y + 96x^2y^2 + 256xy^3 + 256y^4$

5 $64x^6 - 192x^5y + 240x^4y^2 - 160x^3y^3 + 60x^2y^4 - 12xy^5 + y^6$

6 $243x^5 - 810x^4y + 1080x^3y^2 - 720x^2y^3 + 240xy^4 - 32y^5$

7 $\dfrac{1}{16}a^4 + a^3b + 6a^2b^2 + 16ab^3 + 16b^4$

8 $128 - 448x + 672x^2 - 560x^3 + 280x^4 - 84x^5 + 14x^6 - x^7$

9 $\dfrac{1}{a^8} + \dfrac{8}{a^6} + \dfrac{28}{a^4} + \dfrac{56}{a^2} + 70 + 56a^2 + 28a^4 + 8a^6 + a^8$

10 $a^{10} - 5a^6 + 10a^2 - \dfrac{10}{a^2} + \dfrac{5}{a^6} + \dfrac{1}{a^{10}}$

11 $256x^8 + 1024x^7y + 1792x^6y^2 + 1792x^5y^3 \ldots$

12 $x^{15} - 45x^{14}y + 945x^{13}y^2 - 12285x^{12}y^3 \ldots$

13 $729x^6 + 2916x^5y + 4860x^4y^2 + 4320x^3y^3 \ldots$

14 $a^{12} + 12a^{11}b + 66a^{10}b^2 + 220a^9b^3 + 495a^8b^4 \ldots$

15 $a^{11} - 11a^{10}b + 55a^9b^2 - 165a^8b^3 + 330a^7b^4 \ldots$

16 $a^{14} - 28a^{13}b + 364a^{12}b^2 - 2912a^{11}b^3 + 16016a^{10}b^4 \ldots$

17 $1 + 16b + 120b^2 + 560b^3 + 1820b^4 + 4368b^5 \ldots$

18 $x^{10} - 10x^8 + 45x^6 - 120x^4 + 210x^2 - 252 \ldots$

19 $210b^6a^4$

20 $5376a^3b^6$

Exercise 5.3

1 $1 - x + x^2 - x^3 + x^4 \ldots$

2 $1 - 5b + 15b^2 - 35b^3 \ldots$

3 $3(1 + 4b + 10b^2 + 20b^3 \ldots)$ or $3 + 12b + 30b^2 + 60b^3 \ldots$

4 $2(1 - 9x + 54x^2 - 270x^3 + 1215x^4 \ldots)$ or
$2 - 18x + 108x^2 - 540x^3 + 2430x^4 \ldots$

5 $1 + x - 0.5x^2 + 0.5x^3 \ldots$ **6** $1 - x - 0.5x^2 - 0.5x^3 \ldots$

7 $1 - \dfrac{1}{3}x - \dfrac{1}{9}x^2 \ldots$ **8** $1 + 0.75x - 0.844x^2 \ldots$

9 $1 - x + 1.5x^2 - 2.5x^3 \ldots$ **10** $1 + 0.5x + 0.625x^2 + 0.9375x^3 \ldots$

11 $1 - \dfrac{1}{3}x + \dfrac{2}{9}x^2 \ldots$ **12** $1 - 2x + 6x^2 - 20x^3 + 70x^4 \ldots$

13 $2 + 6x + 27x^2 + 135x^3 \ldots$ **14** $\dfrac{1}{2} - \dfrac{1}{16}x + \dfrac{3}{256}x^2 - \dfrac{5}{2048}x^3 \ldots$

15 $\dfrac{3}{16}\left(1 - 2x + \dfrac{5}{2}x^2 - \dfrac{5}{2}x^3 \ldots\right)$

Exercise 5.4

1 $2x - 2x^2 + 2x^3 - 2x^4 + 2x^5 \ldots$ **2** $1 - \dfrac{1}{2}x^2 + x^3 \ldots$

3 $1 - 2x + \dfrac{5}{2}x^2 - 4x^3 \ldots$ **4** $-\dfrac{3}{16} + \dfrac{9}{16}x - \dfrac{27}{32}x^2 + \dfrac{15}{16}x^3 \ldots$

5 $1 - x + \dfrac{7}{2}x^2 \ldots$

Exercise 5.5

1 $+5\%$ **2** -9% **3** $+0.5\%$ **4** -6%

5 -10% **6** $+6\%$ **7** $+4.5\%$ **8** $-4\%, +5.5\%$

9 $+1.75\%$ **10** -4%

11 $+2\%$, no change   **12** 3.75%

13 $+2.5\%, +3\%$ or $+2\%$ **14** i) $+4\%$, ii) -4.5%

15 $+4.5\%$

Chapter 6

Exercise 6.1

1 1.649 **2** 4.462 **3** $0.3\bar{6}$ **4** $0.0\bar{6}$

5 0.607 **6** 3.297 **7** 2.231 **8** $0.18\bar{3}$

9 0.2 **10** 0.455

Exercise 6.2

1 $1 + 4x + 8x^2 + \dfrac{32}{3}x^3 \ldots$

2 $1 + \dfrac{x}{2} + \dfrac{x^2}{8} + \dfrac{x^3}{48} \ldots$

3 $1 - 3x + 4.5x^2 - 4.5x^3 + 3.375x^4 \ldots$

4 $2 - 8x + 16x^2 - \dfrac{64}{3}x^3 + \dfrac{64}{3}x^4 - \dfrac{256}{15}x^5 \ldots$

5 $4x + 2x^2 + \dfrac{x^3}{2} + \dfrac{x^4}{12} \ldots$

6 $1 + 4.5x + 2.125x^2 + 0.521x^3 \ldots$

7 $2 - 2.25x^2 - 2.25x^3 \ldots$

8 $2 + 9x^2 + 6.75x^4 \ldots$

9   i) $1 + 2x^2 + \dfrac{2}{3}x^4 + \dfrac{4}{45}x^6 \ldots$

     ii) $2x + \dfrac{4}{3}x^3 + \dfrac{4}{15}x^5 \ldots$

10   i) $1 - \dfrac{x^2}{2!} + \dfrac{x^4}{4!} - \dfrac{x^6}{6!} \ldots + j\left(x - \dfrac{x^3}{3!} + \dfrac{x^5}{5!} \ldots\right)$

     ii) $1 - \dfrac{x^2}{2!} + \dfrac{x^4}{4!} - \dfrac{x^6}{6!} \ldots - j\left(x - \dfrac{x^3}{3!} + \dfrac{x^5}{5!} \ldots\right)$

     iii) $1 - \dfrac{x^2}{2!} + \dfrac{x^4}{4!} - \dfrac{x^6}{6!} \ldots$

     iv) $j\left(x - \dfrac{x^3}{3!} + \dfrac{x^5}{5!} \ldots\right)$

Exercise 6.3

1 $4e^{4x}$ **2** $-e^{-x}$ **3** $-2e^{-x}$ **4** $-2e^{-2x}$

5 $3e^{3x}$ **6** $\dfrac{1}{2}e^{x/2}$ **7** $-\dfrac{1}{2}e^{-x/2}$ **8** $-4e^{-4x}$

9 $2e^x$ **10** $6e^{2x}$

Chapter 7

Exercise 7.1

1 $-3 + 8x$ **2** $28t^3 - 8$

3 $-10.5x^{2.5} + 8x^{-3}$ **4** $6x - 3.5x^{-4.5} + 4.2x^{-5.2}$

5 $-9.6t^{-2.6} - 4.8t^{-3.4}$ **6** $-33.6t^{-3.4}$

7 $-40x^{-3.5} - 30x^{-2.5}$ **8** $3 + 2x^{-2}$

9 $-7x^{-4.5} - 9x^2$ **10** $-1.5x^{-1.5} + 2$

11 $-3x^{-4} + x^{-2} + 1.75x^{-0.5}$

12 $\dfrac{8}{3}x^{-2/3} + \dfrac{1}{24}x^{-7/8}$

13 $\dfrac{3}{2}x^{1/2} - \dfrac{2}{3}x^{-1/3}$

14 $2.8t^{-0.8} - 4.8t^{-0.6}$

15 $2t^{-1/2} + \dfrac{1}{8}t^{-3/2}$

Exercise 7.2

1 $2\cos 2\theta$

2 $-2\sin(2x + 3)$

3 $-8\cos(3 - 2\theta)$

4 $4\cos\theta\sin^3\theta$

5 $-\dfrac{1}{2}\sin x(\cos x)^{-1/2}$

6 $-\dfrac{1}{2}\cos x(\sin x)^{-3/2}$

7 $4\cos(4\theta - 7) + 4\sin(7 - 4\theta)$

8 $-4\sin x\cos x$ or $-2\sin 2x$

9 $2\cos\theta\left(\dfrac{1}{\sqrt{\sin\theta}} - 1\right)$

10 $-9\cos(6 - 3x)\sin^2(6 - 3x)$

Exercise 7.3

1 $\dfrac{9}{9x - 11}$

2 $\dfrac{18}{9x - 11}$

3 $\dfrac{18}{9x + 4}$

4 $\dfrac{8x - 3}{4x^2 - 3x + 6}$

5 $\dfrac{2}{2x + 3} - \dfrac{1}{4 - x}$

6 $\dfrac{3}{3x - 7} - \dfrac{2x}{x^2 + 1}$

7 $\dfrac{1}{2x}$

8 $-\dfrac{\sin x}{\cos x}$ or $-\tan x$

9 $\dfrac{1}{2x}$

10 $-\dfrac{1}{2x}$

Exercise 7.4

1 $12e^{4x}$

2 $12e^{4x+1}$

3 $-5e^{-2x}$

4 $-8e^{-2x}$

5 $-8e^{-(2x+5)}$ or $-8e^{-2x-5}$

6 $-22.5e^{1-3x}$

7 $7.5 + 2e^{4+x}$

8 $30x^2e^{2x^3}$

9 $12\cos\theta.e^{4\sin\theta}$

10 $\cos\theta - 2\sin\theta.e^{\cos\theta}$

Exercise 7.5

1 $8(2x - 5)^3$

2 $6(2x - 5)^2$

3 $20(2x - 5)^3$

4 $-5(2x - 5)^{-2}$

5 $-4(2x - 5)^{-3}$

6 $3.5(1 + 7x)^{-1/2}$

7 $-24x(1+4x^2)^{-3/2}$

8 $-\dfrac{3}{2}(4-x)^{1/2}$

9 $15(1-2x)(1+x-x^2)^4$

10 $-6(4+3x^2)(4x+x^3)^{-3}$

Exercise 7.6

1 $3(2x+1)e^{2x}$

2 $x^2+2+(3x^2+2)\ln 3x$

3 $\theta\cos\theta+\sin\theta$

4 $2(2-5x-6x^2)$

5 $(3x^3+3x^2+12x+4)e^{3x}$

6 $2\theta\cos\theta-(4+\theta^2)\sin\theta$

7 $\cos^2\theta-\sin^2\theta$ or $\cos 2\theta$

8 $3(1-4\theta\sin\theta)e^{4\cos\theta}$

9 $\left(\dfrac{3x}{3x+1}+\dfrac{1}{2}\ln(3x+1)\right)\dfrac{1}{\sqrt{x}}$

10 $\dfrac{3(\sqrt{x}-7)}{3x+1}+\dfrac{1}{2\sqrt{x}}\ln(3x+1)$

Exercise 7.7

1 $-\dfrac{(1+x)}{e^x}$

2 $\dfrac{(x+1)\cos x-\sin x}{(x+1)^2}$

3 $\dfrac{1-\ln x}{x^2}$

4 $-\dfrac{(\sin t+\cos t)}{e^t}$

5 $\dfrac{1}{(1-x)^2}$

6 $-\dfrac{1}{\sin^2\theta}$ or $-\operatorname{cosec}^2\theta$

7 $\dfrac{4(x-4)e^{2x}}{(2x-7)^2}$

8 $\dfrac{x^2-2x-4}{(x-1)^2}$

9 $\dfrac{1}{(x-2)^2}\left(\left(\dfrac{x-2}{2+x}\right)-\ln(2+x)\right)$

10 $\dfrac{-\theta(\sin\theta+\cos\theta)-(1+\cos\theta-\sin\theta)}{\theta^2}$

Exercise 7.8

1 $4\sin\theta\cos\theta$ or $2\sin 2\theta$

2 $2(x+1)e^x$

3 $\dfrac{1-\ln(1+x)}{(1+x)^2}$

4 $\dfrac{1-4x}{e^x}$

5 $\dfrac{\theta(\cos\theta-\sin\theta)-(\sin\theta+\cos\theta)}{2\theta^2}$

6 $-2\sin(2\theta-\pi)+\cos(\theta-\pi)$

7 $\dfrac{2}{x}+1-\ln 2x$

8 $(2\sin x+\cos x)e^{2x}$

9 $\dfrac{4}{2x+3}$

10 $\dfrac{\sec^2 t-\tan t}{e^t}$

11 $\theta(2\cos(\theta+2\pi)-\theta\sin(\theta+2\pi))$

12 $-2\sin\theta.e^{2\cos\theta}$

13 $-7(5+2x)^{-3/2}$

14 $\dfrac{2-\ln x}{2x^{3/2}}$

15 $-\sin\theta$

Exercise 7.9

1 $2(e^{4x} - e^{-4x})$, i) -806.9, ii) 0, iii) 806.9

2 11.0

3 $-\dfrac{\cos\theta}{\sin^2\theta}$, 1.414

4 -2.55

5 3.5

6 $\dfrac{\sin\theta}{\cos^2\theta}$, 3.46

7 m^2, $2\pi(2r + 0.45)$, change in area with respect to a change in radius, 8.48 m, 5.65 m, 14.14 m

8 $2 - (12\sin 12t + 0.25\cos 12t)e^{-0.25t}$, 6.85, $-2 + (12\cos 12t - 0.25\sin 12t)e^{-0.25}$, 5.99

9 $\dfrac{W(a - x)}{a}$, 28.1

10 $\dfrac{-2(3\sin(3t + \pi/3) + \cos(3t + \pi/3))}{e^t}$, -0.76

Chapter 8

Exercise 8.1

1 $-10\sin x\cos x$, $10(\sin^2 x - \cos^2 x)$ or $-10\cos 2x$

2 $8(4x + 5)^{-1}$, $-32(4x + 5)^{-2}$

3 $12e^{3x-2}$, $36e^{3x-2}$

4 $6(3 + 2t)^2$, $24(3 + 2t)$

5 $\dfrac{6x}{3x^2 + 2}$, $\dfrac{6(2 - 3x^2)}{(3x^2 + 2)^2}$

6 $2xe^{x^2}$, $2(2x^2 + 1)e^{x^2}$

7 $2\cos\theta - 2\theta\sin\theta$, $-2(2\sin\theta + \theta\cos\theta)$

8 $-\tan\theta$, $-\sec^2\theta$

9 $\dfrac{3}{2}(x + 5)^{1/2}$, $\dfrac{3}{4}(x + 5)^{-1/2}$

10 $\dfrac{\cos\theta - \sin\theta}{e^\theta}$, $\dfrac{-2\cos\theta}{e^\theta}$

Exercise 8.2

1 $294(3 + 7x)$, 1911

2 -5

3 $(2 + 6x)e^{3x}$, $6(2 + 3x)e^{3x}$, 61.67, 241.94, $-0.8\overline{3}$

4 $6\cos\theta.e^{3\sin\theta}$, $6(3\cos^2\theta - \sin\theta)e^{3\sin\theta}$, 6, 18

5 $\dfrac{1 - \ln(x - 1)}{(x - 1)^2}$, $\dfrac{-3 + 2\ln(x - 1)}{(x - 1)^3}$, 0.26, -0.65

Exercise 8.3

1 $-5 + 4t - 12t^3$, $4 - 36t^2$

2 $10\left(\dfrac{3t}{1 + 3t} + \ln(1 + 3t)\right)$, $30\left(\dfrac{2 + 3t}{(1 + 3t)^2}\right)$

3 $(2 + 2t + t^2)e^t$, $(4 + 4t + t^2)e^t$

4 $1.37\cos\left(\dfrac{\pi}{4}t + 2.75\right)$, $-1.08\sin\left(\dfrac{\pi}{4}t + 2.75\right)$, 0.62

5 $1 - t + 3t^2$, $-1 + 6t$

Exercise 8.4

1 i) 26 m, ii) 24 ms^{-1}, 28 ms^{-1}, 4 ms^{-1}

2 i) 2.09 ms^{-1}, ii) 1.62 ms^{-1}, -1.18 ms^{-2}

3 $49(1 - e^{-2t})$, tends to 49 ms^{-1}

4 $t = 8$ min, return trip takes 8 min, -1.22 km/min, $2.\bar{6}$ min

5 $0.36\left(\dfrac{-10}{(t - 24)^2} - \dfrac{100}{(t - 24)^3}\right)$, -0.0053 m/hour, 1.73 m/hour

Chapter 9

Exercise 9.1

1 2.5   **2** 1.5   **3** $\dfrac{1}{2}$, 1   **4** -0.74, 1.24

5 $0.9\bar{3}$, 2

Exercise 9.2

1 $-\dfrac{1}{3}$, min   **2** $-\dfrac{1}{4}$, max   **3** -1.5, max; 2, min

4 $-\dfrac{7}{6}$, max; $\dfrac{1}{2}$, min   **5** $-\sqrt{2}$, max; $\sqrt{2}$, min

Exercise 9.3

1 $y_{min} = -1.125$, $(0.75, -1.125)$

2 $y_{max} = 9$, $(-1, 9)$

3 $y_{max} = 28$, $(-2, 28)$; $y_{min} = -80$, $(4, -80)$

4 $y_{max} = 9.26$, $(-1.1\bar{6}, 9.26)$; $y_{min} = 0$, $(0.5, 0)$

5 No max/min

6 $y_{max} = 34$, $(-1.4, 34)$; $y_{min} = 2$, $(0.2, 2)$

7 $y_{min} = 0$, $(0, 0)$

8 $y_{min} = -0.37$, $(-1, -0.37)$
9 $y_{max} = 1.10$, $(0, 1.10)$
10 $y_{max} = 0.35$, $(0.46, 0.35)$; $y_{min} = -6.6 \times 10^{-4}$, $(3.61, -6.6 \times 10^{-4})$

Exercise 9.4

1   Yes, $(0, 4)$ **2**   No   **3**   Yes, $(-2, 0)$
4   Yes $(1.5, 0)$ **5**   Yes, $(4, 3)$

Exercise 9.5

1   2500 Nm

2   Max at $\left(\dfrac{\pi}{2}, 1\right)$, min at $\left(\dfrac{3}{2}\pi, -1\right)$; max at $(0, 1)$, min at $(\pi, -1)$,

   max at $(2\pi, 1)$; max at $\left(\dfrac{3}{4}\pi, \sqrt{2}\right)$, min at $\left(\dfrac{7}{4}\pi, -\sqrt{2}\right)$

3   Yes, 2.58 s, max at $(2.58, 172°)$
4   £5963
5   1, $(0, 1)$, -1 has no real square root, $(0, 0)$
6 $0.2x^3 + x^2 - 3x - 10$, 1.12 tonnes
7   0.18 where $x = 0.607$
8 $h^2 + r^2 = 0.4^2$, $h = (0.16 - r^2)^{1/2}$, min $= 0$, max $= 0.026\,\text{m}^3$
9 x, $2000 - 2x$, $3000 - 2x$; $x(2000 - 2x)(3000 - 2x)$; 392 mm, 1215 mm,
   2215 mm, $1.056 \times 10^9\ \text{mm}^3$
10 $h = \dfrac{12 - r^2}{r}$, $50.27\,\text{m}^3$, $r = 2\,\text{m}$, $h = 4\,\text{m}$

Chapter 10

Exercise 10.1

1 $\dfrac{x^8}{4} - \dfrac{4}{7}x^7 + c$ **2** $-\dfrac{5}{3}t^{-3} - 3t^{-1} + c$

3 $-\dfrac{5}{3}x^{-3} + 3x^{-1} + c$ **4** $-\dfrac{1}{2}x^{-3} - x^5 + c$

5 $\dfrac{3}{4}x^4 + 2x^2 + c$ **6** $\dfrac{4}{5}x^{2.5} - \dfrac{2}{3}x^{1.5} + c$

7 $\dfrac{8}{3}x^{3/2} + c$ **8** $\dfrac{18}{5}t^{5/3} + c$

9 $\dfrac{4}{7}x^{7/4} - 2x + c$ **10** $2.86t^{1.4} + 1.6\overline{6}t^{1.2} + c$

11 $4(x^{1.5} + x^{0.5}) + c$ **12** $-0.8x^{-0.5} + 0.4x^{0.5} + c$
13 $4t^{1.5} - 16t^{0.5} + c$ **14** $2.6\overline{6}x^{1.5} - 0.3\overline{3}x^{0.5} + c$
15 $x^{0.5} + x^2 + c$

Exercise 10.2

1	0.5	**2**	10.37	**3**	2.875	**4**	$-1208.\overline{5}$
5	68	**6**	54.6	**7**	6.03	**8**	21.68
9	1.03	**10**	0.79	**11**	8.74	**12**	0.27
13	4.42	**14**	45.39	**15**	1.68		

Exercise 10.3

1 $-\dfrac{1}{2}\cos 2x + c$

2 $-2\cos 2x - \dfrac{1}{2}\sin 2x + c$

3 $\dfrac{1}{2}\tan 6x + c$

4 $-2\cos \dfrac{1}{2}\theta + c$

5 $4\left(\sin \dfrac{1}{2}t - \cos \dfrac{1}{4}t\right) + c$

6 $-0.25\cos 2x + 4\sin \dfrac{1}{4}x + c$

7 $3\theta + 8\sin \dfrac{1}{2}\theta + c$

8 $2\tan 3x - 4\cos \dfrac{1}{2}x + c$

9 $5(x - \tan 2x) + c$

10 $-\dfrac{1}{6}\cos 3x - 6\cos \dfrac{1}{2}x - \dfrac{1}{2}\sin 2x + c$

Exercise 10.4

1	-2.85	**2**	-3.50	**3**	0.42	**4**	8.49
5	0.27	**6**	6.99	**7**	-0.13	**8**	0.054
9	-0.315	**10**	2.47				

Exercise 10.5

1 $\dfrac{1}{2}(e^{2x} - e^{-2x}) + c$

2 $2e^x + 3e^{-x} + c$

3 $4(e^{x/4} + e^{x/2}) + c$

4 $\dfrac{5}{3}e^{3t} + c$

5 $e^{2x+1} - 2.5e^{1-2x} + c$

6 -11.10

7 2.66

8 -49.03

9 11.75

10 4.53

Exercise 10.6

1	2.43	**2**	0.25	**3**	0.71	**4**	8.32
5	3.77						

Exercise 10.7

1	10.81 unit^2	**2**	12.64 unit^2	**3**	4.39 unit^2	**4**	4.85 unit^2
5	6.04 unit^2						

Exercise 10.8

1	3.05	**2**	2.09	**3**	0.54	**4**	1.06
5	3.82						

Exercise 10.9

1	2.12	**2**	2.86	**3**	2.83	**4**	2.12
5	1.22						

Exercise 10.10

1	10 ms^{-1}	**2**	7.5 N	**3**	0.49 V
4	94.0 N	**5**	$110 \times 10^3 \text{ Nm}^{-2}$	**6**	1.41
7	5625 W	**8**	0, 3.54 A	**9**	35.36 mA

10 $4\sin^2 \omega t$, $2(1 - \cos 2\omega t)$, $4(1 - \cos 2\omega t)^2$, $6 - 8\cos 2\omega t + 2\cos 4\omega t$, 2.45 W

Chapter 11

Exercise 11.1

1	19.525	**2**	4.33	**3**	0.785	**4**	16.16
5	2.958	**6**	2.142	**7**	0.355	**8**	0.265
9	21.50	**10**	0.132				

Exercise 11.2

1	8.08	**2**	0.5	**3**	1.48	**4**	1.5
5	0.637	**6**	i) 3, ii) -2 , iii) 9.375, iv) -2.5
7	0.318	**8**	0.1507 ms^{-1}	**9**	1.932 N	**10**	125 W

Exercise 11.3

1	0.265	**2**	5.03 V	**3**	1.916	**4**	0, 1.768 A
5	53.03 mA						

Chapter 12

Exercise 12.1

1 $\dfrac{4}{x+3} + \dfrac{2}{x-1}$

2 $\dfrac{3}{x-7} - \dfrac{3}{x+5}$

3 $\dfrac{4}{x-4} - \dfrac{1}{x-1}$

4 $\dfrac{3}{x+2} + \dfrac{4}{x-1}$

5 $\dfrac{2}{x+6} + \dfrac{0.5}{x-3}$

6 $\dfrac{2.5}{x+3} + \dfrac{1.5}{x}$

7 $\dfrac{2}{x+1} + \dfrac{3}{x-1}$

8 $\dfrac{4}{x+3} + \dfrac{2}{x+1} + \dfrac{5}{x-4}$

9 $-\dfrac{2}{x+5} + \dfrac{3}{x-1} + \dfrac{1}{x}$

10 $-\dfrac{1}{x+3} + \dfrac{2}{x-3} + \dfrac{4}{x}$

Exercise 12.2

1 $-\dfrac{4}{(x+1)^2} + \dfrac{2}{x+1} + \dfrac{1}{x-5}$

2 $-\dfrac{0.5}{(x+3)^2} + \dfrac{1}{x+3} - \dfrac{1}{x-4}$

3 $\dfrac{3.5}{(x-2)^2} + \dfrac{4.5}{x-2} - \dfrac{1}{x}$

4 $\dfrac{3}{x+3} + \dfrac{1}{(x+1)^2} + \dfrac{2}{x+1}$

5 $\dfrac{5}{(2x-1)^2} + \dfrac{1}{2x-1} + \dfrac{1}{x}$

Exercise 12.3

1 $\dfrac{4x}{x^2+1} - \dfrac{3}{x-1}$

2 $\dfrac{x}{x^2+2} + \dfrac{5}{x}$

3 $\dfrac{4}{x+1} - \dfrac{50}{x^2+10}$

4 $\dfrac{2x-3}{x^2-x+2} - \dfrac{2}{x}$

5 $\dfrac{2x+4}{x^2+2x+4} + \dfrac{1}{x-3}$

Exercise 12.4

1 $4 - \dfrac{7}{x+3} + \dfrac{6}{x-2}$

2 $5 + \dfrac{3}{x-1} + \dfrac{2}{x}$

3 $2x+4 - \dfrac{2}{x+1} + \dfrac{3}{x-1}$

4 $3x - \dfrac{2x}{x^2+4} - \dfrac{6}{x}$

5 $5 + \dfrac{x}{x^2+2} + \dfrac{4}{x-2}$

Exercise 12.5

1 $2\ln(x-1) + 4\ln(x+3) + c$

2 $3\ln(x+2) + 4\ln(x-1) + c$

3 $2\ln\left(\dfrac{x}{10-x}\right) - \dfrac{20}{x} + c$

4 $2\ln(x^2+1) - 3\ln(x-1) + c$

5 $5\ln x + \tfrac{1}{2}\ln(x^2+2) + c$

6 $5x + 2\ln x + 3\ln(x-1) + c$

7 12.11   **8** 3.98

9 10.20   **10** 12.92

11 1.93   **12** 1.78

13 8.92   **14** 6.61

15 13.43

Exercise 12.6

1 10.26 m **2** 18.0 N **3** 0.41 ms^{-2} **4** 1.72, 1.76
5 4.299 × 10^{-3} J

Chapter 13

Exercise 13.1

1 6.13 **2** −2.571 **3** −2.943 **4** 16.161
5 −0.023 **6** 2.142 **7** 0.066 **8** −164.8
9 21.50 **10** 39.495

Exercise 13.2

1 −0.785 **2** 5.111 **3** 0.879 **4** 49.21
5 1.773

Exercise 13.3

1 0.355 **2** 0.191 **3** 0.041 **4** 492.4
5 108.1

Exercise 13.4

1 120.6 **2** 109.7 J **3** −2.013 V **4** 1.92 N
5 4.66 unit^2

Chapter 14

Exercise 14.1

1 $\frac{1}{2}\ln(x^2 + 1) + c$

2 $\frac{1}{6}(x^2 + 5)^3 + c$

3 $3\ln(x^2 + 6) + c$

4 $\frac{(2x^2 + 5)^{-2}}{-2} + c$

5 $\frac{1}{16}(1 + 2x^2 - x^4)^4 + c$

6 $(t^2 - 7)^{1/2} + c$

7 $-\frac{1}{4}\ln(2 - x^2) + c$

8 $\frac{4}{3}(4 + 3x^2)^{3/2} + c$

9 $\frac{(1 + \sin\theta)^3}{3} + c$

10 $\frac{(e^t + e^{-t})^4}{4} + c$

Exercise 14.2

1	1.558	**2**	25.48	**3**	0.109	**4**	7.938
5	−0.321	**6**	114	**7**	3.872	**8**	0.270
9	0.137	**10**	0.504				

Exercise 14.3

1	1.571	**2**	0.524	**3**	$\dfrac{4}{9}\pi$	**4**	0.785
5	0.459	**6**	1.45	**7**	0.524	**8**	0.236
9	0.175	**10**	1.172				

Exercise 14.4

1	0.881	**2**	1	**3**	2.732	**4**	0.055
5	0.096						

Exercise 14.5

1 $\dfrac{1}{2}\theta + \dfrac{1}{4}\sin 2\theta + c$　　　　**2** $-\cos t + c$

3 $\dfrac{1}{2}\tan 2\theta - \theta + c$　　　　**4** $-\dfrac{1}{3}\cos^3 x + \dfrac{1}{5}\cos^5 x + c$

5 $-\dfrac{1}{10}\cos 5\phi - \dfrac{1}{2}\cos \phi + c$　　　　**6** $-\dfrac{1}{12}\sin 6\theta + \dfrac{1}{4}\sin 2\theta + c$

7	$-0.08\overline{3}$	**8**	0.397	**9**	2.856	**10**	0.957
11	0.250	**12**	−0.500	**13**	0.190	**14**	0.018
15	0.091						

Exercise 14.6

1　2π unit^2, 8π unit^2　　　　**2**　$\dfrac{25}{4}\pi$ unit^2, area in the first, positive, quadrant

3	1.207 J	**4**	3.826×10^3 gallons
5	0.174 s	**7**	π unit^2
8	$1.366\,k$, $15.020\,k$	**9**	169.6×10^3
10	136 s		

Chapter 15

Exercise 15.1

1	i)	x,	ii)	y,	iii)	1,	iv)	1
2	i)	x,	ii)	y,	iii)	2,	iv)	1
3	i)	t,	ii)	y,	iii)	1,	iv)	2

4	i) x,	ii) y,	iii) 2,	iv) 3
5	i) x,	ii) y,	iii) 3,	iv) 1
6	i) x,	ii) y,	iii) 3,	iv) 1
7	i) t,	ii) Q,	iii) 2,	iv) 1
8	i) t,	ii) x,	iii) 3,	iv) 1
9	i) x,	ii) y,	iii) 1,	iv) 2
10	i) θ,	ii) y,	iii) 1,	iv) 1

Exercise 15.2

1 Straight line, $y = -2x + c$, $y = -2x + 9$

2 $y = 2x^2 + c$, $y = 2x^2 - 1$

3 $\dfrac{dy}{dx} = -x + 4$, $y = -\dfrac{1}{2}x^2 + 4x + c$, $y = -\dfrac{1}{2}x^2 + 4x - 3$

4 Sine curve, $y = \sin\theta + c$, i) $y = \sin\theta$, ii) $y = 1 + \sin\theta$

5 $y = x^3$

Exercise 15.3

1 $y = \dfrac{3}{2}x^2 + x + 2$

2 $y = -\dfrac{1}{3}x^3 + 2x^2 + 2x - \dfrac{2}{3}$

3 $y = x^2 + \dfrac{4}{x} - 2.5$

4 $y = -4x - \dfrac{1}{x} + 7$

5 $y = -2\sin t - 3\cos t + t + 8$

Exercise 15.4

1 $y = 3e^{2x}$

2 $y = 163.8e^{-4t}$

3 $y = 16.3e^{x/2}$

4 $Q = 0.5e^{3t}$

5 $y = 1.14e^{9x/16}$

Exercise 15.5

1 $y^3 = 1.5(x^2 - 16)$

2 $y^2 + 2y - x^2 - 2 = 0$

3 $\sin y = \sin x - 2$

4 $y = 4x$

5 $y = e^{1-\cos x}$

6 $y^2 = 2\ln(x^2 + 1) + 16$

7 $y^2 + 2y - x^2 + 2x = 0$

8 $y = x - 2$

9 $y^2 = 2(x + \ln x + 1)$

10 $e^y = 2e^x - e$

Exercise 15.6

1 $\dfrac{dy}{dx} = 4x$, $y = 2x^2 + 3$

2 $\dfrac{dv}{dt} = -9.8$, $v = 25 - 9.8t$

3 $Q = 2400$ W

4 $\dfrac{dW}{dx} = kx$, $W = 250x^2$

Exercise 15.7

1 $\dfrac{dy}{dx} = 3y$, $y = 6e^{3x-6}$

2 $5\dfrac{dv}{dt} = kv$, $v = 10e^{-0.03t}$

3 $T = 15 + 25e^{-0.03t}$

4 0.014, 0.022

5 $\dfrac{dN}{dt} = kN$, $N = N_0 e^{-0.365t}$

Exercise 15.8

1 $\dfrac{dy}{dx} = 2\dfrac{x}{y}$, $y^2 = 2x^2 + 1$

2 $\ln\left(\dfrac{m}{4.5}\right) = \dfrac{k}{2}\left(\dfrac{1}{(1+t)^2} - 1\right)$, 0.236

3 $\dfrac{dp}{dV} = -1.395\dfrac{p}{V}$, $pV^{1.395} = C$, $\dfrac{p_1}{p_0} = 0.38$

4 0.246

5 $kx^2 - 19.6mx - 108m = 0$

Chapter 16

Exercise 16.1

1 12 unit^2

2 17.3 unit^2

3 4.39 unit^2

4 4.05 unit^2

5 Symmetry

Exercise 16.2

1 6.25 unit^2

2 14.14 unit^2

3 11 unit^2

5 16 unit^2

Exercise 16.3

1 4 unit^2, (1.60, 2.29)

2 63 unit^2, (3.04, 14.61)

3 (1.57, 0.39)

4 1.718 unit^2, (0.58, 0.93)

5 (0.5, 1.2)

6 (1.27, 0)

7 5.3 unit^2, (0.75, 2.4)

8 20 unit^2, (2.6$\bar{3}$, 3.2$\bar{6}$)

9 (2, 3.4)

10 (0.75, 4)

Exercise 16.4

1 0.495 m from O perpendicular to $CDOAB$

2 0.76 m on line bisecting AD and BC

3 At D

4 0.1$\bar{3}$ m from EA on line bisecting EA and DB

5 0.48 m on line bisecting AD and BC

6   0.26 m from both AB and AF
7   0.21 m from $ABCD$ perpendicular to $ABCD$
8   136 mm from AF, 211 mm from AB
9   (a) 0.68 m horizontally from A, 0.35 m from AB;   (b) 0.87 m
10 (a) At O,   (b) At O,   (c) 0.97 m from lower horizontal tangent,
        (d) 0.97 m from low tangent whose diameter bisects $\angle COB$

Chapter 17

Exercise 17.1

1 20.11 unit^3	**2** 117.8 unit^3	**3** 4.935 unit^3
4 6.283 unit^3	**5** 1.571 unit^3	**6** 642.8 unit^3
7 15.71 unit^3	**8** 7.189 unit^3	**9** 29.98 unit^3
10 8.516 unit^3		

Exercise 17.2

1 $y = -1.5x + 3$, cone, 18.85 unit^3
2   16.76 unit^3
3   113.1 unit^3, symmetry
4 $y = -1.5x + 3$, cone, different radii and heights, make the intercepts on the axes the same
5 $y = \frac{2}{3}x$, 502.7 unit^3, cone, 100.5 unit^3, 402.1 unit^3

Exercise 17.3

1 8.38 unit^3	**2** 13.40 unit^3
3 (0, 0) and (1, 1), 0.94 unit^3	**4** 28.48 unit^3
5 (0, 0) and (2, 8), 143.6 unit^3	

Exercise 17.4

1 (1.75, 0)	**2** (3.34, 0)	**3** (0.66, 0)	**4** (0.5, 0)
5 (0, 3.51)			

Exercise 17.5

1 0.65 m	**2** 0.34 m	**3** 0.80 m
4 1.49	**5** 0.58 m	**6** 0.16 m
7 0.32 m	**8** 0.63 m from vertex	**9** 0.84 m
10 0.82 m		

Chapter 18

Exercise 18.1

1   $I_{XX} = 1360$ cm^4, $I_{YY} = 2000$ cm^4 2 $I_{XX} = 2720$ cm^4, $I_{YY} = 4000$ cm^4
3   $I_{XX} = 6315$ cm^4, $I_{YY} = 9515$ cm^4 4 $I_{XX} = 4708$ cm^4
5   $I_{XX} = 5218$ cm^4 6 $I_{XX} = 2641$ cm^4
7   $I_{XX} = 3170$ cm^4 8 $I_{YY} = 191$ cm^4
9   $I_{XX} = 383$ cm^4, $I_{YY} = 383$ cm^4 10 $I_{YY} = 1832$ cm^4

Exercise 18.2

1   $I_{YY} = 1114$ cm^4 2 $I_{XX} = 2958$ cm^4 3 $I_{YY} = 1446$ cm^4
4   $I_{YY} = 8716$ cm^4 5 $I_{XX} = 4058$ cm^4

Exercise 18.3

1   $I_{QQ} = 3360$ cm^4 2 $I_{QQ} = 3.136\pi \times 10^4$ cm^4
3   $I_{QQ} = 15830$ cm^4 4 $I_{QQ} = 766$ cm^4
5   $I_{QQ} = 7456$ cm^4

Index